PROGRESS IN COLLOID & POLYMER SCIENCE

Editors: F. Kremer (Leipzig) and G. Lagaly (Kiel)

Volume 107 (1997)

Analytical Ultracentrifugation IV

Guest Editors:

R. Jaenicke and H. Durchschlag (Regensburg)

STEINKOPFF
DARMSTADT

Springer-Verlag Berlin Heidelberg GmbH

ISBN 978-3-662-16050-3   ISBN 978-3-7985-1656-4 (eBook)
DOI 10.1007/ 978-3-7985-1656-4

ISSN 0340-255 X

Die Deutsche Bibliothek –
CIP-Einheitsaufnahme

Progress in colloid & polymer science. –
Darmstadt : Steinkopff ; New York :
Springer
    Früher Schriftenreihe
Vol. 107. Analytical ultracentrifugation
IV. – 1997
Analytical ultracentrifugation IV / guest ed.:
Jaenicke and Durchschlag. – Darmstadt :
Steinkopff, 1997
    (Progress in colloid & polymer science ;
    Vol. 107)
    ISBN 978-3-662-16050-3

© Springer-Verlag Berlin Heidelberg 1997
Originally published by Dietrich Steinkopff Verlag
GmbH & Co. KG, Darmstadt in 1997
Softcover reprint of the hardcover 1st edition 1997

Chemistry Editor:
Dr. Maria Magdalene Nabbe;
Production: Holger Frey, Ajit Vaidya.

Typesetting and Copy-Editing:
Macmillan Ltd., Bangalore, India

Progr Colloid Polym Sci (1997) V
© Steinkopff Verlag 1997

PREFACE

The 10th Symposium on Analytical Ultracentrifugation took place at the University of Regensburg on March 13th and 14th, 1997. More than 100 participants, among them 30 speakers, met at the Institute of Biophysics and Physical Biochemistry (Faculty of Biology and Preclinical Medicine) to discuss recent results and developments in the fields of Macromolecular Chemistry, Biophysical Chemistry, and Polymer Science from the perspective of ultracentrifugal analysis. In the present volume of Progress in Colloid & Polymer Science selected topics from the lectures and posters have been collected in order to give an overview over the present state of the art in ultracentrifugation at the turning point from the preferably manual to the fully computerized methodology.

It has been a festive occasion to host the ultracentrifuge community in Regensburg again after a pause of 12 years. Clearly, the 10th Symposium had to be put into an appropriate historical frame. Insiders know that it was Cassius and not Marcus Aurelius, the founder of Regensburg, who inspired Svedberg to develop the ultracentrifuge, but there are some hidden connections between the venerable 27 years of history of the University and Svedberg's method: many of the grand old men from the heroic age of biophysical chemistry have been frequent guests (and even guest professors) in Regensburg, to mention only three: Heini Eisenberg, Max A. Lauffer, and Howard K. Schachman. Apart from these highlights in the grey everyday life of the faculty, there are more historical monuments. In the Institute of the hosts, the veteran of the Spinco E's in this country still survives, and two more of the old-timer generation have been in use around the clock for 25 years. The user (himself close to the age of Svedberg's ingenious first air-driven machine) enjoys his hobby-dinosaurs, and the co-organizer of the Symposium shares his enthusiasm. Not that both would not like to do comparative studies with the XL-A or XL-I, but some surprise has to be left for the next Regensburg Symposium, perhaps after another 12 years.

As in the case of the previous nine meetings, the 1997 Symposium attracted researchers from universities, research institutes, and industrial companies. The highlights which gave the 10th meeting its jubilee flavor were the keynote lectures given by Professors Donald J. Winzor and Allen P. Minton. They not only set the frame and the standards, but also gave the whole conference a very stimulating atmosphere. The organizers would like to thank both speakers for accepting the invitation and contributing to this special issue. Inviting speakers from Australia and the United States was only possible thanks to the financial support from a number of sponsors:

BASF AG, Ludwigshafen
Beckman Instruments GmbH, Munich
Biacore AB, Freiburg
Merck KGaA, Darmstadt
Pharmacia Biotech Europe GmbH, Freiburg
Regensburger Universitätsstiftung Hans Vielberth

Their generosity is gratefully acknowledged. We owe special thanks to the Regensburger Universitätsstiftung Hans Vielberth; without its help we would not have been able to organize this meeting.

The present volume of Progress in Colloid & Polymer Science contains contributions on the following topics:

> Technical Innovations
> Innovations in Data Analysis
> Modeling
> Biological Systems
> Ions and Polyelectrolytes
> Polymers, Colloids, Supramolecular Systems
> Emulsions, Gels, and Dispersions
> Interacting Systems and Assemblies

We hope that the rapid publication of the Proceedings helps to enhance the awareness of our colleagues in Physical Chemistry, Biophysics, Biochemistry and Polymer Science that a new era in the characterization of macromolecules has started with the recent renaissance of analytical ultracentrifugation in many fields of research. In this context we should like to thank the editor Dr. M. M. Nabbe for her support in preparing this volume.

Regensburg, July 1997

R. Jaenicke
H. Durchschlag

Progr Colloid Polym Sci (1997) VII
© Steinkopff Verlag 1997

CONTENTS

Progr Colloid Polym Sci (1997) 107:1–10
© Steinkopff Verlag 1997

P.R. Wills
M.P. Jacobsen
D.J. Winzor

Direct analysis of sedimentation equilibrium distributions reflecting macromolecular interactions

P.R. Wills
Department of Physics
University of Auckland
Auckland, New Zealand

M.P. Jacobsen · Prof. D.J. Winzor (✉)
Centre for Protein Structure, Function
and Engineering
Department of Biochemistry
University of Queensland
Brisbane, Queensland 4072, Australia
E-mail: winzor@biosci.uq.edu.au

Abstract This review of the characterization of protein interactions by sedimentation equilibrium emphasizes procedures that entail direct determination of the thermodynamic activity of the smallest species contributing to the concentration distribution for that constituent. This approach, which has been regarded as an Australasian oddity for over two decades, is first illustrated by evaluating the association constant for α-chymo-trypsin dimerization by the original omega analysis and subsequent refinements thereof. Notable in that regard is the introduction of the psi function, which has evolved from its omega counterpart. Application of the corresponding approach to sedimentation distributions for mixtures of ovalbumin and cytochrome c is presented to illustrate the potential of the psi function for characterizing interactions between dissimilar macromolecular reactants. Also discussed is the means by which these direct analyses of sedimentation equilibrium distributions afford realistic allowance for effects of thermodynamic nonideality on the statistical-mechanical basis of excluded volume.

Key words Protein self-associa-tion – heterogeneous association – sedimentation equilibrium – thermo-dynamic nonideality

Introduction

For many years analytical ultracentrifugation was the major source of information on the heterogeneity and molecular weight of macromolecules. In the field of protein chemistry, however, that role has largely been taken over by gel electrophoretic and gel chromatographic techniques, whereupon the main emphasis of analytical ultra-centrifugation has been transferred to the determination of equilibrium constants for protein interactions. The traditional molecular weight approach to ultracentrifugal ana-lysis influenced initial studies, which were based on the concentration dependence of weight-average molecular weights determined from sedimentation equilibrium distri-butions [1–3]. Such procedures still prevail [4–6], but more direct methods have also been developed [7–9] for analyzing the sedimentation equilibrium distribution, which is effectively a record of total solute concentration $\bar{c}(r)$, as a function of radial distance, r.

The original approach to the characterization of pro-tein self-association by direct analysis of the equilibrium concentration distribution entailed, for ideal systems, evaluation of the equilibrium concentration of monomer throughout the distribution, and hence deduction of the equilibrium constant(s) by combining that parameter with the total concentration at each radial distance [7, 10, 11]. Despite its simplicity, the popularity of that procedure has not extended beyond Australasian shores. Instead, focus has been centred on the alternative direct analysis [8],

2

P.R. Wills et al.
Studies of interactions by sedimentation equilibrium

which is based on a nonlinear least-squares minimization procedure to obtain the best-fit description of $\bar{c}(r)$–r distribution as the sum of exponential expressions for the oligomeric species (monomer, dimer, etc.) postulated to be present: model-dependence is thus part of the analysis from the outset. Although this simulation-based procedure is suitable for quantifying ideal solute interactions, the intricacies of the algorithms used to achieve a best-fit solution have deterred any attempt to incorporate realistic allowance for the effects of thermodynamic nonideality. The reader is referred to a review by Johnson and Straume [12] for a discussion of the numerical and statistical aspects of the analysis incorporated into the software programs based on this procedure.

In this presentation we return to the Australasian approach, which has the advantage that it has already accommodated realistic allowance for the complexities of thermodynamic nonideality in the evaluation of equilibrium constants for protein self-association [9, 13, 14]. Although incomplete, the corresponding characterization of interactions between dissimilar macromolecular reactants by sedimentation equilibrium is at an advanced stage of development.

The omega and psi functions for studies of protein self-association

The omega function, $\Omega(r)$, is an experimental parameter defined by the relationship

$$\Omega(r) = [\bar{c}(r)/\bar{c}(r_F)]\exp[\phi M_1(r_F^2 - r^2)] , \qquad (1a)$$

$$\phi = (1 - \bar{v}\rho_s)\omega^2/(2RT) \qquad (1b)$$

in which $\bar{c}(r)$ and $\bar{c}(r_F)$ are the respective total weight-concentrations of protein at radial distance r and selected reference radial position r_F. The exponent term is calculated from the molecular weight of monomer, M_1, and

the protein partial specific volume $\bar{v}$, which is assumed to apply to monomer as well as all oligomeric species. R is the universal gas constant and T the temperature of a sedimentation equilibrium experiment conducted at angular velocity ω. For many years the remaining parameter (ρ_s) in Eq. (1b) was considered erroneously to be the solution density [13]; but has now been identified as the density of solvent [14, 15]. On the basis that the sedimentation equilibrium distribution of the molar thermodynamic activity of monomer, $z_1(r)$, is given by

$$z_1(r) = z_1(r_F) \exp[\phi M_1(r^2 - r_F^2)] , \qquad (2)$$

the exponential term in Eq. (1a) may be replaced by the thermodynamic activities of monomer at the respective radial positions. Thus Eq. (1a) becomes

$$\Omega(r) = [\bar{c}(r)z_1(r_F)]/[\bar{c}(r_F)z_1(r)] . \qquad (3)$$

In the limit of infinite dilution, the total protein concentration becomes synonymous with the thermodynamic activity of monomer [$\bar{c}(r) \rightarrow z_1(r)M_1$ as $\bar{c}(r) \rightarrow 0$], whereupon it follows that

$$\lim_{\bar{c}(r) \rightarrow 0} \Omega(r) = \Omega_0 = z_1(r_F)M_1/\bar{c}(r_F) . \qquad (4)$$

The thermodynamic activity of monomer at the selected reference radial position may thus be obtained from the value of the ordinate intercept, Ω_0, of the dependence of $\Omega(r)$ upon $\bar{c}(r)$; and hence at all other radial positions throughout the distribution by means of Eq. (2).

This procedure for evaluating $z_1(r_F)M_1$ is illustrated in Fig. 1a, which summarizes results from five sedimentation equilibrium experiments on α-chymotrypsin in acetate–chloride buffer, pH 3.9, I 0.16 [10]. From the ordinate intercept, Ω_0, of 0.32, for a reference concentration, $\bar{c}(r_F)$, of 1.87 g/l, $z_1(r_F)M_1$ is calculated to be 0.60 g/l. Combination of this value with that of the reference radial position (r_F) associated with $\bar{c}(r_F)$ in each sedimentation equilibrium experiment then allows the experimental definition of the

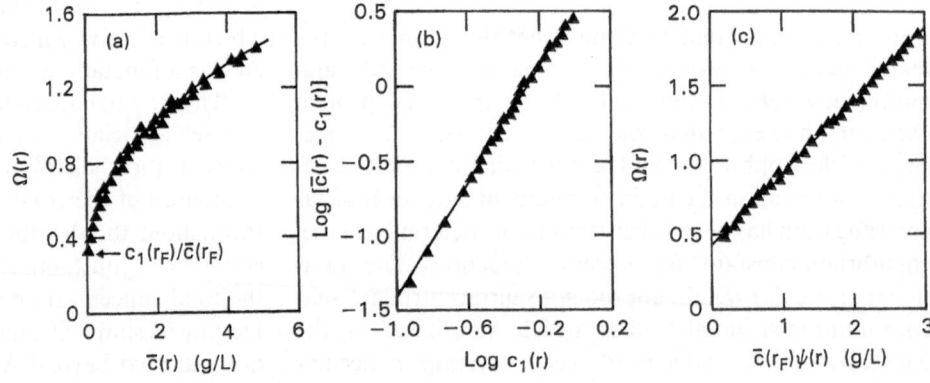

Fig. 1 Characterization of α-chymotrypsin dimerization (pH 3.9, I 0.2) by omega analysis. (a) Evaluation of the thermodynamic activity of monomer, $z_1(r_F) = \Omega_0\bar{c}(r_F)/M_1$, by the original procedure for a reference concentration, $\bar{c}(r_F)$, of 1.87 g/l. (b) Plot of the consequent results in terms of Eq. (5) to obtain the association constant. (c) Revised analysis of the same sedimentation equilibrium distributions to obtain Ω_0 and X_2 via Eq. (9). Adapted from [10, 19]

Progr Colloid Polym Sci (1997) 107: 1–10
© Steinkopff Verlag 1997

relationship between $z_1(r)$ and $\bar{c}(r)$ by means of Eq. (2). Provided that assumed thermodynamic ideality is a justifiable approximation over the range of $\bar{c}(r)$ examined [16], $z_1(r)M_1$ may be regarded as $c_1(r)$, whereupon the results may be tested for conformity with two-state self-association by analysis in terms of the logarithmic form of the law of mass action for such a system:

$$\log c_n(r) = \log[\bar{c}(r) - c_1(r)] = \log X_n - n\log c_1(r) . \quad (5)$$

From the suggested plot of $\log[\bar{c}(r) - c_1(r)]$ versus $\log c_1(r)$, Fig. 1b, the self-association of α-chymotrypsin under these conditions is described adequately by reversible dimerization ($n = 2$) and an association constant (X_2) of 3.5 l/g [10].

An appealing feature of the omega analysis is its independence upon any model of self-association: the thermodynamic activity of monomer throughout the sedimentation equilibrium distribution(s) is obtained without need of any assumptions about the stoichiometry of self-association. Indeed, even the product ϕM_1 may be regarded as an experimental parameter that is determinable from the limiting slope, as $\bar{c}(r) \to 0$, of $\ln \bar{c}(r)$ versus r^2 in a high-speed [17] sedimentation equilibrium experiment. However, as noted by Morris and Ralston [18], a drawback of the omega analysis in this form is the extent of reliance placed upon the accuracy of the curvilinear extrapolation to obtain Ω_0 as the ordinate intercept in Fig. 1a. That defect has now been overcome [19] by writing Eq. [2] in the form

$$c_1(r) = \Omega_0 \bar{c}(r_F)\psi(r) , \quad (6a)$$

$$\psi(r) = \exp[\phi M_1(r^2 - r_F^2)] \quad (6b)$$

and expressing $\bar{c}(r)$ in terms of $c_1(r)$ and the dimerization constant,

$$\bar{c}(r) = c_1(r) + X_2[c_1(r)]^2 . \quad (7)$$

Combination of these two equations gives

$$\Omega(r) = \Omega_0 + X_2 \Omega_0^2 \bar{c}(r_F)\psi(r) \quad (8)$$

which signifies a linear dependence of $\Omega(r)$ upon $\bar{c}(r_F)\psi(r)$ for a reversibly dimerizing system such as α-chymotrypsin (Fig. 1c).

The modified omega analysis has the advantage that Ω_0 and X_2 are evaluated concurrently, but this benefit has been gained at the expense of model independence of the analysis. Having forsaken that attribute of the original omega analysis, a better approach is to eliminate the omega function altogether by writing Eq. (7) as

$$\bar{c}(r) = c_1(r_F)\psi(r) + X_2[c_1(r_F)\psi(r)]^2 + \cdots \quad (9)$$

and using nonlinear least-squares analysis to determine $c_1(r_F)$ and X_2 from the dependence of $\bar{c}(r)$ upon $\psi(r)$. Adaptation of this psi analysis [20] to accommodate multi-state self-association merely requires the addition of higher-order terms in $[c_1(r_F)\psi(r)]$ to the right-hand side of Eq. (9). Such polynomial curve-fitting of $\bar{c}(r)$ as a function of $\psi(r)$ avoids a shortcoming of the omega procedure, for which the two experimental parameters, $\Omega(r)$ and $\bar{c}(r)$, are both dependent variables. Because the measurement of radial distance is essentially error free, $\psi(r)$ may be regarded as an independent variable. Consequently, Eq. (9), or an expanded version thereof, meets the analytical requirement that a dependent variable, $\bar{c}(r)$, be expressed as a function of a statistically independent variable $\psi(r)$.

Allowance for effects of thermodynamic nonideality

In situations where the range of solute concentration covered by the sedimentation equilibrium distribution(s) exceeds that commensurate with assumed thermodynamic ideality, Eq. (7) must be rewritten [20] as

$$\bar{c}(r) = M_1\{z_1(r)/\gamma_1(r) + X_2[z_1(r)]^2/\gamma_2(r)\} \quad (10)$$

where $\gamma_1(r)$ and $\gamma_2(r)$ denote the respective activity coefficients of monomer and dimer at radial distance r. Thus, irrespective of the method used for the determination of $z_1(r)$, estimates of the activity coefficients of both states of the protein are required.

For studies of nonideal protein self-association by sedimentation equilibrium the activity coefficients are usually assumed to be given by the expression [1, 21] $\gamma_1 = \exp[iBM_1\bar{c}(r)]$, where B is an empirical constant. However, this Adams–Fujita approach is unrealistic in that it leads to self-cancellation of the activity coefficient ratio in the expression for the apparent equilibrium constant defined in terms of species concentrations [Eq. (11)].

$$X_2 = \{c_2(r)/[c_1(r)]^2\}\{\gamma_2(r)/[\gamma_1(r)]^2\}$$

$$= X_2^{app}\{\gamma_2(r)/[\gamma_1(r)]^2\} . \quad (11)$$

The inference that the apparent equilibrium constant, X_2^{app}, equates with the true thermodynamic parameter (X_2) is, at best, extremely unlikely. Furthermore, it is at variance with considerations of the effects of thermodynamic nonideality on the statistical-mechanical basis of excluded volume [9, 14, 16, 19, 22–25].

The traditional approach to statistical-mechanical evaluation of activity coefficients [16, 19, 22] is illustrated by considering a reversibly dimerizing system for which nonideality is described adequately by confining consideration to nearest-neighbor interactions. On the basis of spherical geometry for both species, the activity coefficients are given by the expressions [14]

$$\gamma_1(r) = \exp\{2B_{11}c_1(r)/M_1 + B_{12}[\bar{c}(r) - c_1(r)]/M_2 + \cdots\},$$
$$(12a)$$

$$\gamma_2(r) = \exp\{2B_{22}[\bar{c}(r) - c_1(r)]/M_2 + B_{12}[c_1(r)]/M_1 + \cdots\},$$
$$(12b)$$

where

$$B_{ii} = 16\pi N_A R_i^3/3$$
$$+ Z_i^2(1 + 2\kappa R_i)/[4I(1 + \kappa R_i)^2] + \cdots, i = 1, 2,$$
$$(13a)$$

$$B_{12} = \tfrac{4}{3}\pi N_A(R_1 + R_2)^3$$
$$+ Z_1 Z_2(1 + \kappa R_1 + \kappa R_2)/[2I(1 + \kappa R_1)(1 + \kappa R_2)]$$
$$+ \cdots.$$
$$(13b)$$

Z_1, Z_2 and R_1, R_2 are the respective net charges (valences) and radii of monomer and dimer, respectively, and N_A the Avogadro's number. The inverse screening length (κ) is calculated in cm^{-1} from the ionic strength (I) by means of the expression $\kappa = 3.27 \times 10^7\sqrt{I}$.

Evaluation of activity coefficients by this means requires knowledge of $c_1(r)$, the concentration of monomer for each $\bar{c}(r)$; and hence the adoption of an iterative approach. The original version of the omega analysis may be used to obtain the dependence of $z_1(r)$ upon $\bar{c}(r)$, from which the dimerization constant may be determined by iterative, nonlinear curve-fitting in terms of Eqs. (10), (12) and (13) [22]. Alternatively, the expression for the modified omega method [Eq. (8)] is readily adapted by the incorporation of activity coefficients to give

$$\Omega(r)\gamma_1(r) = \Omega_0 + X_2\Omega_0^2 \bar{c}(r_F)\psi(r)[\gamma_1(r)/\gamma_2(r)].$$
$$(14)$$

Initially, thermodynamic ideality $[\gamma_1(r) = \gamma_2(r) = 1]$ is assumed to allow first estimates of $c_1(r_F)$ and hence $c_1(r)$ to be obtained for substitution into Eq. (12) to obtain first estimates of $\gamma_1(r)$ and $\gamma_2(r)$ at each $\bar{c}(r)$. Reanalysis of the results in accordance with Eq. (14) then allows iterative refinement of the values of Ω_0 and X_2.

Resort to the above statistical-mechanical approach leads to the expression of species activity coefficients as a series expansion that requires specification of the concentrations of all solute states, and hence the adoption of iterative procedures. On the other hand, Hill and Chen [26] have taken the viewpoint that a reversibly dimerizing protein should be regarded as a single solute system; and have therefore regarded the self-association as another form of nonideality. Adoption of this latter strategy has the advantage of allowing the thermodynamic activity of monomer, $z_1(r)$, to be expressed as a series expansion in

terms of the total concentration of the single solute component, namely,

$$z_1(r)M_1 = \bar{c}(r)\exp\{2B_2\bar{c}(r)/M_1 + \tfrac{3}{2}B_3[\bar{c}(r)/M_1]^2 + \cdots\},$$
$$(15)$$

where the coefficients B_i are defined from the virial expansion for the osmotic pressure Π in terms of base-molar solute concentration, $\bar{c}(r)/M_1$. Specifically,

$$\Pi/(RT) = \bar{c}(r)/M_1 + B_2[\bar{c}(r)/M_1]^2 + B_3[\bar{c}(r)/M_1]^3 + \cdots.$$
$$(16)$$

The use of Eq. (15) to describe nonideality in terms of total solute concentration is certainly reminiscent of the Adams–Fujita approach [21]; but differs therefrom in that the coefficients (B_i) are now defined on rigorous statistical-mechanical grounds as a mixture of equilibrium constants and excluded volume terms:

$$B_2 = B_{11} - X_2M_1/2,$$
$$(17a)$$

$$B_3 = B_{111} - X_2M_1(4B_{11} - B_{12}) + (X_2M_1)^2,$$
$$(17b)$$

where B_{11} and B_{12} remain defined by Eqs. (13a) and (13b), B_{111} being the corresponding second-order term [22]. Because of its ability to provide the dependence of $z_1(r) M_1$ upon $\bar{c}(r)$, the omega analysis is ideally suited to take advantage of this approach, the relevant expression being [20]

$$\Omega(r) = \exp\{2B_2[\bar{c}(r_F) - \bar{c}(r)]/M_1$$
$$+ \tfrac{3}{2}B_3[\bar{c}(r_F)^2 - \bar{c}(r)^2]/M_1^2 + \cdots\}.$$
$$(18a)$$

In principle, magnitudes of the various B_i may then be obtained by nonlinear regression analysis of $[\Omega(r), \bar{c}(r)]$ results in accordance with Eq. (18a) or its logarithmic counterpart,

$$\ln\Omega(r) = 2B_2[\bar{c}(r_F) - \bar{c}(r)]/M_1$$
$$+ \tfrac{3}{2}B_3[\bar{c}(r_F)^2 - \bar{c}(r)^2]/M_1^2 + \cdots.$$
$$(18b)$$

A problem with the Hill–Chen approach is the slow convergence of the polynomial series, except in instances where excluded volume interactions provide the dominant contribution to B_2 and B_3 [22]. However, this difficulty is overcome [20] by retaining the Hill–Chen strategy but expressing $\bar{c}(r)$ as a polynomial in $z_1(r)$:

$$\bar{c}(r)/M_1 = z_1(r)\{1 + 2b_2z_1(r) + 3b_3[z_1(r)]^2 + \cdots\},$$
$$(19)$$

where, for a dimerizing system,

$$b_2 = \tfrac{1}{2}X_2M_1 - B_{11},$$
$$(20a)$$

$$b_3 = 2B_{11}^2 - \tfrac{1}{2}X_2M_1B_{12} - \tfrac{1}{2}B_{111}.$$
$$(20b)$$

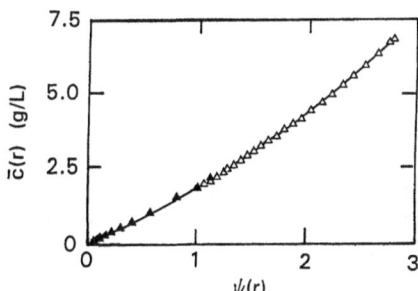

Fig. 2 Allowance for effects of thermodynamic nonideality in sedimentation equilibrium studies of α-chymotrypsin dimerization (pH 4.1, I 0.08). Results are plotted for two experiments (▲, △), together with their best-fit description (——) in terms of Eq. (21) truncated at quadratic term. Adapted from [20]

Substitution of $z_1(r_F)\psi(r)$ for $z_1(r)$ in Eq. (19) then gives

$$\bar{c}(r)/M_1 = z_1(r_F)\psi(r) + 2b_2[z_1(r_F)\psi(r)]^2$$
$$+ 3b_3[z_1(r_F)\psi(r)]^3 + \cdots \quad (21)$$

as the counterpart of Eq. (9) for psi analysis of sedimentation equilibrium distributions reflecting nonideality. Nonlinear least-squares fitting of $\bar{c}(r)/M_1$ as a polynomial in $\psi(r)$ thus provides values of $z_1(r_F)$, b_2, b_3, etc.; and hence of X_2 via Eqs. (13) and (20).

Results from sedimentation equilibrium experiments on α-chymotrypsin (pH 4.1, I 0.08) are summarized in Fig. 2, which refers to analysis with $\bar{c}(r_F) = 1.93$ g/l [20]. Nonlinear least-squares analysis of the results in terms of Eq. (21) truncated at the cubic term yielded an unacceptable description of the data because of inconsistencies between the values of b_2 and b_3, the latter signifying a negative value for the dimerization constant when substituted into Eq. (20b). The solid line in Fig. 2 is the best-fit description in terms of Eq. (21) truncated at the quadratic term and values of $2510(\pm 80)$ l/mol and $59.2(\pm 0.5)$ μM for b_2 and $z_1(r_F)$, respectively. On the basis of a radius of 2.44 nm and a valence of $+10$ for monomeric α-chymotrypsin under these conditions, a value of 309 l/mol is obtained for B_{11} from Eq. (13a). Substitution of these estimates of B_{11} and b_2 into Eq. (20a) then yields a dimerization constant (X_2) of 0.23 (± 0.01) l/g [20], which compares favourably with that of 0.16 l/g inferred from Fig. 1 of [27].

Effects of nonideality arising from the presence of small cosolutes

At this stage attention is turned to effects of thermodynamic nonideality stemming from the presence of a small inert cosolute (M) in sedimentation equilibrium studies performed under conditions where the behaviour of the protein alone would be effectively ideal. Two situations are considered: one in which the protein undergoes reversible dimerization, and the other in which the only effect of the cosolute is on the extent of protein solvation.

An interesting outcome of the omega analysis has been encountered in studies of α-chymotrypsin dimerization under conditions where the nonideality is dominated by the presence of sucrose [28]. Because the molar concentration of the inert cosolute, $c_M(r)/M_M$, greatly exceeds those of monomeric and dimeric enzyme, the expressions for the activity coefficients of the two protein states are simply

$$\gamma_1(r) \approx \exp[B_{1M}c_M(r)/M_M] , \quad (22a)$$

$$\gamma_2(r) \approx \exp[B_{2M}c_M(r)/M_M] . \quad (22b)$$

The fact that $c_M(r)$ is effectively invariant across the entire protein distribution at sedimentation equilibrium justifies the approximation that $\gamma_1(r) \approx \gamma_1(r_F)$, whereupon the counterpart of Eq. (4) becomes

$$\lim_{\bar{c}(r)\to 0} \Omega(r) = c_1(r)/\bar{c}(r_F) . \quad (23)$$

On the basis that Ω_0 defines the concentration of monomer at the reference radial position, the equilibrium constant evaluated by omega analysis (and, by inference, psi analysis) under these circumstances is an apparent value governed by the relationship

$$(X_2)_{app} = X_2\exp[(2B_{1M} - B_{2M})c_M(r)/M_M + \cdots] \quad (24a)$$

$$\approx X_2[1 + (2B_{1M} - B_{2M})c_M(r)/M_M + \cdots] . \quad (24b)$$

Such linear dependence of the measured association constant for α-chymotrypsin dimerization (pH 3.9, I 0.2) is illustrated in Fig. 3a, which also includes the theoretical dependence predicted by Eq. (24b) with values of B_{iM} calculated via Eq. (13b) on the basis of respective radii of 2.44 and 3.07 nm for monomeric and dimeric enzyme [28], the radius of sucrose having been taken as 0.31 nm [15].

The second situation concerns the protein solvation effect of a small cosolute on the measurement of molecular weight of a noninteracting protein by sedimentation equilibrium. In many respects protein solvation is an unfortunate choice of term for an excluded volume phenomenon, the formal equivalence of the protein solvation and excluded volume treatments having been established in relation to the stabilization of protein structure by inert solutes [29, 30]. Use of the statistical-mechanical concept of excluded volume rather than that of protein solvation [31–34] has led to the development of a simple and economical method of protein molecular weight measurement in buffers supplemented with high concentrations of cosolute [35].

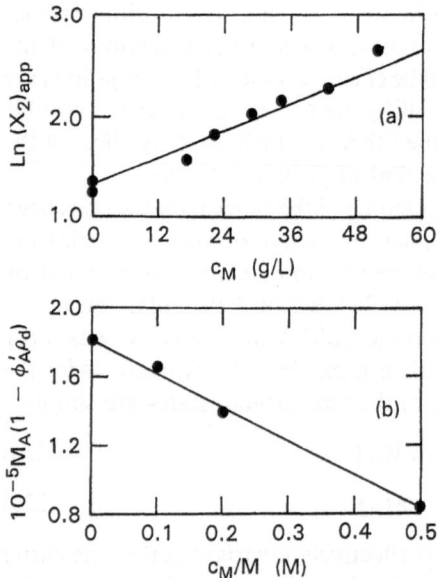

Fig. 3 Effects of thermodynamic nonideality arising from the presence of a high concentration of a small cosolute. (a) Enhancement of the extent of α-chymotrypsin dimerization by sucrose, the results being plotted according to Eq. (24). (b) Determination of the second virial coefficient, B_{AM}, for the excluded volume interaction between sucrose and thyroglobulin, the results being plotted in accordance with Eq. (25). Adapted from [28] and [35], respectively

Casassa and Eisenberg [36] provided the initial solution to the problem by establishing that analysis of sedimentation equilibrium distributions for a protein in the presence of a small cosolute could be simplified to that for the standard single-solute system by prior dialysis of the protein solution against buffer supplemented with cosolute. However, although this action allowed the cosolute to be regarded as part of the solvent, it also introduced the problem of evaluating ϕ'_A, the apparent partial specific volume of protein, from the difference between the densities of dialyzed protein solutions and those of the corresponding diffusates [33, 34]. The apparent partial specific volume may certainly be determined in this manner, but the amount of protein required has rendered its measurement either unfeasible or unattractive.

By commencing with the condition of sedimentation equilibrium for the distribution of protein (A),

$$z_A(r) = z_A(r_F) \exp[M_A(1 - \bar{v}_A\rho_s)\omega^2(r^2 - r_F^2)/(2RT)] \,, \quad (25)$$

where ρ_s is again the density of unsupplemented solvent [9, 15], it has been shown [35] that the concentration distribution is given by the relationship

$$c_A(r) \approx c_A(r_F)\exp\{[M_A(1 - \bar{v}_A\rho_s) - B_{AM}c_M(r_F)$$

$$\times (1 - \bar{v}_M\rho_s)]\omega^2(r^2 - r_F^2)/(2RT) + \cdots\} \,. \quad (26)$$

This equation is to be compared with the expression derived by Eisenberg [37], namely

$$c_A(r) \approx c_A(r_F) \exp[M_A(1 - \phi'_A\rho_d)\omega^2(r^2 - r_F^2)/(2RT) \,, \quad (27a)$$

$$\rho_d = \rho_s + (1 - \bar{v}_M\rho_s)c_M \quad (27b)$$

in which ρ_d is the density of the diffusate (inert solute concentration c_M) with which the protein solution has been brought to dialysis equilibrium prior to ultracentrifugation. Equations (26) and (27) are formally identical with the buoyant molecular weight, $M_A(1 - \phi'_A\rho_d)$, defined correct to first order as

$$M_A(1 - \phi'_A\rho_d) = M_A(1 - \bar{v}_A\rho_s) - (1 - \bar{v}_M\rho_s)B_{AM}c_M(r_F)$$

$$+ \cdots, \quad (28)$$

where $c_M(r_F)$ is effectively c_M, the concentration of cosolute included in the buffer. The molecular weight of the protein may thus be obtained as the ordinate intercept of the limiting linear dependence of buoyant molecular weight upon the concentration of cosolute included in the solvent.

The effect of sucrose on the buoyant molecular weight of bovine thyroglobulin in Tris-chloride buffer (pH 7.5) is summarized in Fig. 3b, which clearly exhibits the linear dependence predicted by Eq. (28). Analysis of the results in those terms yields values of $180\,000$ ($\pm\,9000$) for the buoyant molecular weight and 1500 ($\pm\,300$) l/mol for B_{AM} from the ordinate and slope, respectively. Combination of the former with the reported partial specific volume ($\bar{v}_A$) of $0.723\,\text{ml/g}$ [38] yields a molecular weight of $660\,000$ ($\pm\,33\,000$), an estimate consistent with the accepted value of $670\,000$ for thyroglobulin [39]. Furthermore, the excluded volume, B_{AM}, closely approximates the effective hydrodynamic volume of $1490\,\text{l/mol}$ that is calculated from the sedimentation coefficient of $19.4S$ for thyroglobulin [38, 39].

Although sedimentation equilibrium has been used to evaluate subunit molecular weights of proteins from distributions obtained in the presence of urea or guanidine hydrochloride [40, 41], the need for associated measurement of ϕ'_A by densitometry seems to have been an effective deterrant to widespread use of that method. In view of the above success with evaluating M_A for thyroglobulin in the presence of sucrose, we now consider the situation in which the small cosolute is replaced by a small denaturant such as urea. Clearly, the assumed inertness of cosolute that is implicit in Eq. (28) precludes its application to results in the denaturant concentration range conducive to progressive dissociation of the protein. However, for denaturant concentrations commensurate with complete dissociation of the protein into its constituent subunits, the dependence of buoyant molecular weight upon denaturant concentration should be described by Eq. (28) with M_A re-identified as the subunit molecular weight.

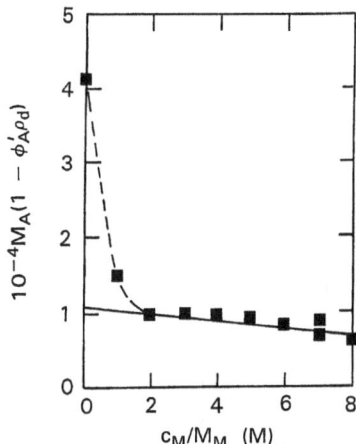

Fig. 4 Determination of the subunit molecular weight of aldolase by sedimentation equilibrium in the presence of urea. Adapted from [35]

The dependence of the buoyant molecular weight of aldolase upon urea concentration [35] is presented in Fig. 4. At low urea concentration the steeper slope can be rationalized in terms of dissociation of the tetrameric enzyme into subunits, whereas the lesser slope at higher denaturant concentration (solid line) reflects the nonideality stemming from excluded volume interactions between denatured subunits and cosolute. Linear regression analysis of results in the latter region signifies an ordinate intercept of 10 900 ($\pm$ 1400) for the buoyant molecular weight of aldolase subunits in urea-free buffer. Combination of this estimate of $M_A(1 - \bar{v}\rho_s)$ with the reported partial specific volume of 0.742 ml/g [42] and the unsupplemented buffer density (ρ_s) yields a subunit molecular weight of 43 500 ($\pm$ 5000), which is a quarter of that, 165 000 ($\pm$ 10 000), for the native enzyme from the other ordinate intercept in Fig. 4 [35].

Characterization of complex formation between dissimilar reactants

Despite the greater biological prevalence of interactions between dissimilar macromolecular reactants, protein self-association has been the dominant phenomenon studied by sedimentation equilibrium. After initial theoretical considerations of the feasibility of employing sedimentation equilibrium for the quantitative characterization of such systems [43], a decade elapsed before the appearance of any reports of experimental attacks on the problem [44, 45]. That counterpart of the omega analysis for characterization of self-association [7] then lay dormant for a further 20 years before its revival in a study of the interaction between lysozyme and chitosan [46]. Mean-

while, the advent of the Beckman XL-A ultracentrifuge had kindled interest in the use of sedimentation equilibrium for characterization of protein-oligonucleotide interactions [47–49] by a protocol stemming from the North American procedure for quantifying protein self-association [8]. Encouraged by the successful development of the psi procedure for direct analysis of solute self-association, we are in the process of examining the feasibility of using a similar approach for the study of interactions between dissimilar solutes.

In order to characterize the interaction between a macromolecular ligand, S, and several sites on an acceptor, A, a mixture of the two reactants is subjected to sedimentation equilibrium at angular velocity ω and temperature T. After attainment of chemical as well as sedimentation equilibrium, the distributions of the various species are given by the expression [cf. Eq. (2)]

$$z_i(r) = z_i(r_F)\exp[\phi_i M_i \omega^2(r^2 - r_F^2)] \,, \tag{29}$$

where $i = A, S, AS, AS_2$, etc. In keeping with the earlier studies [44–49], we shall restrict consideration to thermodynamically ideal systems to allow the thermodynamic activities, $z_i(r)$, to be replaced by molar concentrations, $C_i(r)$. We also restrict consideration to the simplest situation: that in which equilibrium distributions are available for both constituent (total) concentrations, $\bar{C}_A(r)$ and $\bar{C}_s(r)$.

The total concentration of ligand at any radial distance r may be expressed in terms of the corresponding free concentration, $C_s(r)$, and that of free acceptor, $C_A(r)$, as

$$\bar{C}_s(r) = C_s(r) + K_1 C_A(r)C_s(r) + 2K_2 C_A(r)C_s(r)^2 + \cdots , \tag{30}$$

where K_1, K_2, etc., are the respective stoichiometric equilibrium constants for the formation of species AS, AS_2, etc., from acceptor and ligand. We now introduce the psi function [20] for the ligand and acceptor species,

$$\psi_i(r) = \exp[\phi_i M_i(r^2 - r_F^2)], \quad i = A, S \tag{31}$$

to allow Eq. (30) to be written in the form

$$\bar{C}_s(r) = C_s(r_F)\psi_s(r) + K_1 C_A(r_F)C_s(r_F)\psi_A(r)\psi_s(r)$$
$$+ 2K_2 C_A(r_F)[C_s(r_F)]^2 \psi_A(r)[\psi_s(r)]^2 + \cdots . \tag{32}$$

On the grounds that $C_s(r_F)$ and $C_A(r_F)$, as well as the various K_i, are all constant parameters, multivariate nonlinear regression analysis in terms of Eq. (32) with $\psi_s(r)$ and $\psi_A(r)$ as the independent variables has the potential to yield magnitudes of all parameters as curve-fitting coefficients. For a simple 1:1 interaction the evaluation of $C_s(r_F)$, $C_A(r_F)$ and K_1 by this means may well be a reasonable proposition. At this stage we prefer to retain model-independence of the analysis; and therefore divide Eq. (32)

by $\psi_s(r)$ to obtain

$$\bar{C}_s(r)/\psi_s(r) = C_s(r_F) + K_1 C_A(r_F) C_s(r_F) \psi_A(r)$$
$$+ 2K_2 C_A(r_F)[C_s(r_F)]^2 \psi_A(r)\psi_s(r) + \cdots . \quad (33)$$

The concentration of free ligand at the reference radial position may thus be obtained as the ordinate intercept of the dependence of $\bar{C}_s(r)/\psi_s(r)$ upon $\psi_A(r)$, this dependence being linear if complex formation is restricted to 1:1 stoichiometry ($K_2 = 0$).

Determination of the magnitude of $C_s(r_F)$ allows the concentration of free ligand, $C_s(r)$, to be calculated as $C_s(r_F)\psi_s(r)$ throughout the distribution. Furthermore, because the corresponding distribution for total acceptor concentration, $\bar{C}_A(r)$ is also available, the binding function, $v(r)$, can be evaluated as

$$v(r) = [\bar{C}_s(r) - C_s(r)]/\bar{C}_A(r) . \quad (34)$$

Thus, provided that $C_s(r)$ varies sufficiently across a sedimentation equilibrium distribution (or, indeed, a series of equilibrium distributions), a binding curve may be generated by plotting $v(r)$ as a function of $C_s(r)$. Binding constants (K_1, K_2, etc.) are then obtained by nonlinear regression analysis in terms of the expression

$$v(r) = K_1 C_s(r)/[1 + K_1 C_s(r)] + K_2 C_s(r)/[1 + K_2 C_s(r)]$$
$$+ \cdots . \quad (35)$$

Interaction of cytochrome c with ovalbumin (pH 6.3, I 0.03)

The selection of ovalbumin (A) and cytochrome c (S) as the two reactants of a model system reflects the existence a convenient chromophore on the ligand to allow the determination of $C_s(r)$ throughout the sedimentation equilibrium distribution. On the basis of the limited valence data available for cytochrome c [50] and ovalbumin [51], the two reactants should bear net charges of approximately equal magnitude but opposite sign in the pH range 6.0–6.5, thereby ensuring some form of electrostatic interaction between the two reactants at low ionic strength. Furthermore, the fact that the AS complex is thus likely to be effectively uncharged should also render negligible the formation of complexes with greater than 1:1 stoichiometry under the conditions of study (pH 6.3, I 0.03).

Figure 5a presents sedimentation equilibrium distributions recorded at 410 nm (▲) and 280 nm (●) for an interacting mixture of ovalbumin and cytochrome c. Because the former distribution reflects only the cytochrome c constituent, the radial dependence of $\bar{C}_s(r)$ is readily obtained (▲, Fig. 5b). Knowledge of the relative magnitudes of the absorption coefficients for cytochrome c at 280 and

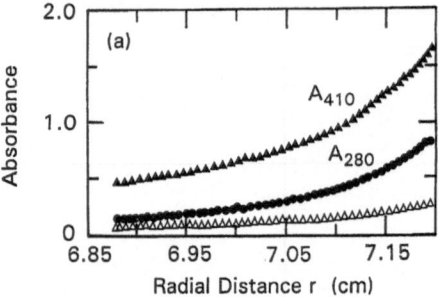

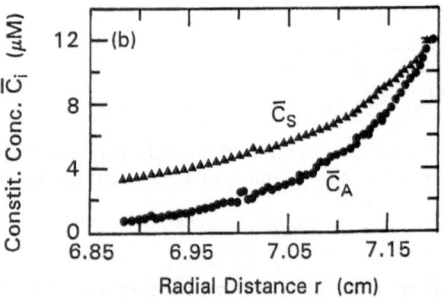

Fig. 5 Determination of the two constituent concentration distributions reflecting sedimentation equilibrium (15000 rpm, 20 °C) of a mixture of ovalbumin (A) and cytochrome c (S). (a) Equilibrium distributions recorded at 410 nm (▲) and 280 nm (●), together with that at 280 nm for the cytochrome c constituent (△). (b) The constituent concentration distributions for ovalbumin (●) and cytochrome c(▲)

410 nm allows the contribution of S constituent to $A_{280}(r)$ to be calculated (△, Fig. 5a), whereupon the remainder is attributed to ovalbumin constituent: the consequent distribution in terms of $\bar{C}_A(r)$ is shown (●) in Fig. 5b.

Analysis of distribution for cytochrome c constituent in terms of Eq. (33) is presented in Fig. 6, where the essentially linear dependence of $\bar{C}_s(r)/\psi_s(r)$ upon $\psi_A(r)$ indicates dominance of 1:1 complex formation between cytochrome c and ovalbumin under these conditions. This essentially linear dependence renders fairly accurate the extrapolation required to obtain $C_s(r_F)$ as the ordinate intercept; and hence the delineation of $C_s(r)$ throughout the distribution as a prelude to evaluating the binding function, $v(r)$, at each radial distance. Results from that sedimentation equilibrium experiment and fourteen others are presented as a binding curve in Fig. 7, where every tenth point has been included. Nonlinear curve-fitting of the entire data set to the rectangular hyperbolic relationship for 1:1 complex formation [Eq. (35) with $K_2 = 0$] yields a binding constant ($\pm 2\,\text{SD}$) of 62000 (± 2000) M^{-1}.

The procedure used to characterize the interaction between cytochrome c and ovalbumin (Figs. 6 and 7) differs markedly from its recent counterparts for analysis of heterogeneous association [48, 49] in that the emphasis is placed on the extraction of interaction parameters from

Progr Colloid Polym Sci (1997) 107:1–10
© Steinkopff Verlag 1997

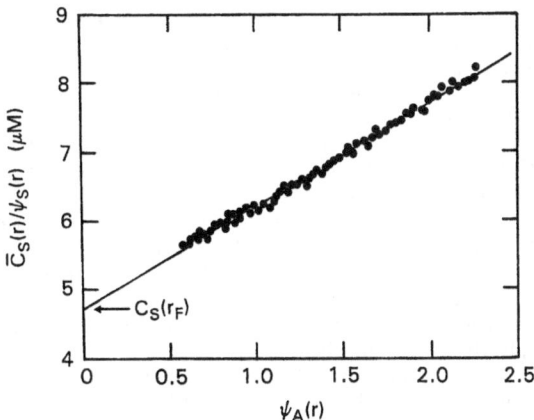

Fig. 6 Determination of the concentration of free cytochrome c from the sedimentation equilibrium distribution for total ligand concentration, $\bar{c}_s(r)$, by means of Eq. (33). The ordinate intercept defines the magnitude of $C_s(r_F)$

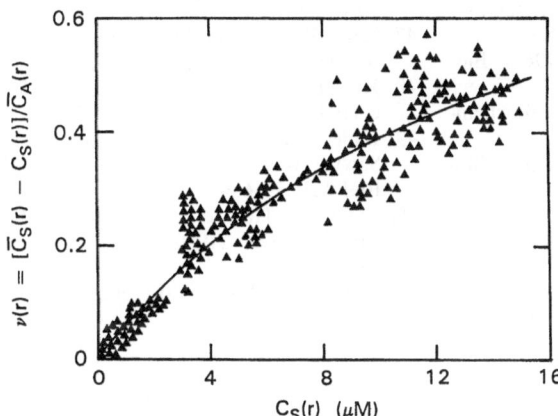

Fig. 7 Binding data (▲) obtained by sedimentation equilibrium for the interaction of cytochrome c (S) with ovalbumin (pH 6.3, I 0.03), together with their best-fit description (——) in terms of a binding constant of $62\,000\,M^{-1}$ for 1:1 complex formation

the experimental distributions rather than on their evaluation by iterative simulation of distributions to seek solutions that match the experimental records. Whereas the earlier analyses are thus model-dependent from the outset, the experimental results presented in Fig. 7 are model-independent. In that regard the considerable scatter of points in Fig. 7 reflects our refusal to adopt the common

practices of (i) applying smoothing programs to data prior to analysis and (ii) incorporating a model-dependent "baseline correction factor" as an additional curve-fitting parameter to decrease the uncertainty inherent in the returned parameters. The reported uncertainty (3%) in K is thus the estimated experimental error in the actual data set rather than in a set that has already been subjected to selective trimming prior to commencing the statistical assessment of random error.

Concluding remarks

The ultimate goal of this research into the characterization of macromolecular interactions by sedimentation equilibrium is the development of simple and rigorous analytical procedures that make realistic allowances for the effect of thermodynamic nonideality on the statistical-mechanical basis of excluded volume. For studies of protein self-association that goal has been achieved: the psi procedure not only provides a very simple and direct means of data analysis but also accommodates rigorous statistical-mechanical allowance for the effects of thermodynamic nonideality [20]. Our corresponding attack on the characterization of interactions between dissimilar macromolecular reactants is well under way. The sedimentation equilibrium study of the ovalbumin-cytochrome c interaction has emphasized the rigour, model-independence and extreme simplicity of the psi analysis for characterizing thermodynamically ideal heterogenous associations. Furthermore, quantitative relationships have already been derived that express the difference between $\bar{C}_s(r)$ and $\bar{C}_A(r)$ as a function of the two independent variables $\psi_s(r)$ and $\psi_A(r)$; but their application awaits the location of suitable numerical procedures for further analytical assessment.

In summary, the general reluctance of the rest of the ultracentrifuge community to take advantage of contributions made by the Australasian contingent to the study of interacting systems by sedimentation equilibrium has been their loss but our gain. That disinterest has provided us with two decades of breathing space in which to assimilate the statistical-mechanical deliberations of Terrell Hill; and hence to advance the study of macromolecular interactions to the stage where allowance for effects of thermodynamic nonideality can be made on a rigorous scientific basis.

References

1. Adams ET, Williams JW (1964) J Am Chem Soc 86:3454–3461
2. Roark DE, Yphantis DA (1969) Ann NY Acad Sci 164:245–278
3. Hoagland VD, Teller DC (1969) Biochemistry 8:594–602
4. Bucci E (1986) Biophys Chem 24:47–52
5. Chatelier RC, Minton AP (1987) Biopolymers 26:507–524
6. Minton AP (1990) Anal Biochem 190:1–6

7. Milthorpe BK, Jeffrey PD, Nichol LW (1975) Biophys Chem 3:169–176

8. Johnson ML, Correia JJ, Yphantis DA, Halvorson HR (1981) Biophys J 36: 575–588

9. Wills PR, Jacobsen MP, Winzor DJ (1996) Biopolymers 38:119–130

10. Tellam R, de Jersey J, Winzor DJ (1979) Biochemistry 24:5316–5321

11. Tellam R, Winzor DJ (1980) Arch Biochem Biophys 201:20–34

12. Johnson ML, Straume M (1994) In: Schuster TM, Laue TM (eds) Modern Analytical Ultracentrifugation. Birkhäuser, Boston, pp 37–65

13. Williams JW, Van Holde KE, Baldwin RL, Fujita H (1958) Chem Revs 58:715–806

14. Wills PR, Winzor DJ (1992) In: Harding SE, Rowe AJ, Horton JC (eds) Analytical Ultracentrifugation in Biochemistry and Polymer Science. Roy Soc Chem, Cambridge, pp 311–330

15. Wills PR, Comper WD, Winzor DJ (1993) Arch Biochem Biophys 300:206–212

16. Ogston AG, Winzor DJ (1975) J Phys Chem 79:2496–2500

17. Yphantis DA (1964) Biochemistry 3:297–317

18. Morris M, Ralston GB (1985) Biophys Chem 23:49–61

19. Jacobsen MP, Winzor DJ (1992) Biophys Chem 45:119–132

20. Wills PR, Jacobsen MP, Winzor DJ (1996) Biopolymers 38:119–130

21. Adams ET, Fujita H (1963) In: Williams JW (ed) Ultracentrifugal Analysis in Theory and Experiment. Academic Press, New York, pp 119–128

22. Wills PR, Siezen RJ, Nichol LW (1980) Biophys Chem 11:71–82

23. Jeffrey PD (1981) In: Frieden C, Nichol LW (eds) Protein–Protein Interactions. Wiley, New York, pp 213–256

24. Harris SJ, Winzor DJ (1988) Arch Biochem Biophys 265:458–465

25. Winzor DJ, Wills PR (1994) In: Schuster TM, Laue TM (eds) Modern Analytical Ultracentrifugation. Birkhäuser, Boston, pp 66–80

26. Hill TL, Chen YD (1973) Biopolymers 12:1285–1312

27. Aune KC, Goldsmith LC, Timasheff SN (1971) Biochemistry 10:1617–1621

28. Shearwin KE, Winzor DJ (1988) Biophys Chem 31:287–294

29. Wills PR, Winzor DJ (1993) Biopolymers 33:1627–1629

30. Winzor DJ, Wills PR (1995) In: Gregory RB (ed) Protein-Solvent Interactions Dekker, New York, pp 483–520

31. Cohen G, Eisenberg H (1968) Biopolymers 6:1077–1100

32. Reisler E, Eisenberg H (1969) Biochemistry 8:4572–4578

33. Lee JC, Timasheff SN (1981) J Biol Chem 256:7193–7201

34. Timasheff SN (1995) In: Gregory RB (ed) Protein-Solvent Interactions, Dekker, New York, pp 445–482

35. Jacobsen MP, Wills PR, Winzor DJ (1996) Biochemistry 35:13 173–13 179

36. Casassa EF, Eisenberg H (1964) Adv Protein Chem 19:287–395

37. Eisenberg H (1976) Biological Macromolecules and Polyelectrolytes in Solution. Clarendon Press, Oxford.

38. Derrien Y, Michel R, Pedersen KO, Roche J (1949) Biochim Biophys Acta 3:436–441

39. Steiner RF, Edelhoch H (1961) J Am Chem Soc 83:1435–1444

40. Lee JC, Timasheff SN (1974) Arch Biochem Biophys 165:268–273

41. Prakash V, Timasheff SN (1981) Anal Biochem 117:330–335

42. Taylor JF, Lowry C (1956) Biochim Biophys Acta 20:109–117

43. Nichol LW, Ogston AG (1965) J Phys Chem 69:4365–4367

44. Nichol LW, Jeffrey PD, Milthorpe BK (1976) Biophys Chem 4:259–267

45. Jeffrey PD, Nichol LW, Teasdale RD (1979) Biophys Chem 10:379–387

46. Cölfen H, Harding SE, Vårum KM, Winzor DJ (1996) Carbohydrate Polym 30:45–53

47. Laue TM, Senear DF, Eaton S, Ross AJB (1993) Biochemistry 32:2469–2472

48. Kim T, Tsuykiyama T, Lewis MS, Wu C (1994) Protein Sci 3:1040–1051

49. Bailey MF, Davidson BE, Minton AP, Sawyer WH, Howlett GJ (1996) J Mol Biol 263:671–684

50. Laue TM, Hazard AL, Ridgeway TM, Yphantis DA (1989) Anal Biochem 182:377–382

51. Creeth JM, Winzor DJ (1962) Biochem J 83:566–574

Progr Colloid Polym Sci (1997) 107:11–19
© Steinkopff Verlag 1997

A.P. Minton

Alternative strategies for the characterization of associations in multicomponent solutions via measurement of sedimentation equilibrium

Dr. A.P. Minton (✉)
Building 8, Room 226
National Institute of Health
Bethesda, Maryland 20892-0830, USA

Abstract Four strategies for the analysis of sedimentation equilibrium of solutions containing two solute components in the context of models for equilibrium association are described:

(1) Direct modeling of the equilibrium gradients of a single experimentally measurable signal measured in each of several samples of differing composition;

(2) Direct modeling of the equilibrium gradients of two (or more) signals measured in each of several samples of differing composition;

(3) Calculation of the independent component concentration gradients in individual samples via analysis of the gradients of two or more signals in the sample, followed by modeling of the component concentration gradients from several samples of differing composition; and

(4) Calculation of the signal-average buoyant molar mass as a function of solution composition, followed by modeling of the composition dependence of the signal-average buoyant molar mass in the context of an association model.

Examples of the application of each strategy are presented and compared, and conclusions are drawn regarding the relative utility of each method for different types of experiments and systems studied.

Key words Self-association – hetero-association

Introduction

The measurement of concentration gradients at sedimentation equilibrium has proven to be a powerful method for the quantitative characterization of biomolecular associations in solution. Until now, both theoretical work and experimental applications have been primarily directed at the detection and analysis of self-association in solutions containing a single sedimenting solute component. However, it is becoming increasingly appreciated that by far the greater number of associations of biological interest fall into the category of hetero-association. During the last ten years, efforts have been made in several laboratories, including our own, to develop robust methods for the detection and quantitative characterization of hetero-associations in solutions containing two or more interacting solute components. The purpose of the present work is to review and compare some of these methods.

The present communication is organized as follows. In the next section, sedimentation equilibrium in a solution of two solute components is formally described in order to review basic concepts and introduce notation to be utilized subsequently. In the subsequent section the concepts of primary and secondary data are introduced and the transformation of primary to secondary data described. This is followed by two sections in which several procedures for modeling primary and secondary data with an equilibrium association model are described. In the final section the relative advantages and disadvantages of the various methods are compared.

Chemical and sedimentation equilibrium in an associating system[1]

Consider a solution containing two solute components A and B. The solution may contain monomeric A and B in equilibrium with one or more homo- and/or hetero-oligomeric species. The molecular species of composition A_iB_j will be denoted by $\{ij\}$, where either i or j, but not both, may be zero,[2] and a property X of species A_iB_j will be denoted by X_{ij}. We shall refer to any experimentally measurable property of an individual species as a *signal*, and the magnitude of one particular signal (out of many possible signals) associated with species $\{ij\}$ will be denoted by $s_{k,ij}$, where k is the index of the signal.[3] In the present work we shall limit ourselves to consideration of those signals, the strength of which is directly proportional to the weight/volume concentration of the contributing species, viz.

$$s_{k,ij} = \alpha_{k,ij} w_{ij} \, . \tag{1}$$

We define the equilibrium association constant for formation of species $\{ij\}$ with respect to w/v concentrations as follows:

$$K_{ij}^w = \frac{w_{ij}}{w_{10}^i w_{01}^j} \, . \tag{2}$$

This equilibrium constant is linearly related to the equilibrium constants defined with respect to molar concentrations and signal units:

$$K_{ij}^c \equiv \frac{c_{ij}}{c_{10}^i c_{01}^j} = \frac{M_{10}^i M_{01}^j}{M_{ij}} K_{ij}^w \, , \tag{3}$$

$$K_{ij}^{(k)} \equiv \frac{s_{k,ij}}{s_{k,10}^i s_{k,01}^j} = \frac{\alpha_{k,ij}}{\alpha_{k,10}^i \alpha_{k,01}^j} K_{ij}^w \, . \tag{4}$$

Equation (4) serves to remind us that $K_{ij}^{(k)}$ varies with the particular measurement (e.g., the wavelength at which absorbance is measured), whereas K_{ij}^w and K_{ij}^c are constant at constant temperature, pressure and other environmental variables such as solution pH and ionic strength.

---

[1] In this section relations originally presented elsewhere [1, 2] are summarized and some additional notation introduced.
[2] Pure monomeric A and pure monomeric B are thus denoted by $\{10\}$ and $\{01\}$, respectively.
[3] For example, the absorbance of $\{ij\}$ at two different wavelengths would be denoted by $s_{1,ij}$ and $s_{2,ij}$.

If all species sediment ideally, then at sedimentation equilibrium

$$w_{ij}(r) = w_{ij}(r_{\text{ref}}) \exp\left[\phi_{ij}(r^2 - r_{\text{ref}}^2)\right]$$

$$= K_{ij}^w w_{10}(r_{\text{ref}})^i w_{01}(r_{\text{ref}})^j \left[\phi_{ij}(r^2 - r_{\text{ref}}^2)\right] \, , \tag{5}$$

where r denotes the radial position, r_{ref} denotes an arbitrarily selected reference position, and

$$\phi_{ij} = \frac{M_{ij}(1 - \bar{v}_{ij}\rho)\omega^2}{2RT} \, , \tag{6}$$

where M_{ij} and $\bar{v}_{ij}$, respectively, denote the molar mass and partial specific volume of $\{ij\}$, ρ the solvent density, ω the rotor velocity, R the molar gas constant, and T the absolute temperature. Ordinarily the partial specific volume of each species is given to a good approximation by the weight average [1]

$$\bar{v}_{ij} = f_{A,ij} \bar{v}_A + f_{B,ij} \bar{v}_B \, , \tag{7}$$

where $\bar{v}_x$ denotes the partial specific volume of pure component X and $f_{X,ij}$ is the mass fraction of component X in species A_iB_j:

$$f_{A,ij} = \frac{iM_{10}}{iM_{10} + jM_{01}} \, , \tag{8a}$$

$$f_{B,ij} = 1 - f_{A,ij} \, . \tag{8b}$$

Combination of Eqs. (6) and (7) yields

$$\phi_{ij} = \frac{(iM_A^* + jM_B^*)\omega^2}{2RT} \, , \tag{9}$$

where $M_A^* \equiv M_{10}(1 - \bar{v}_A\rho)$ and $M_B^* \equiv M_{01}(1 - \bar{v}_B\rho)$. The total concentration of X at each radial position is given by

$$w_X(r) = \sum_{i,j} f_{X,ij} w_{ij}(r) \, . \tag{10}$$

Primary and secondary data

Primary data

The set of primary data acquired from a sedimentation equilibrium experiment consists of the dependence of one or more signals, or experimentally measurable properties of a solution, upon the distance from the axis of rotation in the centrifuge. The signals must be known functions of the solute composition of the solution. The value of a particular signal s_k at radial position r is given by

$$s_k(r) = \sum_{i,j} s_{k,ij}(r) + \delta_k(r) \, , \tag{11}$$

Progr Colloid Polym Sci (1997) 107:11–19
© Steinkopff Verlag 1997

where δ_k is the baseline level of signal measured in the absence of all solute components. To simplify the following analysis, it will be assumed that the baseline may be independently measured (or estimated) and the signal corrected appropriately.

Combination of Eqs. (1) and (11) yields

$$S_k(r) \equiv s_k(r) - \delta_k(r) = \sum_{i,j} \alpha_{k,ij} w_{ij}(r) \ . \tag{12}$$

In the present work we will confine our attention to a particular class of signals, namely those for which individual species signals are a weight-average of component signals

$$\alpha_{k,ij} = \alpha_{k,A} f_{A,ij} + \alpha_{k,B} f_{B,ij} \ , \tag{13}$$

where $\alpha_{k,x}$ is a composition-independent[4] constant of proportionality between the w/v concentration of component X and the contribution of X to the kth signal. As an example, assumption of the validity of Eq. (13) in the case of absorbance as a signal would be conditional on the absence of detectable hyper- or hypochromism in mixtures of A and B. Combination of Eqs. (10), (12) and (13) yields

$$S_k(r) = \alpha_{k,A} w_A(r) + \alpha_{k,B} w_B(r) \ . \tag{14}$$

Secondary data

In the present work we shall consider two useful classes of secondary data. The first is the concentration gradient of an individual component within a two-component solution, i.e., $w_A(r)$ and/or $w_B(r)$. The gradient of a single component may, of course, be obtained trivially from a single signal $S_1(r)$ in the special case that either $\alpha_{1,A}$ or $\alpha_{1,B}$ is zero. This may be accomplished when, for example, only one of the two components contains a chromophore that absorbs in the visible [3], or contains a radiolabel [4].

If both components contribute to each of two measured signals, then one can solve the two simultaneous equations

$$S_1(r) = \alpha_{1,A} w_A(r) + \alpha_{1,B} w_B(r) \ , \tag{15a}$$

$$S_2(r) = \alpha_{2,A} w_A(r) + \alpha_{2,B} w_B(r) \tag{15b}$$

for $w_A(r)$ and $w_B(r)$ as functions of the independently measured α coefficients, $S_1(r)$, and $S_2(r)$, provided that $\alpha_{1,A}/\alpha_{2,A} \neq \alpha_{1,B}/\alpha_{2,B}$ [2]. This method suffers from the disadvantage that if the two signals are not selected with careful attention to the relative magnitudes of the α coefficients, the solution may be very sensitive to quite small errors in the measured quantities. More robust estimates

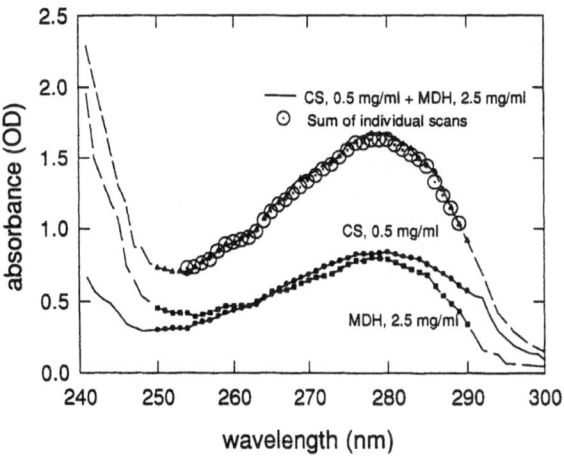

Fig. 1 UV spectra of three solutions containing citrate synthase (CS), malate dehydrogenase (MDH), and a mixture of the two proteins. Curves represent wavelength scans taken in double-sector cells in the XLA analytical ultracentrifuge at a radial position midway between meniscus and base at 1000 rpm, prior to centrifugation to sedimentation equilibrium at higher rotor velocity. The data points represent regular arrays of data obtained at 1 nm increments by filling in absorbance values at skipped wavelengths by linear interpolation. Agreement between the sum of the two scans of individual components and the scan of the solution mixture indicates that Eq. (14) is valid for this system. The program EXCOEF simultaneously fits Eq. (14) to these three sets of data at each wavelength λ to obtain best-fit values of $\alpha_{CS,\lambda}$ and $\alpha_{MDH,\lambda}$.

of $w_A(r)$ and $w_B(r)$ may be obtained from measurement of the gradients of more than two signals per sample, followed by a least-squares solution of a set of more than two linearly independent equations of the form of Eq. (14) [5–7]. The program EXCOEF utilizes wavelength absorbance scans of solution mixtures obtained on the XLA analytical ultracentrifuge to facilitate the calculation of effective extinction coefficients of each of two components. The program TWOCOMP uses the extinction coefficients so obtained to calculate individual component concentration gradients from radial absorbance scans of solution mixtures taken at multiple wavelengths via linear least-squares analysis [8]. Examples of data processing by EXCOEF and TWOCOMP are presented and described in Figs. 1 and 2.

The second type of secondary data to be discussed is the signal average buoyant molar mass, defined as follows:

$$M_k^* \equiv \frac{\sum_{i,j} s_{k,ij} M_{ij} (1 - \bar{v}_{ij} \rho)}{\sum_{i,j} s_{k,ij}} \ . \tag{16}$$

Combination of Eqs. (1), (7), (8), (13), and (16) yields

$$M_k^* = \frac{\sum_{i,j} (\alpha_{k,A} f_{A,ij} + \alpha_{k,B} f_{B,ij}) w_{ij} (i M_A^* + j M_B^*)}{\sum_{i,j} (\alpha_{k,A} f_{A,ij} + \alpha_{k,B} f_{B,ij}) w_{ij}} \tag{17}$$

---

[4]I.e., independent of the concentration and state of association of component X.

Fig. 2 Gradients of individual
components calculated by
TWOCOMP from radial scans
of a mixture of CS (0.5 mg/ml)
and MDH (2.5 mg/ml) at two
wavelengths (triangles), three
wavelengths (squares), and five
wavelengths (circles). Note that
although the scatter of points in
a calculated gradient decreases
with increasing number of
wavelengths analyzed, the basic
shape of the gradient remains
constant.

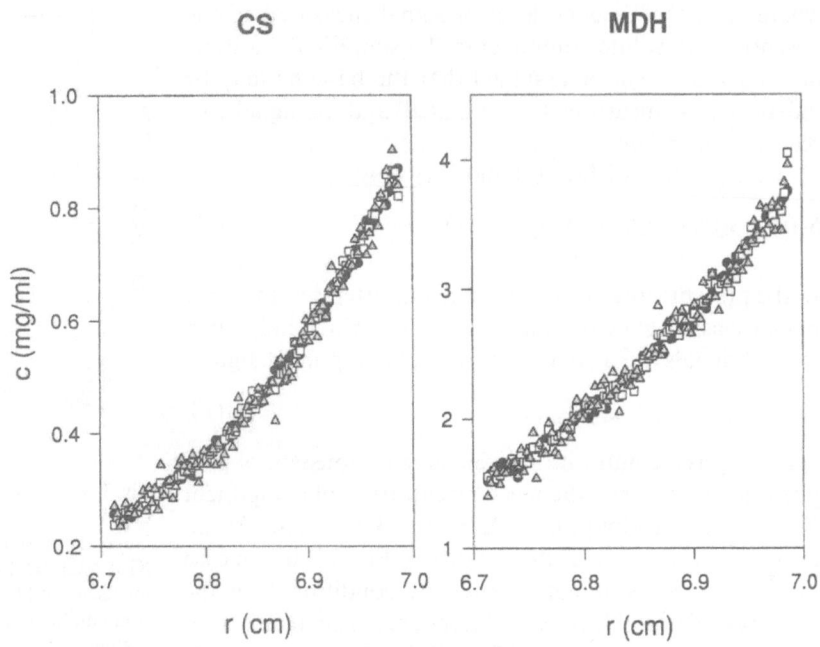

Fig. 2 Gradients of individual components calculated by TWOCOMP from radial scans of a mixture of CS (0.5 mg/ml) and MDH (2.5 mg/ml) at two wavelengths (triangles), three wavelengths (squares), and five wavelengths (circles). Note that although the scatter of points in a calculated gradient decreases with increasing number of wavelengths analyzed, the basic shape of the gradient remains constant.

the value of which is readily calculable given experimentally measured values of $\alpha_{k,A}$, $\alpha_{k,B}$, M_A^*, and M_B^*, together with an association model that specifies the values of w_{ij}. It may be shown [1] that when the solution is thermodynamically ideal with respect to all sedimenting solutes, and Eqs. (7) and (14) are valid, then the signal average buoyant molar mass may be obtained from experiment by

$$M_k^*(r') = \frac{RT}{\omega^2} \frac{1}{r'} \left(\frac{d \ln S_k(r)}{dr} \right)_{r=r'}. \tag{18}$$

It follows from Eq. (18) that, in the most general case, the evaluation of M_k^* requires numerical differentiation of the primary data. Under favorable conditions one can experimentally evaluate $S_k(r)$ and $M_k^*(r)$ over a range of compositions occurring in a single sample between the meniscus and base at sedimentation equilibrium. As described above, transformation of the primary data obtained from multiple signals $S_1(r)$, $S_2(r)$, ... , permits the evaluation of $w_A(r)$ and $w_B(r)$. In this fashion one may obtain the dependence of M_1^*, M_2^*, ... , upon w_A and w_B over the range of solution compositions found in a single sample.

A simple alternative method for obtaining the dependence of M_1^*, M_2^*, ... , upon w_A and w_B is as follows. Under the conditions of many sedimentation equilibrium experiments (e.g., those carried out at lower speeds or using short solution columns), the radial dependence of the composition is so small that M_k^* is almost invariant throughout the cell and may be treated as a single sample-average value

$\langle M_k^* \rangle$ that is characteristic of the loading composition of the solution contained therein. Under such conditions Eq. (18) may be integrated to yield [4]

$$S_k(r) = S_k(r_{\mathrm{ref}}) \exp \left[\frac{\langle M_k^* \rangle \omega^2}{2RT} (r^2 - r_{\mathrm{ref}}^2) \right] \tag{19}$$

and the value of M_k^* corresponding to the known loading composition $\{w_{A,\mathrm{load}}, w_{B,\mathrm{load}}\}$, i.e.,$\langle M_k^* \rangle$, may be obtained via a simple least-squares fit of a straight line to a plot of $\ln S_k(r)$ against r^2.[5] By means of this procedure one may obtain the values of M_1^*, M_2^*, ... for a single (loading) composition. Each additional sample provides an additional composition. An example of the evaluation of M_1^* at three solution compositions, taken from the literature, is presented in Fig. 3.

Although at first glance it may seem that the amount of secondary data obtained from a single experiment by means of this simplified procedure is much smaller than that obtained via the general procedure, one should keep in mind the following points:

1. As many as nine short-column experiments may be run concurrently in a single analytical centrifuge run, and potentially many more samples may be run concurrently using preparative ultracentrifuges in an analytical mode [9, 10].

---

[5] The degree of goodness of the straight line fit will establish whether it is legitimate to approximate M_k^* by a constant $\langle M_k^* \rangle$ across the entire sample cell in a particular experiment.

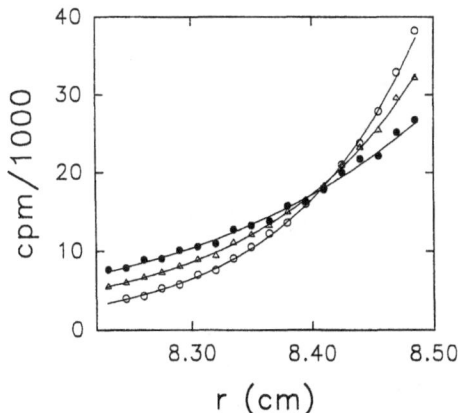

Fig. 3 Gradients of radiolabeled complement subcomponent C1r (3.0 μM) measured in the presence of three concentrations of unlabeled C1s (solid circles, 10 nM; triangles, 1 μM; open circles, 3 μM). Solid curves represent the best fits of Eq. (19) to each of the gradients; the corresponding best fit values of $M_{w,C1r}^*$ are 50.6, 73.4, and 92.8 ($\pm$ 5%) kg/mol. Reproduced with permission from [4].

2. Short-column experiments ordinarily come to equilibrium much more rapidly than long-column experiments, and, depending upon conditions, may be completed in as little as two to three hours [11], permitting two or three analytical centrifuge runs (as many as 27 samples) per day.

3. Each sample may yield multiple signal-average buoyant molar masses $M_1^*, M_2^*, \ldots$, each of which serves as an additional data point in the modeling process (see below).

If a particular signal k' is associated with a single component X (i.e. $\alpha_{k',Y} = 0$ for $Y \neq X$) then it follows from Eq. (17) that *for that particular signal* the signal average buoyant molar mass is equal to the weight average buoyant molar mass of component X:

$$M_{k'}^* = M_{w,X}^*, \tag{20}$$

where

$$M_{w,X}^* \equiv \frac{\sum_{i,j} f_{X,ij} w_{ij} (i M_A^* + j M_B^*)}{w_X}. \tag{21}$$

A further simplification obtains when $\bar{v}_A = \bar{v}_B$. Then to a very good approximation [4]

$$M_{w,X}^* \simeq M_{w,X}(1 - \bar{v}\rho), \tag{22}$$

where $\bar{v} \equiv (\bar{v}_A + \bar{v}_B)/2$ and

$$M_{w,X} = \frac{\sum_{i,j} f_{X,ij} w_{ij} (i M_{10} + j M_{01})}{w_X}. \tag{23}$$

Equation (22) is exact in the limit $\bar{v} = \bar{v}_A = \bar{v}_B$.

Modeling primary data with an association scheme

Using Eq. (5) together with Eqs. (13)–(17), one can calculate $S_k(r)$ for a single sample (or "experiment") as a function of $\alpha_{k,A}$, $\alpha_{k,B}$, M_A^*, M_B^*, a finite set of non-zero K_{ij}^w defining a particular association scheme,[6] and the reference concentrations $w_{10}(r_{ref})$ and $w_{01}(r_{ref})$. The first four of these parameters are independently measurable in principle, whereas the K_{ij} and reference concentrations are not. These latter parameters are termed adjustable or floating parameters.

We now define a chi-square (normalized error) function measuring the overall discrepancy between calculated and observed values for a single experiment

$$\chi^2|_k^E \equiv \sum_{n=1}^{n_E} \left[\frac{S_{k,E}^{meas}(r_n) - S_{k,E}^{calc}(r_n)}{\sigma_{k,E}(r_n)} \right]^2, \tag{24}$$

where E is an index identifying the particular experiment, n is an index of data points within experiment E, $S_{k,E}^{meas}(r_n)$ is the experimentally measured value of S_k for $r = r_n$ in E, $S_{k,E}^{calc}(r_n)$ is the calculated value of S_k for $r = r_n$ in E, and $\sigma_{k,E}(r_n)$ is the standard deviation of the value of $S_{k,E}^{meas}(r_n)$. Least squares modeling of a single experiment is accomplished by minimizing the value of $\chi^2|_k^E$ with respect to variation in the floating parameters.

Consider now a set of experiments conducted on several solutions containing A and B, but at different loading concentrations of each of the solute components. One may simultaneously model all signals from all experiments with the same set of parameters given above, except that a distinctive set of reference concentrations, denoted by $w_{10}(E, r_{ref})$ and $w_{01}(E, r_{ref})$ applies to each experiment. Modeling of the combined data is accomplished by minimization of the compound χ^2 function [1]

$$\chi^2 = \sum_E \sum_k \chi^2|_k^E. \tag{25}$$

Additional experimental information, such as knowledge of the loading concentrations of each component, sample volumes, loading values of S_k and/or independently measured values of the $\alpha_{k,X}$, can provide constraints derived from considerations of conservation of mass [1, 12, 13] or conservation of signal [2]. These constraints may be used to eliminate some or all of the reference concentrations as independently floating parameters, thereby reducing substantially the number of floating parameters and simplifying the modeling process. We strongly recommend that such constraints be employed to eliminate reference

---

[6] For example, the association scheme $A + A \rightleftharpoons A_2$; $A + B \rightleftharpoons AB$; $AB + A \rightleftharpoons A_2B$ would be defined by non-zero values of K_{20}, K_{11}, and K_{21}.

concentrations as floating parameters in the analysis of equilibrium gradients, because the values of reference concentrations obtained via least-squares modeling correlate highly with the value of other parameters such as the K_{ij}^w, rendering the evaluation of equilibrium constants problematic [14].

The computer programs 2C1SFIT and 2CSFIT (2 components, 1 or 2 signal fit) implement conservation of mass constraints to globally model one or two signals from each of several concurrent experiments. Sample results obtained using 2C1SFIT are presented in Fig. 4, and comparable results obtained using 2C2SFIT are presented in Fig. 5.

Modeling secondary data with an association scheme

Individual component concentration gradients, calculated from multiple signal gradients as described in the section on chemical and sedimentation equilibrium, may be modeled with an association scheme using the program 2C2SFIT by setting the proportionality constants $\alpha_{1,1} = \alpha_{2,2} = 1$ and $\alpha_{1,2} = \alpha_{2,1} = 0$, resulting in $S_1(r) = w_1(r)$ and $S_2(r) = w_2(r)$. Results comparable to those shown in Fig. 5 are obtained.[7]

The dependence of the value of M_k^* on the composition of the solution (w_A, w_B) is modeled as follows. Let each discrete solution composition ("data point") for which a value of M_k^* has been obtained be denoted $\{w\}_n$, where n is an index of data points. We define the least-squares error function

$$\chi^2 \equiv \sum_k \sum_n [M_{k,\text{expt}}^*(\{w\}_n) - M_{k,\text{calc}}^*(\{w\}_n)]^2 , \tag{26}$$

where $M_{k,\text{calc}}^*$ is given by either Eq. (17) or Eqs. (20)–(21) as appropriate. The error function χ^2 defined by Eq. (26) is thus a function of four parameters (M_A^*, M_B^*, $\alpha_{k,A}$, and $\alpha_{k,B}$) that are in principle independently determinable, and the indeterminate floating parameters K_{ij} characterizing a particular reaction scheme. A particular association model, specifying a limited number of oligomeric species $\{ij\}$ in equilibrium with the reactant species $\{10\}$ and $\{01\}$, is fit to this set of secondary data by minimizing χ^2 with respect to variation in the K_{ij} and any other undetermined para-

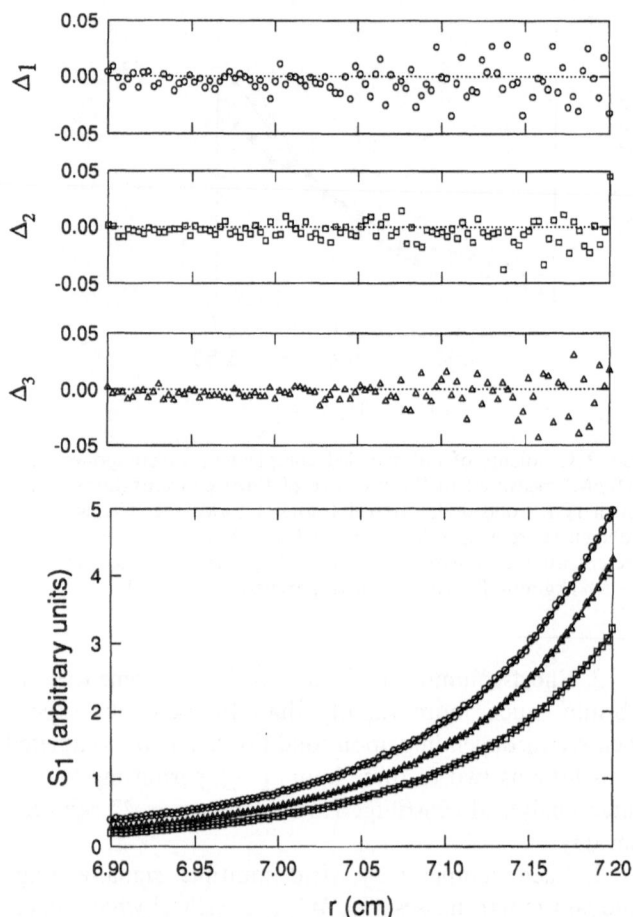

Fig. 4 Direct modeling of gradients of a single experimental signal (S_1) with an association model using 2C1SFIT. Data are simulated using the program 2COMPSIM with the following parameters: $M_A = 90\,000$ g/mol; $M_B = 62\,000$ g/mol; $\bar{v} = \bar{v}_A = \bar{v}_B = 0.73$ ml/g; $\alpha_{1,A} = 1.70$; $\alpha_{1,B} = 0.32$; $T = 20\,°C$; RPM = 10000; $\log K_{11}^w = 0.5$; all other K's negligible. Sample composition $\{w_{A,\text{load}}, w_{B,\text{load}}\}$ (g/l): circles, $\{0.5, 2.5\}$; squares, $\{0.5, 0.5\}$; triangles, $\{0.5, 1.5\}$. Normally distributed pseudo-noise with $\sigma = 0.003 + 0.005 S_1$ is added to the calculated gradients [1]. Global nonlinear least-squares modeling of all three gradients was carried out using conservation of mass constraints, assuming prior knowledge of the loading composition and the loading value of S_1 for each sample, but no assumed prior knowledge of α coefficients, thus eliminating reference concentrations as adjustable parameters. Best fit values of the three floating parameters were: $M_A = 89\,880$ g/mol; $M_B = 61\,240$ g/mol; $\log K_{11}^w = 0.507$. Best-fit gradients are plotted as solid lines through the data; residuals are plotted against r separately for each simulated data set.

meters.[8] An example of modeling the experimentally measured dependence of M_k^* upon w_A and w_B, taken from the literature, is presented in Fig. 6.

---

[7] Alternatively, the program 2COMPFIT globally models a set of component concentration gradients using constraints derived from conservation of signal rather than conservation of mass. Application of this analysis might be preferred when doubts exist as to the validity of the assumption of conservation of mass; i.e., when solute species may be denaturing and precipitating over the time course of the experiment.

---

[8] As pointed out above, if a signal S_k is proportional to the total concentration of only one of the two solute components, M_k^*, and hence χ^2, becomes independent of the α coefficients.

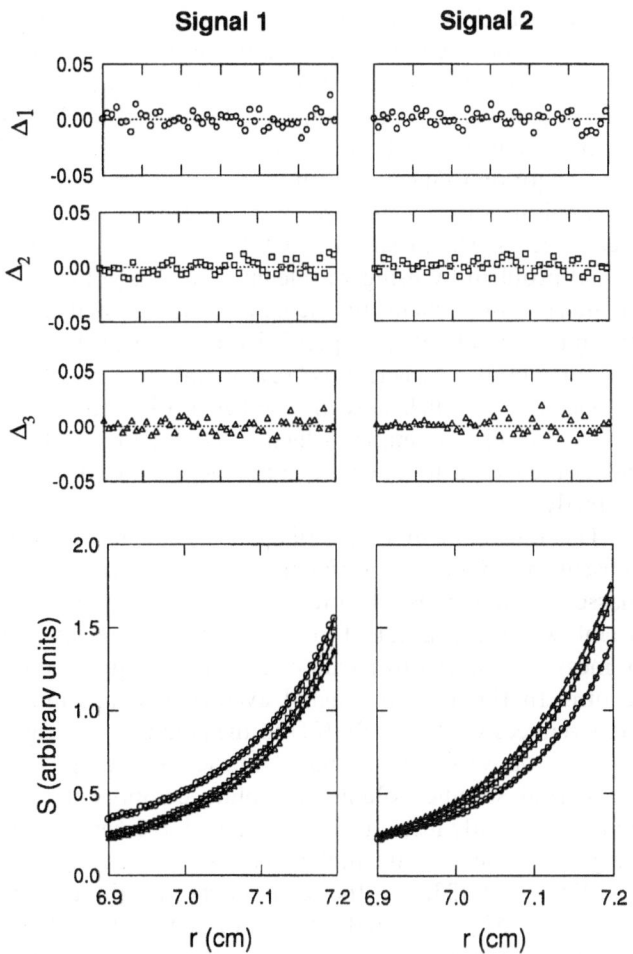

Fig. 5 Direct modeling of gradients of two experimental signals (S_1 and S_2) with an association model using 2C2SFIT. Data are simulated using the program 2COMPSIM with the following parameters: $M_A = 68\,000$ g/mol; $M_B = 34\,000$ g/mol; $\bar{v} = \bar{v}_A = \bar{v}_B = 0.73$ ml/g; $\alpha_{1,A} = 0.5$; $\alpha_{1,B} = 1.0$; $\alpha_{2,A} = 1.0$; $\alpha_{2,B} = 0.5$; $T = 20\,°C$; RPM = 10000; $\log K_{11}^w = 0.5$; $\log K_{12}^w = 0.8$; all other K's negligible. Sample composition $\{w_{A,load}, w_{B,load}\}$ (g/l): circles, $\{0.3, 0.6\}$; squares, $\{0.5, 0.4\}$; triangles, $\{0.6, 0.3\}$. Normally distributed pseudo-noise with $\sigma_k = 0.003 + 0.005\,S_k$ is added to the calculated gradients [1]. Global nonlinear least-squares modeling of all six gradients (three samples, 2 signals/sample) was carried out using conservation of mass constraints, assuming prior knowledge of the loading composition and the loading values of S_1 and S_2 for each sample, but no assumed prior knowledge of α coefficients, thus eliminating reference concentrations as adjustable parameters. The molar masses of the two monomeric components were constrained to their known values (which may be determined from experiments on the individual components). Minimization of the compound χ^2 with respect to variation in the values of all $\log K_{ij}$ for $i, j \leqslant 2$ returned the following best-fit values: $\log K_{20}^w = -11.3$; $\log K_{02}^w = -1.74$; $\log K_{11}^w = 0.476$; $\log K_{21}^w = -3.53$; $\log K_{12}^w = 0.796$; $\log K_{22}^w = -5.81$. Best-fit gradients are plotted as solid lines through the data; residuals are plotted against r separately for each simulated data set.

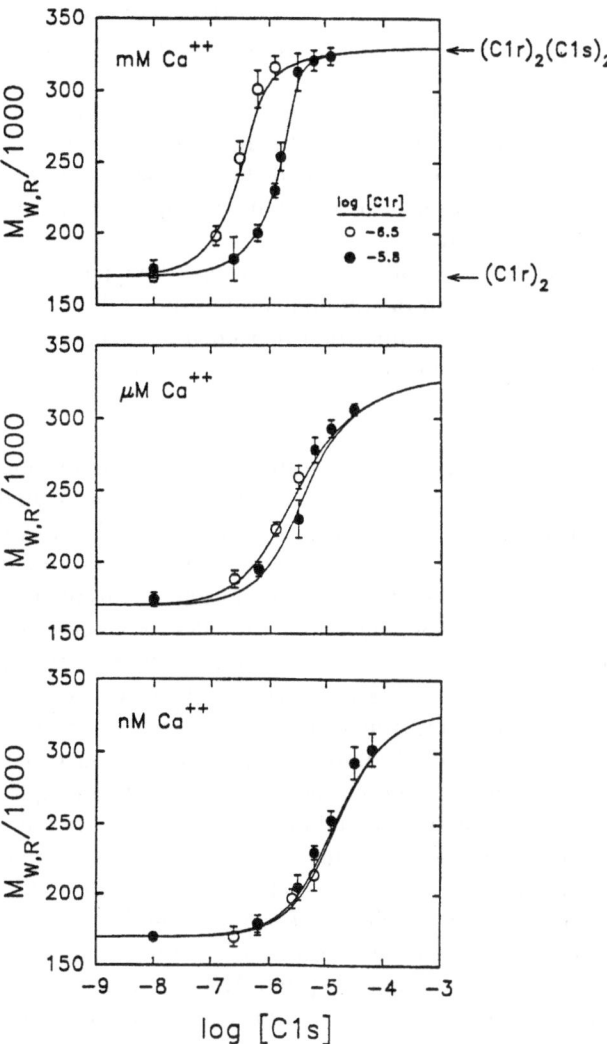

Fig. 6 Dependence of the weight-average molar mass of labeled complement subcomponent C1r [text Eq. (23)] upon the concentrations of C1r and unlabeled complement subcomponent C1s. Individual graphs present data obtained in the presence of three different calcium concentrations. Each graph contains data obtained at two different C1r concentrations and as many as nine different C1s concentrations. Curves represent the calculated best fit of a heteroassociation model of the following form: $(C1r)_2 + C1s \rightleftarrows (C1r)_2(C1s)$; $(C1r_2)(C1s) + C1s \rightleftarrows (C1r)_2(C1s)_2$. Figure is modified from [4]. Details of the model and fitting procedure are given in the cited publication.

Discussion

In the present work we have focused on four methods for the quantitative characterization of association equilibria in ideal solutions containing two sedimenting solute components:

(1) Direct modeling of the equilibrium gradients of a single signal measured in multiple samples of differing composition;

(2) Direct modeling of the equilibrium gradients of two (or more) signals measured in multiple samples of differing composition;

(3) Calculation of the independent component concentration gradients via analysis of the gradients of two or more signals, followed by modeling of the component concentration gradients from multiple samples of differing composition; and

(4) Calculation of the signal-average buoyant molar mass as a function of solution composition, followed by modeling of the composition dependence of the signal-average buoyant molar mass in the context of an association model.

The first three of these methods are similar in concept, and differ primarily in the number of individual signal gradients measured per sample. Of the three, method 1 requires the least data and is the simplest to implement. Under favorable circumstances this method can provide reasonably good estimates of association constants. Method 1 is probably the method of choice when one has prior knowledge of the values of M_A^* and M_B^* (obtainable from experiments conducted on solutions of the individual components), and has a good idea of the correct association scheme. As the number of samples of differing composition and the number of linearly independent signals/sample acquired increase, so does the ability of the investigator to discriminate between alternative possible association schemes. For the example presented in Fig. 5, given two signals from each of three samples, it was possible to evaluate M_A^* and M_B^* approximately, as well as identifying the significant species present ($\{10\}$, $\{01\}$, $\{11\}$, and $\{12\}$) without any prior knowledge whatsoever. Given prior knowledge of the values of M_A^* and M_B^*, modeling of the solution mixture data provided more accurate estimates of $\log K_{11}^w$ and $\log K_{12}^w$ than did modeling a single signal from each of the three samples. Modeling of component gradients rather than signals is advantageous when more than two signals are collected per sample (see, e.g., the data of [7]), and/or when the validity of the assumption of conservation of mass is in doubt (footnote 7).

When two or more signals are collected per sample, calculation and modeling of derived component gradients, as opposed to direct modeling of signals, may also be advantageous to users of the XLA analytical ultracentrifuge for an entirely different but eminently practical reason. When the XLA is programmed to automatically collect radial absorbance scans at two or more wavelengths from each of two or more sample cells in a rotor, the instrument occasionally collects data at an unpredictable wavelength that may differ from the programmed wavelength by up to 2 nm. Moreover, the wavelength at which the scan is actually performed may vary between different sample cells in the same rotor. If an extinction coefficient of one or both solute components varies significantly (i.e., by more than 2% or so) over a wavelength spread of 2 nm, then a lack of wavelength reproducibility can cause systematic errors in the analysis due to the use of significantly inaccurate α coefficients. The program EXCOEF deals with this potential problem by using wavelength scans together with an interpolation scheme to calculate extinction coefficients at 1 nm intervals over a specified range of wavelengths. The extinction coefficient corresponding to the *actual* wavelength at which a scan is taken (rather than that assumed in programming), which is recorded in the data file, is then used to calculate component concentrations correctly.

The fourth method described here, calculation and subsequent modeling of signal-average buoyant molar masses as functions of solution composition, is most useful when the gradient of a single component can be measured directly by means of a label or unique spectral feature. In this case the signal average buoyant molar mass reduces via Eqs. (19)–(21) to the component weight average buoyant molar mass $M_{w,X}^*$. Assuming that the partial specific volumes of solute components can be independently measured or calculated, the component weight-average buoyant molar mass is readily converted via Eqs. (22)–(23) to the component weight-average molar mass $M_{w,X}$. Given the masses of each of the isolated solute components (individual polypeptide chains, nucleic acids, etc.), knowledge of the dependence of $M_{w,X}$ on solution composition provides valuable guidance in the construction of model reaction schemes. For example, the data shown in Fig. 6 indicate, in an entirely model-independent fashion, that in the absence of C1s, C1r is present as a dimer, and that in the presence of saturating amounts of C1s, C1r exists as a heterotetramer with composition $(C1r)_2(C1s)_2$. This is a natural and intuitive guide to the two-step association scheme subsequently used to model the data quantitatively [4]. Additionally, analysis of the dependence of the component weight-average molar mass (or, more generally, the *apparent* component weight-average molar mass) on solution composition may be used, in principle, to characterize hetero-associations in solutions of nonideally sedimenting solutes, although this has so far been done only for simulated model systems [15].

The programs mentioned by name in this report have been written for IBM-compatible personal computers running under DOS, and may be obtained from the author upon request. Those programs (2C1SFIT, 2C2SFIT, 2COMPFIT) that have been used to model either primary or secondary data in the context of equilibrium association

Progr Colloid Polym Sci (1997) 107:11–19
© Steinkopff Verlag 1997

schemes currently allow for the potential presence of the following species: {10}, {01}, {20}, {02}, {11}, {21}, {12}, and {22}. An extended version of 2COMPFIT that allows for the presence of the additional species {30}, {03}, {31}, {13}, {32}, {23}, and {33} has been completed but not fully tested as of the date of writing.

References

1. Hsu C, Minton AP (1991) J Mol Recognition 4:93–104
2. Minton AP (1994) In: Schuster TM, Laue TM (eds) Modern Analytical Ultracentrifugation. Birkhäuser, Boston, pp 81–93
3. Laue TM, Senear DF, Eaton S, Ross JBA (1993) Biochemistry 32:2469–2472
4. Rivas G, Ingham KC, Minton AP (1994) Biochemistry 33:2341–2348
5. Press WH, Flannery BP, Teukolsky SA, Vertterling WT (1986) Numerical Recipes: The Art of Scientific Computing, Cambridge University Press, Cambridge
6. Henry ER, Hofrichter J (1992) Meth Enzymol 210:129–192
7. Lewis MS, Shrager RI, Kim S-J (1994) In: Schuster TM, Laue TM (eds) Modern Analytical Ultracentrifugation. Birkhäuser, Boston, pp 94–115
8. Bailey MF, Davidson BE, Minton AP, Sawyer WH, Howlett GJ (1996) J Mol Biol 263:671–684
9. Minton AP (1989) Anal Biochem 176:209–216
10. Darawshe S, Minton AP (1994) Anal Biochem 220:1–4
11. Laue TM (1992) Beckman Instruments Technical Information Bulletin DS-835
12. Nichol LW, Ogston AG (1965) J Phys Chem 69:4365–4367
13. Lewis MS (1991) Biochem 30:11716–11719
14. Brooks I, Wetzel R, Chan W, Lee G, Watts DG, Soneson KK, Hensley P (1994) In: Schuster TM, Laue TM (eds) Modern Analytical Ultracentrifugation. Birkhäuser, Boston, p 351
15. Chatelier RC, Minton AP (1987) Biopolymers 26:1097–1113

Progr Colloid Polym Sci (1997) 107:20–26
© Steinkopff Verlag 1997

E.K. Dimitriadis
M.S. Lewis

Non-linear curve-fitting methods for data from the XL-A analytical ultracentrifuge

Dr. E.K. Dimitriadis (✉) · M.S. Lewis
Biomedical Engineering and
Instrumentation Program
NCRR/NIH
Building 13, Room 3N17
13 SOUTH DR. MSC 5766
Bethesda, Maryland 20892-5766, USA

Abstract In equilibrium or velocity sedimentation experiments with the XL-A analytical ultracentrifuge it is customary to acquire absorbance data. The least-squares method is most widely used for fitting non-linear models to such collected data. It is here shown that due to the non-Gaussian characteristics of the noise in the absorbance data, the least-squares method is not optimal and introduces a systematic bias to the estimated parameters. This bias can be eliminated by either using the maximum-likelihood method on the absorbance data or otherwise by fitting the intensity data directly. The probability distribution of the noise in the latter is Gaussian and the least-squares estimation is equivalent to maximum likelihood. The methodology for using the intensity data is developed and simulations for a variety of systems are performed.

Key words Ultracentrifugation – Gaussian noise – curve-fitting – least-squares estimation – maximum-likelihood estimation

Introduction

The optical system of the XL-A analytical ultracentrifuge is that of a dual beam spectrophotometer which measures the intensity of transmitted light through the pair of sectors of every cell. One sector contains a buffer solution which is used as reference and which is usually very transparent to the wavelengths of the light used; the other sector contains the same buffer plus one or more species of solutes, such as proteins, peptides or oligonucleotides, whose behavior is to be investigated. Because of instrument noise, the sectors are scanned a number of times at every radial position. The intensity data are then converted to absorbance by taking the base-10 logarithm of the ratio between the reference intensity and the intensity transmitted through the solution sector and the computed absorbance data at every position are averaged. The acquired data correspond to the profile of solute concentrations as a function of radial position in the sector.

To investigate the state of the solution, solute behavior models are assumed and corresponding mathematical models for the expected concentration profiles are written. The efficacy of the assumed models is then tested by attempting to curve-fit the mathematical models to the collected absorbance data. Usually, quantities such as molecular masses or association constants are used as the fitting parameters. The final decision for the most probable behavior of the solute in the particular solution is based on the quality of the fit as described by various statistical measures. The method used most often for curve fitting is that of least squares [1] which seeks estimates of the unknown model parameters so that the sum of squares of the deviations of the actual data from the model curve is minimized. In addition, it is known that optimal least-squares estimation is achieved if the data are weighted by

Progr Colloid Polym Sci (1997) 107: 20–26
© Steinkopff Verlag 1997

the inverse of the variance. Hence, weighted least squares has been the method of choice. This, however, requires an iterative scheme as the optimal weights are not known at the outset. It is known from estimation theory [2] that the above method gives unbiased estimations of the unknown parameters under one of the following two conditions: either the noise probability distribution is Gaussian or else, no statistical information about the characteristics of the noise is available.

Fortunately, the design of the XL-A permits the acquisition of raw intensity data, and if enough samples are collected it should be possible to examine the probability distribution of the noise in some detail. We have performed such measurements and found that while the intensity is contaminated by Gaussian noise, the absorbance noise is not normally distributed. The fundamental statistical principles explaining this observation are presented and the non-Gaussian distribution for the absorbance data is derived. The conclusion is that the least-squares method estimates are statistically biased. Unbiased estimations can be achieved either by fitting the intensity data directly or, since the absorbance noise characteristics are known, by using the maximum likelihood method. The latter is equivalent to least squares only when the noise is Gaussian but otherwise it should give better, more reliable estimates of the model parameters than least squares would give in this case.

Methods

The "method scan" procedure in the XL-A acquisition software was invoked to collect a large number of intensity data at each of a number of radial locations with the purpose of establishing the noise characteristics of the instrument. Four-hole rotors were used where, in each of the three cells one of the sectors was loaded with 185 μl phosphate-buffered saline (PBS) solution and the other with 180 μl of one of three concentrations of equine myoglobin in PBS. It is expected that myoglobin will not self-associate. A total of 99 samples of transmitted intensity were collected from each sector and at about 30 radial positions between the meniscus and the sector bottom. The collected data are shown in Fig. 1 where the scatter is seen to be significant. The collected data were stored and processed by a 486-based data acquisition computer. Histograms of the intensity data, transmitted intensity versus frequency of occurrence, were constructed and attempts were made to fit the data at different radial locations and along the whole solution column with Gaussian distributions. The fit was very good in all cases and an example of the data along with the fitted Gaussian is shown in Fig. 2. It may be argued that this should be expected by virtue of

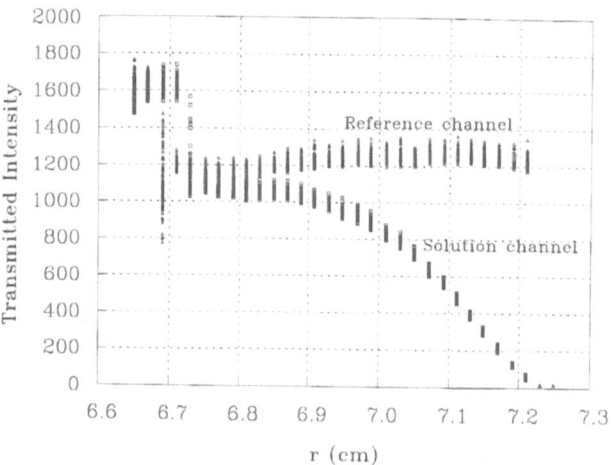

Fig. 1 Raw intensity data collected using the "Method Scan" on the XL-A; 99 data points collected at each radial position

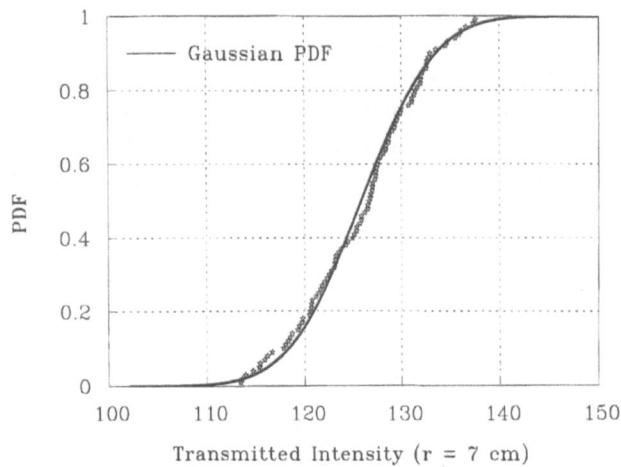

Fig. 2 Probability distribution function (PDF) of intensity data at a single radial position

the Central Limit Theorem [3] since there is a fair number of noise contributors. It was also observed that the Gaussian noise distributions differed along the solution column. In fact, the variances of these distributions appeared to be inversely proportional to the transmitted intensity itself. The noise variance was plotted as a function of intensity and then fitted by a low-order polynomial to quantitatively establish their relationship. Figure 3 shows the experimentally observed variances versus signal strength and the fitted curve which is linear. The availability of such information is needed in order to generate realistic data for computer simulations of any number of associating or nonassociating systems in the ultracentrifuge. The generated intensity data that correspond to the two sectors were

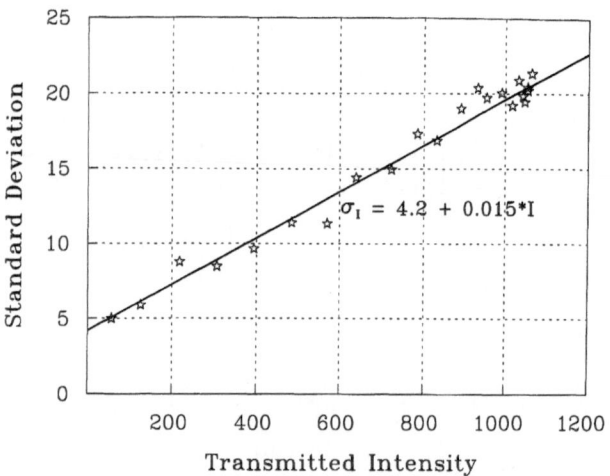

Fig. 3 Noise standard deviation as a function of transmitted intensity signal strength: raw data and least-squares linear fit

then combined to compute absorbances at a large number of radial positions (200–250). Absorbance distribution histograms, however, departed from Gaussian distributions sufficiently to motivate a more detailed investigation into the issues involved from a theoretical, statistical point of view.

Theory

The method of least squares is known to perform optimal data noise filtering [1] or otherwise is known as an unbiased estimator only under the condition that the statistical characteristics of the noise are Gaussian. If this condition is not met and the noise can statistically be described by a different distribution, unbiased parameter estimation can still be performed by using the more general maximum likelihood estimation (MLE) method [3]. Also, it is easy to show that, when the MLE is applied to signals contaminated by Gaussian noise, the mathematical formulation leads to equations that are identical to those obtained by the least-squares approach [1]. Both the MLE method, which needs *a-priori* knowledge of the noise probability distribution, and the least-squares method applied to signals with Gaussian noise have been shown to give parameter estimations whose error covariance asymptotically tends to the theoretical lowest limit achievable, known as the Cramer–Rao lower bound [3]. Application of the least-squares method, however, to signals contaminated by anything but Gaussian noise would give biased estimations with lowest possible bound of the estimated parameter variance larger than the Cramer–Rao bound.

Here, the above-described measurements have ascertained that the noise in the transmitted intensity data from both sectors is indeed Gaussian in distribution. Any linear combination of two Gaussian random variables, x and y, results in a new random variable whose probability distribution is the two-dimensional Gaussian distribution

$$P(x, y) = \frac{U(x)U(y)}{2\pi\sigma_1\sigma_2\sqrt{1-r^2}}$$

$$\times \exp\left[-\frac{1}{2(1-r^2)}\left(\frac{(x-\bar{x})^2}{\sigma_1^2} - \frac{2rxy}{\sigma_1\sigma_2} + \frac{(y-\bar{y})^2}{\sigma_2^2}\right)\right],$$

$$(1)$$

where σ_1 and σ_2 are the standard deviations of x and y, respectively, $\bar{x}$ and $\bar{y}$ are their respective mean values, and r is their correlation coefficient which, for the intensity data, would be zero since the intensities in the two sectors are independent. The unit step functions $U(x)$ and $U(y)$ signify the fact that no negative intensity measurements are possible. The intensity data, however, are not combined linearly when absorbances are computed. In fact, taking the ratio of two zero-mean, Gaussian signals result in a new signal whose noise is distributed according to what is known as the Cauchy distribution which, in this case, can be written as [4]

$$P_z(z) = \frac{\sqrt{1-r^2}\,\sigma_1\sigma_2/\pi}{\sigma_2^2(z - r\sigma_1/\sigma_2)^2 + \sigma_1^2(1-r^2)}, \qquad (2)$$

where $z = x/y$ and the other symbols are as defined above. Yet another variable transformation defined by $w = \log_{10}(z)$ will result in a signal whose noise characteristics should be similar to those of absorbance. The probability distribution of the new variable is in general given by [4]

$$P_w(w) = \sum_{i=1}^{n} \frac{P_z(z(w))}{(\partial w/\partial z)_{z = z_i(w)}}, \qquad (3)$$

where $z = \exp(w)$ is the inverse transformation and z_i are the n real roots of the transformation. Fortunately, in this case there is only one such root. Using the new transformation and Eq. (2) in Eq. (3) leads to the probability distribution of w,

$$P_w(w) = \frac{2\sqrt{1-r^2}\,\sigma_1}{\pi\sigma_2} \frac{e^w}{e^{2w} - 2r(\sigma_1/\sigma_2)e^w + (\sigma_1/\sigma_2)^2} \qquad (4)$$

with a mean value

$$\bar{w} = \frac{2\cos^{-1}(r)}{\pi} \ln\left(\frac{\sigma_1}{\sigma_2}\right) \qquad (5)$$

which, interestingly enough, is not zero unless the variances of the two parameters are identical or the two variables are totally dependent ($r = 1$). Here, $r = 0$, and this simplifies Eq. (4) to

$$P_w(w) = \frac{1}{\pi \cosh(w - \bar{w})} \tag{6}$$

with a distribution mean $\bar{w} = \ln(\sigma_1/\sigma_2)$. The shift in the mean of the transformed variable is a clear indication that the use of the least-squares approach on the absorbance data will introduce a systematic bias in the estimation process. The above exercise, however, has to be modified because the intensity signals are not zero-mean. The procedure is slightly modified from the above and is developed using the two signals defined in Eq. (1) except that they are assumed to be independent ($r = 0$). Two new random variables, X and Y, are defined as the 10-base logarithms of x and y, respectively. Based on Eq. (3), the probability density function of X will be given by

$$f_x(X) = \log(10) 10^x f_x(10^x) \tag{8}$$

with a similar expression for the pair (y, Y), where $f_x(x)$ and $f_y(y)$ are the Gaussian distributions of x and y, respectively. If x and y represent the transmitted intensity signals for the reference and solution sectors, the absorbance signal will have characteristics of a new random variable defined by $z = \log_{10}(x) - \log_{10}(y) = X - Y$ whose probability density is given by the correlation of the two distributions [4],

$$p_z(z) = \int_{-\infty}^{\infty} f_x(Y + z) f_y(Y) \, dY \ . \tag{9}$$

When the equation for the Gaussian distribution and Eq. (8) are introduced into Eq. (9), the indicated integration results in [5]

$$p_z(z) = \frac{\ln(10) 10^z}{2\beta(z)\pi\sigma_1\sigma_2} \exp(-\delta)$$

$$\times \left[1 + \gamma(z)\sqrt{\frac{\pi}{\beta(z)}} \exp\left(\frac{\gamma(z)^2}{4\beta(z)}\right) \right.$$

$$\left. \times \left(1 + \Phi\left(\frac{\gamma(z)}{2\sqrt{\beta(z)}}\right) \right) \right], \tag{10}$$

where

$$\beta(z) = \frac{10^{2z}\sigma_2^2 + \sigma_1^2}{2\sigma_1^2\sigma_2^2}, \quad \gamma(z) = \frac{10^z \bar{x}\sigma_2^2 + \bar{y}\sigma_1^2}{\sigma_1^2\sigma_2^2},$$

$$\delta = \frac{\bar{x}^2\sigma_2^2 + \bar{y}^2\sigma_1^2}{2\sigma_1^2\sigma_2^2} \tag{11}$$

and the special function $\Phi(x)$ is the well-known probability integral or error function [5] defined by

$$\Phi(x) = \frac{2}{\sqrt{\pi}} \int_0^x \exp(-t^2) \, dt \ , \tag{12}$$

When the original variables, x and y, have zero-mean values, β and γ are zero and, as expected, Eq. (10) becomes identical to Eq. (6). Computations using Eqs. (10) and (11) indicate that the shift of the mean is still a function of the ratio of the two variances, its value is effectively independent of the mean values of x and y and the non-zero mean values of x and y lead to a sharpening of the distribution of z. As discussed above, the variances in both sectors are functions of the signal strengths and, from the fitted equation in Fig. 2, it is seen that the standard deviation can exhibit up to a fourfold variation, which implies a non-negligible shift of the distribution mean.

The cost function for the general MLE problem is the conditional joint probability distribution of the measurements at all the radial locations given the values of the parameters. For random noise, the measurements at the different radial locations are independent and the cost function is simply the product of the probability distributions at each measurement point. This cost function has to be maximized. If the noise probability distribution at a measurement point is Gaussian, it is easy to show that maximization of the cost function leads to the same algorithm as does least squares. When the noise is not Gaussian, one needs to go directly to the MLE method using the known noise distribution. Here the distribution is given by Eq. (10) and the MLE cost function to be maximized becomes

$$P(A(\mathbf{r})/f_A(\mathbf{r}; \boldsymbol{\alpha})) = \prod_{i=1}^{n} p_z(A(r_i); f_A(r_i; \boldsymbol{\alpha})) \ , \tag{13}$$

where $\mathbf{r}$ is the vector of radial locations, $A(r_i)$ is the absorbance measured at location r_i, f_A is the function to be fitted to the absorbance data and $\boldsymbol{\alpha}$ is the vector of the unknown parameters in the model. The scheme requires explicit expression of the probability density function in terms of its mean value which for the expression in Eq. (10) is not possible in closed form. Instead, the method could be developed numerically. It is convenient to maximize the logarithm of this function, which is equivalent to maximizing the function itself. The scheme results in a set of equations for the parameters a_i, $i = 1, m$:

$$\frac{\partial}{\partial a_i} \ln(P(A(\mathbf{r})/f_A(\mathbf{r}; \boldsymbol{\alpha}))) = 0 \ , \quad i = 1, m \ . \tag{14}$$

For fitting absorbance data this approach should give better results than the weighted least-squares method. Implementation of the method is as easy as the least squares with the additional advantage that the iteration required for computing optimal weights for the least-squares method is avoided.

Thus, in view of the noise characteristics in the ultracentrifuge and the mathematical manipulations associated with parameter estimation, the user is given two choices. The first choice is to implement the MLE method to fit the usual absorbance data and the second choice is to acquire intensity data and to fit these directly using the least-squares approach. This second approach is complicated by the fact that there are no models for the intensity data in the two sectors. Instead, there is a model for a combination of the two data sets that is a transformation of the model for the absorbance data:

$$I_s(\mathbf{r}) = I_0(\mathbf{r})10^{f_A(\mathbf{r};\mathbf{a})} \qquad (15)$$

where subscripts s and 0 denote the solution and reference sectors, respectively. Two methods were developed here to resolve the problem of the lacking models for the two intensities. These are described below:

Method 1: *Spline smoothing of the reference sector intensity*: Cubic splines were used to construct a smooth curve that optimally fits the reference intensity data. The smoothed intensity is constructed as the sum of scaled cubic spline functions [6, 7]. The scaling coefficients are computed for optimal smoothing using the weighted least-squares method. This computation has to be performed initially in the analysis and the scaling coefficients stored for use when fitting the data in the solution sector based on the model in Eq. (10). A problem with the spline smoothing of the reference data is that some correlations in the noise at neighboring locations are implied. This should not be a major problem, especially if optimal smoothing splines are used which combine the least squares with minimization of curvature errors.

Method 2. *Simultaneous fitting of the data from both sectors in higher-dimensional space*: A new four-dimensional function is constructed from Eq. (10) as follows:

$$G_I(i_s, i_0, r) = i_s - i_0 10^{f_A(r;\mathbf{a})}, \qquad (16)$$

where the new variables represent the two intensities. Given the intensity data and the vector of radial locations the vector of unknown parameters α can be estimated so that the deviation of the function G_1 from zero is minimized in the least-squares sense. This method has the theoretical advantage that the data in the reference sector are treated as uncorrelated while the spline method implies some correlation.

Based on the above considerations, a series of simulations were performed to demonstrate and compare the behavior of the various approaches to the problem of estimating equilibrium parameters for a single non-associating species and in a variety of interacting systems in equilibrium experiments in the ultracentrifuge. Obviously, analysis of actual data from the centrifuge would not be useful in the context of this effort.

Results

In all the simulations, reference sector intensity data were constructed by first fitting actual data from the reference sector using cubic splines as discussed in the Methods section. The scaling coefficients were stored and used as benchmark to generate reference sector noisy data for all subsequent simulations. In all simulations, experiments with two cells were assumed with a 2:1 concentrations ratio in the solution sectors. In the following, the method of spline smoothing of the reference sector data will be designated SSLS for Spline Smoothing-Least Squares, the multi-dimensional data fitting method MDLS for Multi-Dimensional-Least Squares and the usual least squares directly on the absorbance data AWLS for Absorbance Weighted Least Squares. The computer package MLAB, by Civilized Software, Inc. (Bethesda, MD, USA) was used in all cases.

Monomer solution

A solution of myoglobin was assumed for the first simulation. It is known that myoglobin does not self-associate and has a compositional molecular mass of 16 952 Da. The molecular mass M, the cell bottom concentration c_b, and a baseline error offset e_b were the fitting parameters in the model given by

$$f_A(r; M, c_b, e_b) = c_b \exp(AM(r^2 - r_b^2)) + e_b, \qquad (17)$$

where $f_A(r)$ is total solute concentration, expressed as absorbance at 280 nm, as a function of radial position r_b; $A = (1 - \bar{v}\rho)\omega^2/2RT$ where $\bar{v}$ is the compositional partial specific volume [8], ρ the solvent density, ω is the rotational speed in rad/sec, R the gas constant and T the absolute temperature. Intensity data were generated using the nominal molecular mass and with zero-mean Gaussian noise, with variance according to Fig. 2, added. Ten samples were generated and then averaged for each radial position. The resulting intensity values were used to compute absorbance profiles for the two sectors. Curve-fitting was performed using the three methods described above and starting with the same initial guesses for all methods. The results for the molecular mass were as follows: 16 971 Da for SSLS, 16 963 Da for MDLS and 17 758 Da for AWLS. The quality of fit was excellent in all cases. It is clear that the first two methods perform better since their error is of the order of 0.1%, while the error for ALS is of the order of 5%, which is not negligible. Equally important is the measurement of $1 - \bar{v}\rho$ term for a monomeric species for which the molecular mass is reliably known. This is often done to correct compositional values of $1 - \bar{v}\rho$

for each species before analyzing data from an association experiment.

Monomer–dimer association

A macromolecule with the same molecular mass as myoglobin was assumed to self-associate forming a dimer. The fitting parameters for this case were the cell bottom monomer concentrations the baseline errors and the natural logarithm of the equilibrium association constant $\ln(k_{12})$ which was given on the absorbance scale. The appropriate model for this association is

$$f_A(r; \ln(k_{12}), c_b, e_b) = c_b \exp(g(r)) + c_b^2 \exp(\ln(k_{12})$$
$$+ 2g(r)) + e_b , \qquad (18)$$

where $g(r) = AM_1(r^2 - r_b^2)$ and the remaining symbols are defined above. For data generation, three values of the association constant were used: $\ln(k_{12}) = 2, 4$ and 8 covering a wide range of association strengths. Excellent quality fits were achieved in all cases. Using the first two methods the error in estimating $\ln(k_{12})$ was consistently less than 1%. Fitting the absorbance data, the error was of the order of 3% for the weakest association, 1.5% for the intermediate and 1% for the strongest. Note, however, that the actual error in the association constant itself is roughly proportional to the product of k_{12} with the magnitude of the estimation error for its logarithm. Therefore, the approximate percentage error for k_{12} for the weakest association is 6% (2×0.03); for the intermediate strength of association the error is also 6% (4×1.5), and it is 8% (8×0.01) for the strongest association. The possible estimation errors are greater here than in the single species analysis since, with effectively the same amount of information (same number of data points) estimation is performed in a higher-dimensional parameter space. For this reason and because it has been observed that quite often the optimization surface gradients are very small over an extensive area around the optimal point, reaching true convergence is more critical. Here, every effort was made to achieve true convergence, however, by making the convergence criteria increasingly more stringent. It is certain that the above computed errors are systematic, and due to the biased nature of the least-squares estimator as it is applied to absorbance data.

Two interacting proteins forming a heterodimer

Two macromolecules, α and β, with molecular mass ratio of $3:1$ in a solution with $1:1$ molar concentration ratio were assumed to undergo one-to-one association forming a heterodimer. The natural logarithm of the equilibrium association constant $\ln(k_{\alpha\beta})$, the concentrations at the cell bottom of the two monomers, $c_{b,\alpha}$ and $c_{b,\beta}$ and the baseline offset $e_{\alpha\beta}$ were used as fitting parameters. The association model was written as

$$f_A(r; \ln(k_{\alpha\beta}), c_{b,\alpha}, c_{b,\beta}, e_{\alpha\beta}) = c_{b,\alpha} \exp(g_\alpha(r))$$
$$+ c_{b,\beta} \exp(g_\beta(r)) + c_{b,\alpha} c_{b,\beta} \exp(\ln(k_{\alpha\beta}) + (g_\alpha(r)$$
$$+ g_\beta(r))) + e_{\alpha\beta} , \qquad (19)$$

where $g_\alpha(r) = A_\alpha M_\alpha(r^2 - r_b^2)$ and $g_\beta(r) = A_\beta M_\beta(r^2 - r_b^2)$. Again three values of the natural logarithm of the equilibrium association constant $\ln(k_{\alpha\beta}) = 2, 4$ and 8 were used for data generation. Equilibrium association simulations were again performed and the estimated parameters were compared to the data generation parameter values. It was systematically observed that the SSLS method gave the best results. For the weakest association the greatest error in the value of $\ln(k_{\alpha\beta})$ resulted when using the AWLS and it was about 5%. For the intermediate association strength the corresponding estimation error was of the order of 3% and for the strongest of about 2%. A final computation of interest is shown in Fig. 4 where the PDF of the residues obtained after fitting the simulated data for the interacting system. The PDFs were then fitted by both Gaussian functions and by the new PDF given by Eq. (10). It is clear that the new PDF describes the noise much more closely and that explains the greater errors of the least-squares estimates from absorbance as opposed to intensity data.

Fig. 4 Absorbance data fitting residual PDF: raw data, fitted Gaussians and log-ratio PDFs

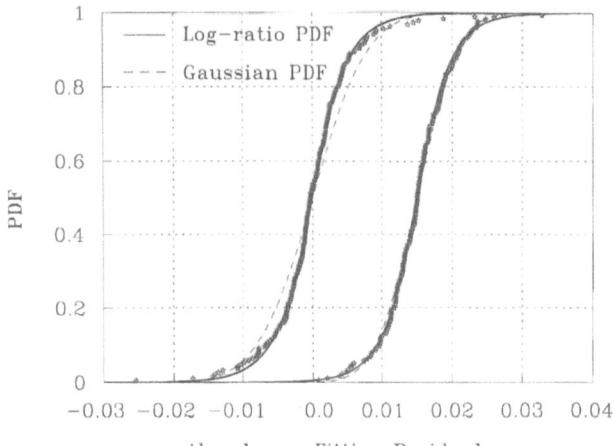

Discussion

It has been shown that the noise probability distribution in the usual absorbance data collected in an XL-A ultracentrifuge is not Gaussian. The noise in the transmitted intensity, however, is Gaussian. The simple non-linear transformation between the two forms of the data allows for the non-Gaussian distribution to be analytically derived in closed form. The form of the new distribution is not as convenient as one might desire for easy implementation of the MLE method. Its implementation, therefore, has been postponed to the future. This method is known to perform unbiased estimation for the case when the statistical characteristics of the noise are known. It has also been observed that the variance of the Gaussian noise in the intensity data is proportional to the signal strength itself. The probability distribution is not an explicit function of this variance, however, and this means that maximum likelihood may be implemented without regard to the actual noise variations along the radius. Here, this variation was measured for the purpose of generating realistic centrifuge data for simulations. In general, of course, clear understanding of the noise characteristics and of the noise sources is the necessary prerequisite for the long-term implementation of more powerful and robust estimation methods.

The performed simulations show that a small but clear advantage exists in using intensity data to fit the desired parameters. The improvement in accuracy is always welcome but it would be even more so in certain cases which experience from future use of the methods may reveal. For example, it might prove critical in cases where it is difficult to choose between two competing models either by the quality of the fit or by the thermodynamic analysis. This issue becomes even more important when parameter estimation errors are compounded with deviations between compositional and actual parameters of the macromolecules.

There is an additional advantage in using intensity data that has to do with the truncation of absorbance data required to avoid cell bottom effects and non-linear photocathode response at very high absorbance values. Additionally, the bias introduced in the analysis by the application of least-squares estimation on the absorbance data is maximized near the cell bottom where the variance ratio in the two sectors is largest, thus introducing a maximal distribution mean shift away from zero. There is a minimal need for truncation at the cell bottom when using intensity data, since the data from the solution sector are the least noisy in that region and abberant points are easily identified. In contrast, however, absorbance data at the cell bottom are the most noisy because of the transformation from intensities to absorbance. This condition leads to uncertainty as to where truncation should be implemented.

References

1. Radhakrishna RC (1965) Linear Statistical Inference and its Applications. Wiley, New York
2. Eykhoff P (1974) System Identification: Parameter and State Estimation. Wiley, New York
3. Manukian EB (1986) Modern Concepts and Theorems of Mathematical Statistics. Springer Berlin
4. Papoulis A (1991) Probability, Random Variables, and Stochastic Processes, 3rd ed. McGraw-Hill Book, New York
5. Gradshteyn IS, Ryszhik IM (1980) Tables of Integrals, Series and Products, 4th ed. Academic Press, New York
6. Unser M, Aldroubi A, Eden M (1993) IEEE Trans Signal Process **41** (2):821
7. Unser M, Aldroubi A, Eden M (1993) IEEE Trans Signal Process **41**(2):834
8. Fujita, H (1975) Foundations of Ultracentrifugal Analysis. Wiley, New York

Progr Colloid Polym Sci (1997) 107:27–35
© Steinkopff Verlag 1997

J. Behlke
O. Ristau

Rapid molecular mass determination by sedimentation velocity experiments and direct fitting of the concentration profiles

Prof. Dr. J. Behlke (✉) · O. Ristau
Max-Delbrueck Center for Molecular
Medicine
Robert-Roessle-Straße 10
13122 Berlin, Germany

Abstract We present a method for the direct molecular mass determination from sedimentation velocity experiments. It is based on a non-linear least-squares fitting procedure of the radial concentration profiles and simultaneous estimation of the sedimentation coefficient and the ratio of sedimentation/diffusion coefficients considering approximate solutions of the Lamm equation. Different model functions from Faxén as well as Archibald type derived by Fujita [4] were used to describe the sedimentation behavior of macromolecules during the centrifugation. By means of a computer program, LAMM sedimentation and diffusion constants of some proteins were determined. The method presented here appears to be useful for the rapid molecular mass determination of proteins larger than 10 kDa. One of the equations of the Archibald type is also suitable for substances of low molecular mass of about 1 kDa. The model function neither requires the existence of a plateau region nor a meniscus region free of solute.

Key words Sedimentation coefficient – diffusion coefficient – molecular mass – proteins – non-linear least-squares fit

Introduction

Analytical ultracentrifugation is a powerful tool for molecular mass determination of macromolecules (see, e.g., refs. [1, 2]). In addition to the meniscus depletion sedimentation equilibrium technique proposed by Yphantis [3], sedimentation velocity experiments may also be used to obtain the molecular mass (M) of macromolecules. The latter method, however, normally yields the sedimentation coefficient (s), which in order to calculate the molecular mass has to be combined with the diffusion coefficient (D) of the sample: the latter is usually obtained from overlay experiments using a synthetic boundary cell. s and D determined simultaneously are from high-speed sedimentation experiment, and by using the analytical functions developed by Fujita [4, 5] to describe the form of the sedimentation velocity boundary. These functions are approximate solutions of the Lamm equation [6] for the ultracentrifuge, mostly of the Faxén type [7]. The equations given by Fujita [4, 5] can be used to fit the concentration profiles of a sedimentation velocity run by a non-linear least-squares technique to estimate sedimentation and diffusion coefficients. The use of such equations with one or more error functions is time-consuming and requires a powerful computer. Recently, Philo [8, 9] and we [10] have reported first results using Fujita equations [4, 5] applicable to the conventional double-sector cell as well as the synthetic boundary cell. To establish the efficacy of the Fujita equations we have analyzed concentration profiles generated by the finite-element method [11].

The aim of the current study is to present sedimentation and diffusion coefficients obtained from sedimentation velocity experiments on different small globular proteins. Application of this technique to characterize more extended proteins or oligopeptides is also discussed.

Material and methods

Sperm-whale myoglobin and hen-egg lysozyme were from SERVA, Heidelberg and cytochrome c from MERCK, Darmstadt, FRG.

Sedimentation velocity runs were performed using a Beckman XL-A ultracentrifuge equipped with UV absorption optics. Experiments were carried out in either conventional double-sector or synthetic boundary cells.

In addition to experimental curves, noise-free data simulated by the finite-element method [11, 12] were also used in the analysis. Radial concentration profiles calculated for a sedimentation coefficient of 2 S and a diffusion coefficient of 1×10^{-6} cm^2/s in simulated experiments at 42 000 or 50 000 rpm (synthetic boundary cells) and 50 000 rpm (conventional cells) were based on either 800 or 1600 data points between $r_m = 6.4$ cm and $r_b = 7.2$ cm. The accuracy of the curves was determined by a simulation time increment dt of 0.5 s and radial steps dr of 0.0005 cm.

To estimate sedimentation and diffusion coefficients concentration profiles were fitted directly to the five different equations given by Fujita [4, 5]. A previously described computer program [10], LAMM, that runs under MS-DOS 6.0 on a Pentium PC, was used. It can read up to 18 XL-A data files for simultaneous and global fitting. Additional parameters such as the elapsed time, $\omega^2 t$, and rotor speed are taken from the data header. All parameters of the model functions as well as a baseline (offset) can be estimated or held constant at the initial value. The initial boundary positions and the part of the profiles to be included in the fitting procedure for each experimental record were determined by manual trimming of the data set. The initial values for all other parameters were calculated by the program. The program accepts only absorption values smaller than 3.2 OD. The five model functions presented by Fujita [4, 5] for sector-shaped cells are the following.

Synthetic boundary (Eq.(2.127) of Fujita [5]):

$$c = \frac{c_0 e^{-\tau}}{2} [1 - \mathrm{erf}(\xi)]$$

$$\xi = \frac{z - \tau}{2\sqrt{\varepsilon\tau}}, \quad x = \left(\frac{r}{r_0}\right)^2, \quad \tau = 2\omega^2 st \quad (1)$$

$$\varepsilon = \frac{2D}{s\omega^2 r_0^2}, \quad z = \ln(x) .$$

In this equation, c_0 denotes the loading concentration, r_0 the liquid–liquid meniscus position, $\mathrm{erf}(\xi)$ the error function, D the diffusion constant, ω the angular velocity, and s the sedimentation constant.

Synthetic boundary experiments with allowance for s-c dependence (Eq.(2.191) of Fujita [4]):

$$c = \frac{c_0 e^{-\tau}}{1 - \lambda} \frac{(1 - \mathrm{erf}(p))e^{p^2}}{(1 - \mathrm{erf}(p))e^{p^2} + \sqrt{1 - \lambda}(1 + \mathrm{erf}(\xi))e^{\xi^2}} ,$$

$$\xi = \frac{1 - \sqrt{x(1 - \zeta)}}{\sqrt{\varepsilon\zeta}}, \quad p = \frac{\xi - \alpha\sqrt{\dfrac{\zeta}{\varepsilon}}}{\sqrt{1 - \lambda}}, \quad \lambda = \alpha\zeta ,$$

$$\zeta = 1 - e^{-\tau}, \quad \varepsilon = \frac{2D}{s_0\omega^2 r_0^2}, \quad s_c = s_0\left(1 - \alpha\frac{c}{c_0}\right), \quad (2)$$

In the monograph of Fujita [5] only the concentration gradient of this function is described. The symbol x has the same meaning as in Eq. (1). s_0 is the Svedberg constant at zero concentration. If the concentration-dependent parameter α becomes zero, this function is identical with Eq. (2.128) of ref. [5] for synthetic boundary experiments.

Standard double-sector cell of infinite length (Eq. (2.167) of Fujita [5]):

$$c = \frac{c_0 e^{-\tau}}{2}\left\{1 - \mathrm{erf}(\xi) + \frac{\sqrt{2\varepsilon}\sinh(\tau/2)}{x^{1/4}[1 + (xe^{-\tau})^{1/4}]\sqrt{\pi}}e^{-\xi^2}\right\} .$$

$$(3)$$

All symbols have the same meaning as in Eq. (2). Function (3) accounts for a special case of initial condition for the synthetic boundary cell. If the boundary moves sufficiently away from the meniscus, Eq. (3) can be employed for the analysis of experiments in conventional double-sector cells.

Improved expressions for a double-sector cell with infinite length

The following equation by Fujita [4, 5, 13] includes a parameter (H) suggested by Holladay [14] to improve the description of the boundary form:

$$c = \frac{c_0 e^{-\tau}}{2}\left\{1 - \mathrm{erf}(p) - 2\sqrt{\frac{\tau}{\pi\varepsilon}}e^{-p^2} + \left(1 + \frac{\tau + \ln x}{\varepsilon}\right)\right.$$

$$\left. \times \left[1 - \mathrm{erf}\left(\frac{\tau + \ln x}{\sqrt{4\varepsilon\tau}}\right)\right]e^{(\ln x/\varepsilon)}\right\}, \quad p = \frac{\tau - \ln x}{\sqrt{4\varepsilon\tau}} . \quad (4)$$

In the improved case ε may be modified to ε/H. The exact validity of Eq. (4) is limited to the earlier steps of sedimentation experiments where the ratio $x = (r/r_0)^2$ in the Lamm equation should be close to one. This restriction applies also for the equations developed for the synthetic boundary cell. In the case of Eq. (4) the limitation can be decreased by introducing a time-dependent parameter.

Written in dimensionless variables, the Lamm equation reads

$$\frac{\partial \theta}{\partial \tau} = \varepsilon e^{-z} \frac{\partial^2 \theta}{\partial z^2} - \frac{\partial \theta}{\partial z}, \quad \theta = e^{\tau} \frac{c}{c_0}, \quad z = \ln(x), \tag{4a}$$

In the Fujita and MacCosham [13] expression the factor $\exp(-z)$ is replaced by unity. However, to a first approximation this factor must fall below 1.0 for the later stages of sedimentation, therefore, leading to an overestimation of ε. After solving Eq. (4a) we have to correct ε by a time-dependent factor H, which approximately compensates for the decrease of e^{-z}:

$$H = \frac{1}{1 + (e^{\tau} - 1)p}. \tag{4b}$$

The additional parameter p amounts to about 0.5; it corresponds to about one-half of the moving boundary (solute front) at the reduced time τ. However, its value can also be estimated using the fitting procedure.

Double-sector cell with finite length (Eq. (2.280) of Fujita [4]): This expression[1] allows for solute accumulation at the cell base as well as solute depletion at the meniscus:

$$\frac{c}{c_0} = \frac{e^{-\tau}}{2} [\mathrm{erf}(p_{22}) - \mathrm{erf}(q_2)] + \frac{1}{2a} \{2(1 + a)e^{-\lambda(X-x)}$$

$$- 2e^{-\lambda(1-a)(X-x)-\tau} - (1 + a)[(1 - \mathrm{erf}(q_{11}))e^{-\lambda(X-x)}$$

$$- (1 - \mathrm{erf}(p_1))e^{-\lambda(1-x)}] - e^{-\tau}[(1 - \mathrm{erf}(p_2))e^{\lambda(1-a)(x-1)}$$

$$- (1 - \mathrm{erf}(q_{22}))e^{-\lambda(1-a)(X-x)}]\}, \tag{5}$$

$$p_1 = \frac{(1 + a)\sqrt{\tau}}{2\sqrt{a}} + \frac{\lambda\sqrt{a}}{2\sqrt{\tau}}(x - 1),$$

$$p_2 = \frac{(1 - a)\sqrt{\tau}}{2\sqrt{a}} + \frac{\lambda\sqrt{a}}{2\sqrt{\tau}}(x - 1),$$

$$p_{22} = \frac{(1 - a)\sqrt{\tau}}{2\sqrt{a}} + \frac{\lambda\sqrt{a}}{2\sqrt{\tau}}(X - x),$$

$$q_2 = \frac{(1 - a)\sqrt{\tau}}{2\sqrt{a}} - \frac{\lambda\sqrt{a}}{2\sqrt{\tau}}(x - 1),$$

$$q_{11} = \frac{(1 + a)\sqrt{\tau}}{2\sqrt{a}} - \frac{\lambda\sqrt{a}}{2\sqrt{\tau}}(X - x),$$

$$q_{22} = \frac{(1 - a)\sqrt{\tau}}{2\sqrt{a}} - \frac{\lambda\sqrt{a}}{2\sqrt{\tau}}(X - x),$$

$$a = \frac{2\varepsilon}{1 + X}, \quad \lambda = \frac{1}{\varepsilon}, \quad X = \left(\frac{r_2}{r_0}\right)^2, \quad x = \left(\frac{r}{r_0}\right)^2,$$

---

[1] Note, the original Eq. (5) in Fujita's monograph [4] contains a misprint. At the exponent of the last term in the fourth row the factor λ is neglected.

Here r_0 is the air–liquid position and r_2 the cell base. All other symbols have the same meaning as before. Equation (5) is not of the Faxén type because the cell length is finite (Archibald type). The integration variable z (Eqs. (6) and (7)) is held at 0.5 the mean value for the integration. Therefore, when using this formula very reasonable results in the fitting procedure can be expected in the central part of the concentration distribution curve or the plateau region.

$$z = \frac{x - 1}{x_b - 1}, \quad 0 \leqslant z \leqslant 1, \quad x_b = \left(\frac{r_b}{r_0}\right)^2, \tag{6}$$

$$\frac{\partial \theta}{\delta \tau} = \frac{\varepsilon}{(x_b - 1)} \left[\frac{z(x_b - 1) + 1}{x_b - 1}\right] \frac{\partial^2 \theta}{\partial z^2}$$

$$+ \left[\frac{\varepsilon}{(x_b - 1)} - \frac{z(x_b - 1) + 1}{x_b - 1}\right] \frac{\partial \theta}{\partial z}. \tag{7}$$

For a successful fit it was found to be useful to take into account only a few (not all) data points of the sharply rising part near the bottom. The reason for that is the low precision of the radial steps which lead to some uncertainty in absorbance data in this region. Best results for the measurement of sedimentation and diffusion coefficients were obtained when using a special final absorbance for the scans. This value was determined approximately as the maximal slope of the last concentration profile which was chosen as 0.05 OD/mm. After the first fitting procedure the program determines the slope and the final optical density and eliminates all data points with higher values. In a second run, the program calculates the relevant values for sedimentation and diffusion coefficients. In contrast to the fit procedures using other model functions, the present approach leads to results which do not depend on the length of the plateau region.

From the estimated sedimentation and diffusion coefficients, the molecular mass of the macromolecule under investigation is calculated from the Svedberg equation

$$M = \frac{sRT}{D(1 - \rho\bar{v})}, \tag{8}$$

where R is the gas constant, T the absolute temperature, ρ the solvent density and $\bar{v}$ the partial specific volume.

Numerical methods

The time-consuming part of the fitting functions is the numerical evaluation of the error functions. Fortunately, a very effective method developed by Gautschi [15] can be used to decrease the computing time for the error functions. The derivatives of the fitting functions are calculated algebraically to save computer time. The fitting algorithm is a "damped least-squares" procedure according to Levenberg [16] in the version given by Wynne and Wormel [17]. Standard deviations of the parameters were calculated in the simple linear model version. For most cases this appears to be sufficient because the variance in the results be comparing different experiments is often larger than the allowed confidence interval for one scan. The fitting procedure took 25 s. CPU time on a pentium PC (100 MHz) to analyze 12 data files (3300 data points).

Results

Fitting of noise-free synthetic data

The ability of the five model functions to fit simulated radial concentration distributions obtained, e.g. for a sedimentation coefficient of 2 S and a diffusion coefficient of 1×10^{-6} cm^2/s using the finite-element method [11, 12] was tested. In addition to the sedimentation and diffusion coefficients, the loading concentration (c_0) and the radius position at the meniscus (r_0) or cell base (r_2) were also estimated. The baseline was held at zero.

Table 1 Estimated parameters from Claverie simulation

Equation no.	s [S]	D [10^7 cm^2/s]	c_0	r_0[cm]	α
(1)	1.99_1	9.85_2	1.000_5	6.697_2	—
(2)	1.99_2	10.00_1	0.999_9	6.697_2	—
(2)[a]	1.79_4	8.50_0	0.994_2	6.697_1	0[b]
(2)[a]	2.01_4	10.09_0	1.001_6	6.697_4	0.1114

Note: The following predetermined parameters were chosen: sedimentation coefficient $s = 2$ S; diffusion coefficient, $D = 10 \times 10^{-7}$ cm^2/s; loading concentration of solute, $c_0 = 1$; initial boundary position, $r_0 = 6.697_5$ cm; the concentration-dependent parameter, $\alpha = 0$ or 0.2 (for indicated equation); speed, 50 000 rpm (synthetic boundary) using different equations
[a] $\alpha = 0.2$ (predetermined)
[b] α held at 0

Synthetic boundary data

From the results (Table 1) we can conclude that Eqs. (1) and (2) yield very accurate values for sedimentation and diffusion coefficients as well as the loading concentration and the meniscus position. Moreover, Eq. (2) allows us to analyze the concentration dependence of sedimentation coefficients. However, only the values for s and D were obtained as expected, whereas the concentration-dependent parameter α deviates considerably.

Double sector cell data

In the fitting attempts of synthetic data for double-sector cells, Eq. (5) for a finite as well as the Faxén solutions, Eqs. (1)–(4), in the case of infinite cell length have been considered. The results (Table 2) obtained for fitting radial concentration distributions calculated for 2 S and 1×10^{-6} cm^2/s clearly demonstrate that Eqs. (3) and (4) yield reliable sedimentation coefficients but the diffusion coefficients are too small by about 5–7%. A modification of Eq. (4) as given in ref. [10], see Methods section, lead to more accurate results with respect to the expected diffusion coefficient. In contrast to the Eqs. (1)–(4), Eq. (5) fits the radial concentration profiles from the meniscus to the cell base and includes also the boundary at the bottom. However, because of the limited accuracy of the Archibald solution (Eq. (5)), in this region, it is necessary to cut all values above a profile slope higher than 0.05 A/mm. This value has been obtained empirically. After estimating the slope in a first fit the second fit is carried out with a reduced data set. This procedure enables us to obtain reliable s/D ratios and therefore more precise molecular masses. However, the data for s and D are slightly too low. As has been pointed out earlier [10], Eq. (5) does not allow an optimal fit near the meniscus. When removing the first part of the traces prior to fitting, the expected s and D values can be recovered.

Table 2 Estimated parameters from Claverie simulation

Equation no.	s [S]	D [10^{-7} cm^2/s]	c_0	r_0[cm]	r_2[cm]	p
(3)	2.01_2	9.33_4	1.001	6.398_0	—	—
(4)	1.99_1	9.47_7	1.000	6.399_8	—	0[a]
(4a)	1.98_8	9.88_0	0.999	6.400_1	—	0.4119
(5)	1.95_7	9.77_0	0.996	6.391_7	7.197_3	—
(5)[b]	1.99_8	9.99_7	0.995	6.382_7	7.197_6	—

Note: The following predetermined parameters were used: $s = 2$ S; $D = 1 \times 10^{-6}$ cm^2/s; $c_0 = 1$; initial boundary position, $r_0 = 6.4$ cm; bottom radius, $r_2 = 7.2$ cm; speed, 50 000 rpm; using different equation
[a] p held at 0
[b] Omitting data points $< r = 6.49$ cm

Progr Colloid Polym Sci (1997) 107:27–35
© Steinkopff Verlag 1997

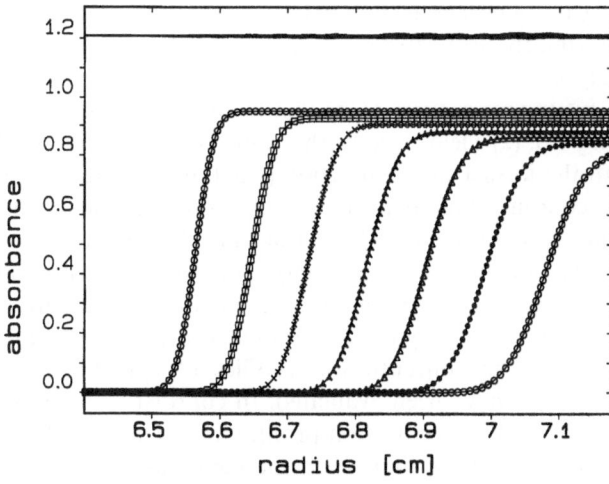

Fig. 1 Claverie simulation for $s = 6$ S and $D = 1 \times 10^{-7}$ cm^2/s, loading concentration $c_0 = 1$ and 50 000 rpm (conventional double sector cell). Delay time 40 min; $\Delta t = 10$ min. Fit using Eq. (4a). Residuals given in 16-fold amplification show only very small deviations. For a clear representation only each second record is shown. Estimated values: $c_0 = 1.0005$; $s = 5.999$ S; $D = 1.001 \times 10^{-7}$ cm^2/s; $p = 0.5$

As mentioned by Holladay [18], the approximate solutions of the Lamm equation should yield reliable data only for globular proteins. Radial concentration profiles for the extremely elongated myosin with a sedimentation coefficient of 6 S and a diffusion coefficient of 1×10^{-7} cm^2/s have also been simulated (Fig. 1) and the traces were fitted using Eqs. (3)–(5). According to the results given in Table 3 we can conclude that the model equations might be applicable for elongated macromolecules.

Most of the fitting functions used in the present study are of the Faxén type and require a plateau region in the traces. In case of small biomolecules (for example oligopeptides) this is difficult to achieve during a sedimentation velocity experiment. For those substances one should use Eq. (5) which is of the Archibald type. Here a plateau region is not necessary. As shown in Fig. 2 the radial concentration distribution curves obtained for a sedimentation coefficient of 0.25 S and a diffusion coefficient of 3×10^{-6} cm^2/s using the finite-element method could be fitted very well using Eq. (5).

Altogether, testing the five model functions on synthetic data sets under different conditions, we can conclude that the most accurate values for sedimentation and diffusion coefficients especially of small proteins (10–20 kDa) can be obtained when fitting the data sets using Eqs. (1) or (2) and for standard double-sector cells with small restrictions by Eqs. (4) and (5). Equation (3) results in reliable sedimentation data but the diffusion coefficients were found to be underestimated. To obtain correct values for both parameters, strict meniscus depletion is necessary

Table 3 Estimated parameters from Claverie simulation

Equation no.	s [S]	D [10^7 cm^2/s]	c_0	r_0[cm]	r_2[cm]
(3)	5.958	1.0019	1.005	6.4000	—
(4)	5.943	1.0233	0.9789	6.3956	—
(4a)	5.999	1.0196	1.0010	6.4000	—
(5)	5.952	1.0590	0.9995	6.3789	7.2004
(5)[a)]	5.971	1.0003	1.0005	6.3999	7.2000

Note: The following predetermined parameters were used: $s = 6$ S; $D = 1 \times 10^{-7}$ cm^2/s loading concentration of solute $c_0 = 1$; initial boundary position $r_0 = 6.4$ cm, bottom radius 7.2 cm; speed 40 000 rpm.
[a)] Fit with truncated data sets

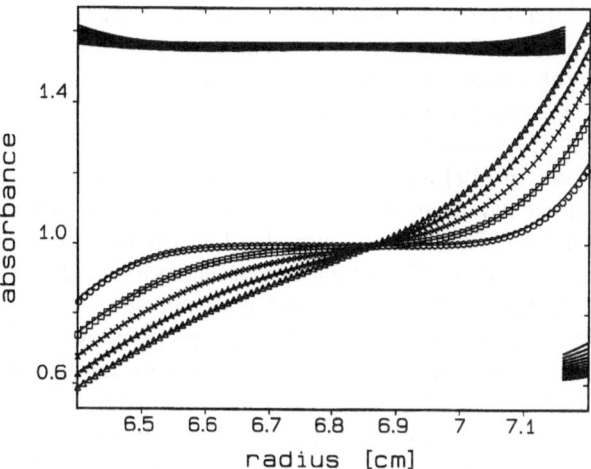

Fig. 2 Claverie simulation for $s = 0.25$ S, $D = 3 \times 10^{-6}$ cm^2/s, $c_0 = 1$ and 50 000 rpm (double-sector cell). Fitting procedure by Eq. (5). Estimated values: $c_0 = 1.0016$, $s = 0.247$ S, $D = 2.997 \times 10^{-6}$ cm^2/s, $r_0 = 6.390$ cm, $r^2 = 7.191$ cm

which demands larger column height, and higher speed of at least 60 000 rpm or lower diffusion coefficients. For the analysis of substances with a molecular mass of about 1 kDa, Eq. (5) yields reliable data.

Fitting of experimental curves

Because the size and shape of small proteins with a molecular mass of 10–20 kDa are of a special interest for many biochemical laboratories, we have analyzed such proteins as cytochrome c, myoglobin and lysozyme by the different model functions. Because the factor α (Eq. (2)) was estimated inaccurately in noise-free synthetic data (see Table 1) this parameter was not considered in the analysis of the experimental curves. When neglecting α, Eq. (2) becomes identical with Eq. (1) and was not considered in the treatment of experimental curves. The substances were tested under nearly identical conditions (protein concentration, buffer composition, speed and column height in the cells). Depending on the equations used for analyzing the concentration profiles, different regions of the curves have been included. On average, 10 different traces of each experiment were used in the calculation of sedimentation and diffusion coefficients.

Cytochrome c

Figure 3 demonstrates a synthetic boundary experiment of cytochrome c. It was analyzed from the position of increasing concentration at radius values of about 6.6 cm up to the plateau region which can be recognized easily in the first profiles. Using Eq. (1) and curves recorded at three different wavelengths the following data were estimated: $s_{20,w} = 1.68_5 \pm 0.00_2$ S and $D_{20,w} = (11.84 \pm 0.07)10^{-7}$ cm^2/s. Considering a partial specific volume of $\bar{v} = 0.713$ cm^3/g [19] and a solvent density $\rho = 1.003$ g/cm^3, the molecular mass of cytochrome c was determined as 12176 ± 87. This value is somewhat lower than the theoretical one with 12330 obtained from the amino acid composition [20] and including the heme moiety.

In order to obtain clear plateau regions in synthetic boundary experiments for low molecular mass substances, it is important to take the first records immediately after reaching the velocity speed. However, sometimes in cases of overlaying runs, the first traces are slightly disturbed

and cannot be merged for the later stages of the sedimentation run. In most cases these data are useless for a further analysis.

Fitting procedures of synthetic boundary experiments using Eq. (2) yield exactly the same results as obtained with the former model function (Eq. (1)) when neglecting the concentration dependence of sedimentation coefficients because the model functions are then identical.

The sedimentation velocity runs in conventional double-sector cells are less sensitive against disturbances compared to the synthetic boundary experiments. For analyzing the concentration profiles of cytochrome c, Eqs. (3)–(5) have been evaluated. In contrast to attempts using Eq. (1), concentration profiles for Eq. (3) should be taken after a delay time when the meniscus position has sedimented partially free of soluted material. Equations (4) and (5) can be used for profiles which are recorded immediately after reaching the final speed. This is possible because both equations take into account the boundary at the meniscus position. In contrast to Eq. (4), Eq. (5) considers also the steep increasing concentration profile at the cell base. According to our experience it is advisable to consider data points up to a maximum absorbance as given in the methodical part and shown in Fig. 4. According to the different regions involved in the analysis, smaller deviations are expected and were observed also in the analysis of runs with cytochrome c (see Table 4). The curves in this region are characterized by a large back diffusion of sedimented material, but they can be fitted well by Eq. (5). Contrary to the synthetic boundary experiments in the

Fig. 3 Radial concentration distributions (symbols) and fitted data (curves) for 0.4 mg/ml cytochrome c in phosphate buffer, pH 7.4, at 38000 rpm (synthetic boundary experiment). The boundary shapes were obtained from Eq. (1) resulting in $s_{20,w} = 1.68_5$ S and $D_{20,w} = 11.84 \times 10^{-7}$ cm^2/s. From the 11 records involved in the fitting only each second is demonstrated

Fig. 4 Plot of experimental and calculated profiles of cytochrome c (conditions as in Fig. 3) at 20°C and 50000 rpm. The model function, Eq. (5), was used for fitting. Again each second data file is shown. Estimated parameters: $s_{20,w} = 1.62_8$ S, $D_{20,w} = 11.40 \times 10^{-7}$ cm^2/s

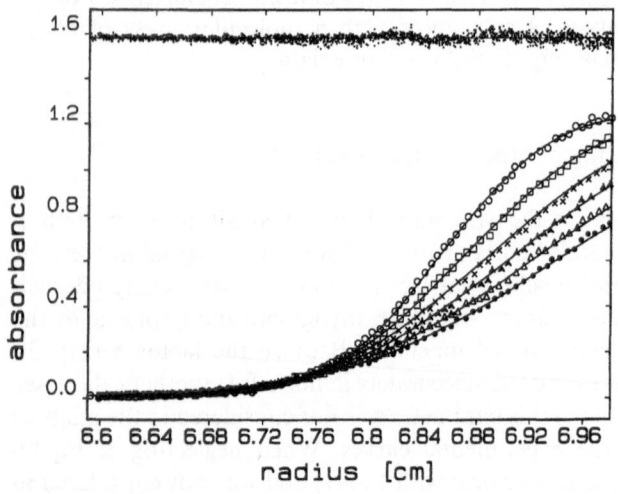

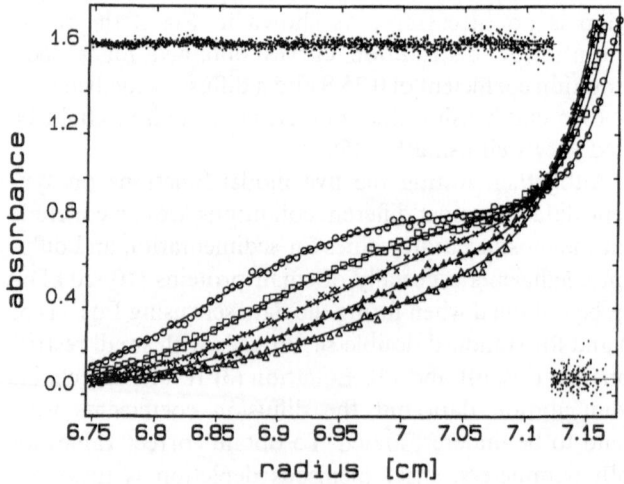

Table 4 Sedimentation and diffusion coefficients of cytochrome c, lysozyme and myoglobin derived by fitting the concentration distribution profiles using Eqs. (1)–(5), as well as molecular mass data and the deviations from the expected values

Protein	Equation No[a]	s [S]	$D[10^7 \text{ cm}^2/\text{s}]$	M_{sD}^b	$\Delta M(\%)^{c}$
Cytochrome c	(1)	$1.69_5 \pm 0.00_3$	11.82 ± 0.10	12 269	− 0.5
Cytochrome c	(3)	$1.69_5 \pm 0.00_5$	11.01 ± 0.17	13 172	+ 6.8
Cytochrome c	(4)	$1.67_0 \pm 0.00_4$	11.35 ± 0.20	12 589	+ 2.1
Cytochrome c	(4a)	$1.67_2 \pm 0.00_4$	11.66 ± 0.19	12 269	− 0.05
Cytochrome c	(5)	$1.62_1 \pm 0.00_3$	11.60 ± 0.08	11 956	− 0.3
Cytochrome c	(5)[d]	$1.66_9 \pm 0.00_5$	11.75 ± 0.11	12 153	− 0.15
Myoglobin	(1)	$1.92_5 \pm 0.00_7$	10.39 ± 0.23	17 655	− 1.2
Myoglobin	(3)	$1.99_0 \pm 0.00_9$	10.02 ± 0.29	18 925	+ 6.0
Myoglobin	(4)	$1.95_2 \pm 0.00_3$	10.14 ± 0.18	18 438	+ 3.2
Myoglobin	(4a)	$1.95_8 \pm 0.00_2$	10.48 ± 0.15	17 803	− 0.3
Myoglobin	(5)	$1.88_7 \pm 0.00_2$	10.07 ± 0.11	17 856	− 0.02
Myoglobin	(5)[d]	$1.98_6 \pm 0.00_5$	10.69 ± 0.14	17 703	− 0.9
Lysozyme	(1)	$1.90_6 \pm 0.00_2$	11.90 ± 0.15	14 478	+ 1.1
Lysozyme	(3)	$1.88_3 \pm 0.00_4$	10.57 ± 0.21	16 112	+ 12.5
Lysozyme	(4)	$1.86_3 \pm 0.00_2$	10.82 ± 0.10	15 563	+ 8.7
Lysozyme	(4a)	$1.86_5 \pm 0.00_1$	11.42 ± 0.11	14 762	+ 3.1
Lysozyme	(5)	$1.79_4 \pm 0.00_2$	11.25 ± 0.03	14 414	+ 0.7
Lysozyme	(5)[d]	$1.85_0 \pm 0.01_2$	11.63 ± 0.11	14 379	+ 0.4

[a] Equation (4a) is the improved model function (4)

[b] $\rho = 1.003$ g/ml, $\bar{v} = 0.713$ mg/g [20] for cytochrome c: $\bar{v} = 0.742$ ml/g [22] of myoglobin; $\rho = 1.0005$ g/ml, $\bar{v} = 0.730$ ml/g [24] for lysozyme

[c] Deviation from the expected molecular mass: 12 330 for cytochrome c [19], 17 860 for myoglobin [21] and 14 316 for lysozyme [23]

[d] Fitting of the truncated profiles by Eq. (5)

runs with conventional standard double-sector cells in the first records the plateau is already beginning to disappear. In spite of such behavior most of the radial concentration profiles, including the last ones which look like the earlier stage of a sedimentation equilibrium run can be fitted reasonably well. When comparing the different equations used in the fitting procedure with respect to the molecular mass calculated from the estimated sedimentation and diffusion coefficients of cytochrome c (Table 4) the most reliable data were obtained using Eqs. (4)–(5). Presumably, the quality of the results depends on the experimental conditions. Although in synthetic boundary experiments the boundary is visible immediately after overlay this process is accompanied often by microdisturbances and therefore causes noise. A further disadvantage of this technique is the difficulty to estimate the exact diffusion time, which differs somewhat from the time derived from the $\omega^2 t$ term of the data header. Another reason for deviations of the results from the theoretical molecular mass derived from the amino acid composition is the partial specific volume. However, an independent parameter for the efficiency of the model equations seems to be the reproducibility of the data obtained by the fitting procedures. In this respect, Eq. (5) seems to yield the most reliable data because it considers the largest radial concentration range for analysis.

Myoglobin

In the same manner as mentioned above we analyzed the sedimentation behavior of sperm whale myoglobin. In comparison with data of overlaying experiments the results of conventional sedimentation runs are more precise (see Fig. 5 and Table 4) and in better agreement with the theoretical molecular mass of 17 860 Da calculated from the amino acid composition [21] including the heme moiety. This is valid also for such concentration profiles with a lower signal to noise ratio. As demonstrated for cytochrome c the uncertainty in the determination of s and D derived from synthetic boundary runs based on the inaccuracy of the diffusion time. Both parameters (s and D) are influenced in the same manner by the measured (or estimated) diffusion time. Therefore, the calculated molecular masses differ less than the sedimentation or diffusion coefficients from which they are obtained.

Lysozyme

In contrast to the two proteins discussed above, lysozyme is a protein without a cofactor. Similar to the other substances the concentration distribution curves were analyzed by the different model equations to obtain the

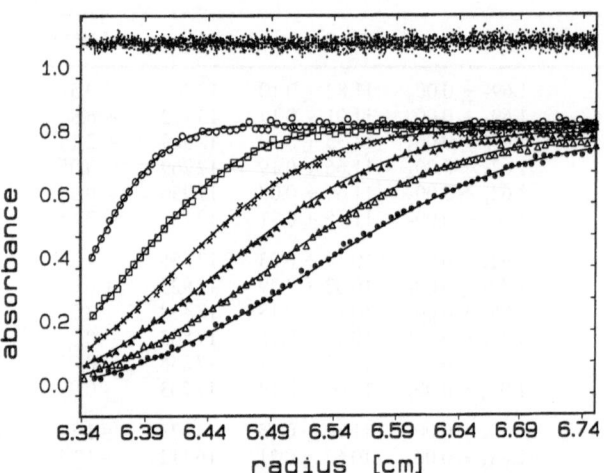

Fig. 5 Sedimentation velocity experiment of sperm whale myoglobin in phosphate buffer, pH 7.4, at 20°C and 50000 rpm (conventional double-sector cell) fitted by Eq. (4a). The following results were obtained: $s_{20,w} = 1.95_6$ S; $D_{20,w} = 11.44 \times 10^{-7}$ cm²/s. Only each second record is shown

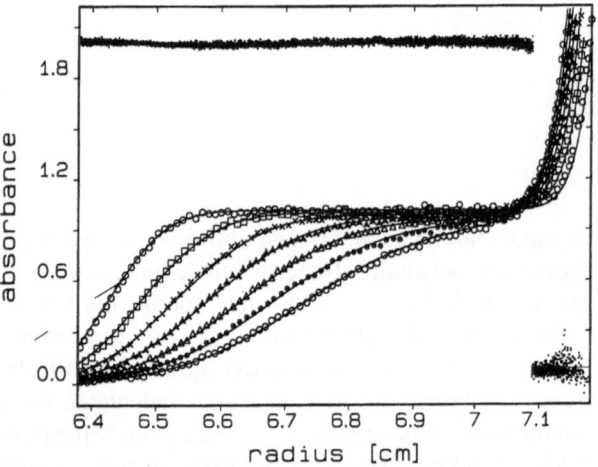

Fig. 6 Plot of experimental and calculated profiles of 0.4 mg/ml hen-egg lysozyme dissolved in phosphate buffer, pH 7.4, at 20°C and 50000 rpm (conventional double-sector cell). Although Eq. (5) accounts for both boundaries the fit in the meniscus region is not optimal. Estimated values: $s = 1.79_5$ S; $D = 11.24 \times 10^{-7}$ cm²/s. When omitting the first data points up to 6.49 cm the following results were obtained: $s_{20,w} = 1.85_0$ S; $D_{20,w} = 11.57 \times 10^{-7}$ cm²/s

hydrodynamic data. The results are summarized in Table 4 reflecting the usefulness of the model functions to get reliable data also for the small protein. As demonstrated before, Eq. (5) yields the most precise results

with respect to the theoretical molecular mass of 14316 Da [23] especially when using truncated data sets (see Fig. 6).

Discussion

The sedimentation equilibrium technique has been considered as the ultracentrifugation method of choice to determine the precise molecular masses of macromolecules. The combined sedimentation/diffusion experiments in two runs and at different speeds appear to be elaborate yielding less reliable results because of some uncertainties in the overlaying technique. The possibility of direct molecular mass determination from sedimentation velocity runs based on approximate solutions of the Lamm equation [6] is a merit of Fujita [4, 5, 13], who has developed different model functions describing radial concentration profiles of high-speed sedimentation experiments. For many years the fitting procedure of such traces by using the above-mentioned equations with up to six error functions seemed to be computationally too demanding for practical applications. However, due to effective methods for numerical evaluation of error functions given by Gautschi [15] and more powerful computers we are now able to use all the five equations from Fujita [4, 5] for the simultaneous estimation of sedimentation and diffusion coefficients as a routine method. These procedures permit the rapid molecular mass determination of substances which are sensitive to increasing temperature and which would degradate in long time sedimentation runs. Because of possible microdisturbance in overlay experiments we prefer sedimentation velocity runs with the conventional double-sector cells. With respect to the usefulness of the different model functions to get reliable results, Eqs. (3)–(5) can be recommended. From our experience, application of Eq. (3) requires higher columns. In contrast, a modified Eq. (4) and particularly Eq. (5) are suitable for experiments with smaller column height. Furthermore, the rotor speed should not be too high. In future work, we hope to eliminate the missing capability of Eq. (5) to fit the meniscus region of the concentration profiles by a small empirical alteration suggested for the improvement of Eq. (4).

Acknowledgement The authors are grateful to thank Dr. Walter Stafford (Boston, USA) for some Claverie simulations. The skillful technical assistance of Mrs. Bärbel Bödner is gratefully acknowledged.

Progr Colloid Polym Sci (1997) 107: 27–35
© Steinkopff Verlag 1997

References

1. Harding SE, Rowe AJ, Horton JC (eds) (1992) Analytical Ultracentrifugation in Biochemistry and Polymer Science. The Royal Society of Chemistry, Cambridge, UK
2. Schuster TM, Laue TM (eds) (1994) Modern Analytical Ultracentrifugation. Birkhäuser, Boston
3. Yphantis DA (1964) Biochemistry 3: 297
4. Fujita H (1962) Mathematical Theory of Sedimentation Analysis. Academic Press, New York
5. Fujita H (1975) Foundations of Ultracentrifugal Analysis. Wiley, New York
6. Lamm O (1929) Ark. Mat. Astr. o. Fys. 21B(2):1
7. Faxén H (1929) Ark. Mat. Astr. o. Fys. 21B(3):1
8. Philo JS (1994) In: Schuster TM, Laue TM (eds), Modern Analytical Ultracentrifugation. Birkhäuser, Boston, p 156
9. Philo JS (1997) Biophys J 72:435
10. Behlke J, Ristau O (1997) Biophys J 72:428
11. Claverie JM, Dreux H, Cohen R (1975) Biopolymers 14:1685
12. Cox DJ, Dale RS (1981) In: Frieden C, Nichol LW (eds), Protein–Protein Interaction. Wiley, New York, p 173
13. Fujita H, MacCosham VJ (1959) J Chem Phys 30:291
14. Holladay LA (1979) Biophys Chem 10:187
15. Gautschi W (1969) Comm ACM 12:635
16. Levenberg K (1944) Quart Appl Math 2:164
17. Wynne CG, Wormell PMJH (1963) Appl Opt 2:1233
18. Holladay LA (1980) Biophys Chem 11:303
19. Timchenko AA, Denesyuk AI, Fedorov BA (1981) Biofizika 26:32
20. Margoliash E, Smith EL, Kreil G, Tuppy H (1961) Nature 192:1121
21. Edmundson AB (1965) Nature 205:883
22. Behlke J, Wandt I (1973) Acta Biol Med Germ 31:383
23. Canfield RE (1963) J Biol Chem 238:2698
24. Schausberger A, Pilz I (1977) Makromolekulare Chem 178:211

Progr Colloid Polym Sci (1997) 107:36–42
© Steinkopff Verlag 1997

H. Cölfen
D.J. Winzor

A computer program based on the psi function for model-independent analysis of sedimentation equilibrium distributions reflecting macromolecular interactions

Dr. H. Cölfen (✉)
Max-Planck-Institute for Colloid and
Interface Research
Colloid Chemistry
Kantstraße 55
14513 Teltow, Germany
E-mail: coelfen@castor.mpikg-
teltow.mpg.de

D.J. Winzor
Centre for Protein Structure, Function
and Engineering
Department of Biochemistry
University of Queensland
Brisbane, Queensland 4072, Australia

Abstract A computer program, written in BASIC, has been developed for the characterization of either solute self-association or the interaction between macromolecular acceptor and ligand. Simplicity and model independence of the procedure to evaluate the free concentration of smallest macromolecular species are features which are illustrated by analysis of simulated Rayleigh distributions based on α-chymo-

trypsin dimerization, and of simulated absorbance records of the distribution based on 1:1 interaction between trimethylamine dehydrogenase (acceptor) and electron transferring flavoprotein (ligand).

Key words Protein self-association – heterogeneous association – sedimentation equilibrium – psi analysis

Introduction

The direct analysis of sedimentation equilibrium distributions for solutes undergoing either self-association or interaction with a second solute has followed different pathways in the northern and southern hemispheres. By far the more common method of attack has entailed the use of highly iterative simulation procedures to obtain the best-fit description of the experimental solute distribution(s) in terms of a postulated model for the interaction [1–4]. On the other hand, the Australasian procedures [5–8] tend to be based on model-independent determination of the thermodynamic activity (concentration) of the smallest macromolecular reactant throughout the sedimentation equilibrium distribution. An advantage of this direct approach is the simplicity of the mathematical manipulations involved in such quantitative characterization of macromolecular interactions – a simplicity that allows its application without a specific computer program. However, in the belief that the method will gain wider acceptance by such action, we have now written a program for evaluating the concentration distributions of monomer and smaller reactant from sedimentation equilibrium

distributions on self-associating and heterogeneously associating systems respectively.

Basis of the procedure

The psi procedure [8] stems from earlier observations by Nichol and coworkers [5–7] that the sedimentation equilibrium distribution for a reacting system is amenable to analysis in terms of the thermodynamic activity of the smallest macromolecular species throughout the distribution. For simplicity of presentation, we shall restrict consideration to systems for which assumed thermodynamic ideality is a reasonable approximation – a simplification which allows the identification of thermodynamic activities with molar concentrations. However, the ramifications of thermodynamic nonideality, for which allowance can also be made, are considered briefly in the discussion.

Solute self-association

For a solute undergoing ideal self-association the sedimentation equilibrium condition for the weight-concentration distribution of each oligomeric species, $c_i(r)$, namely,

$$c_i(r) = c_i(r_F)\psi_i(r) \,, \tag{1a}$$

$$\psi_i(r) = \exp[M_i(1 - \bar{v}_i\rho)\omega^2(r^2 - r_F^2)/2RT] \,, \tag{1b}$$

may be combined with the mass-conservation requirement for the total solute concentration, $c(r)$,

$$c(r) = c_1(r) + c_2(r) + \cdots \tag{2}$$

to give the expressions

$$c(r) = c_1(r_F)\psi_1(r) + c_2(r_F)\psi_2(r) + \cdots, \tag{3a}$$

$$c(r)/\psi_1(r) = c_1(r_F) + c_2(r_F)[\psi_2(r)/\psi_1(r)] + \cdots. \tag{3b}$$

Concentrations of the various oligomers at radial distance r are thus expressed as the product of the corresponding concentration at a selected reference radial position, $c_i(r_F)$, and the psi function, Eq. (1b). M_i and $\bar{v}_i$ denote the molar mass and partial specific volume, respectively, of the species ($i = 1$, monomer; $i = 2$, dimer, etc.) subjected to centrifugation at angular velocity ω and absolute temperature T, ρ is the solvent density [9, 10] and R the universal gas constant.

The free monomer concentration at the reference radial position, $c_1(r_F)$, may thus be determined as the ordinate intercept of the dependence of $c(r)/\psi_1(r)$ upon $[\psi_2(r)/\psi_1(r)]$, irrespective of the mode of self-association. Furthermore, the monomer concentration at any radial distance, $c_1(r)$, may then be calculated by means of Eq. (1) to provide the experimenter with the free monomer concentration throughout the distribution. Evaluation of the association equilibrium constants(s), X_i, then entails consideration of the total concentration in terms of the requisite polynomial in $c_1(r)$,

$$c(r) = c_1(r) + X_2[c_1(r)]^2 + X_3[c_1(r)]^3 + \cdots \tag{4}$$

which expresses the concentration of each oligomeric species as the product of an association constant and the monomer concentration raised to the appropriate power.

Interaction between dissimilar reactants

Adaptation of the above procedure to a thermodynamically ideal interaction involving two dissimilar reactants is readily accomplished by noting that the omega [5–7] and hence psi [11] procedures can be applied to the smaller macromolecular reactant (ligand S) to obtain its free concentration throughout the distribution. Provided that separate distributions may be determined for the acceptor (A) and ligand (S) constituents, the total molar concentration of S constituent at radial distance r, $\bar{C}_S(r)$, may be written as

$$\bar{C}_S(r) = C_S(r_F)\psi_S(r) + C_{AS}(r_F)\psi_{AS}(r) + \cdots, \tag{5a}$$

$$\bar{C}_S(r)/\psi_S(r) = C_S(r_F) + C_{AS}(r_F)[\psi_{AS}(r)/\psi_S(r)] + \cdots, \tag{5b}$$

where $C_S(r_F)$ and $C_{AS}(r_F)$ denote the respective concentrations of free S and AS complex at the reference radial position, r_F. $C_S(r_F)$ is thus determined as the ordinate intercept of the dependence of $\bar{C}_S(r_F)/\psi_S(r)$ upon $[\psi_{AS}(r)/\psi_S(r)]$. Evaluation of $C_S(r)$ as $C_S(r_F)\psi_S(r)$ throughout the distribution then allows determination of the binding function [12]

$$\nu(r) = [\bar{C}_S(r) - C_S(r)]/\bar{C}_A(r) \tag{6}$$

and hence the association constant(s) by standard binding curve analysis [13].

For situations in which the only available sedimentation distribution is in terms of total absorbance, $A(r)$, the counterpart of Eqs. (5a) and (5b) must be written in the form

$$A(r) = A_S(r_F)\psi_S(r) + A_A(r_F)\psi_A(r)$$
$$\quad + A_{AS}(r_F)\psi_{AS}(r) + \cdots, \tag{7a}$$

$$A(r)/\psi_S(r) = A_S(r_F) + A_A(r_F)[\psi_A(r)/\psi_S(r)]$$
$$\quad + A_{AS}(r_F)[\psi_{AS}(r)/\psi_S(r)] + \cdots, \tag{7b}$$

where $A_i(r_F)$ denotes the absorbance due to free species i at the reference radial position r_F. The absorbance due to free S at the reference radial position may thus be obtained as the ordinate intercept of the dependence of $A(r)/\psi_S(r)$ upon $[\psi_A(r)/\psi_S(r)]$, whereupon advantage may be taken of Eq. (1) to obtain $A_S(r)$ throughout the sedimentation equilibrium distribution. Subtraction of $A_S(r)$ from $A(r)$ then leads to a revised total absorbance distribution, $A'(r)$, in which acceptor (A) is the smallest solute species. By reasoning analogous to that used above, we may thus write

$$A'(r) = A_A(r_F)\psi_A(r) + A_{AS}(r_F)[\psi_{AS}(r)/\psi_A(r)] + \cdots \tag{8a}$$

$$A'(r)/\psi_A(r) = A_A(r_F) + A_{AS}(r_F)[\psi_{AS}(r)/\psi_A(r)] + \cdots \tag{8b}$$

whereupon $A_A(r_F)$ may be obtained as the ordinate intercept of the dependence of $A'(r)/\psi_A(r)$ upon $[\psi_{AS}(r)/\psi_A(r)]$. Knowledge of the sedimentation equilibrium distributions in terms of $A(r)$, $A_S(r)$ and $A_A(r)$ then allows evaluation of the binding constant K as

$$K = [\varepsilon_A\varepsilon_S/\varepsilon_{AS}][A(r) - A_S(r) - A_A(r)]/[A_A(r)A_S(r)], \tag{9}$$

where the ratio of molar absorption coefficients (ε_i) is introduced to define the binding constant in traditional units of M^{-1}. The magnitude of K is thus determined from the slope of the dependence of $[A(r) - A_S(r) - A_A(r)]$ upon $A_S(r)A_A(r)$ [11].

The psi computer program

A program has been written in BASIC for the analysis of either solute self-association or solute interaction with a dissimilar reactant. The development of a common program for both types of system re-emphasizes the model independence of the above psi approach [8, 11], which has evolved from its omega counterpart [5–7]. Because the program evaluates the concentration (thermodynamic activity) distribution for the smallest macromolecular species contributing to a sedimentation equilibrium pattern, it applies equally well to determination of the monomer distribution for a self-associating system or of the free ligand distribution for a heterogeneously associating system. As indicated above, the distribution of free acceptor may also be determined by the corresponding analysis of the revised distribution obtained by subtracting that for free ligand. This information, together with that for total concentrations(s), is stored in ASCII format – ready for importation as data files into any standard nonlinear regression package for analysis of their interdependence in terms of the appropriate interaction model(s).

Details of the computer program for analysis of a Rayleigh interference scan of the sedimentation equilibrium distribution for an interacting system are summarized as a flow chart in Fig. 1, where subscripts 1 and 2 refer to the smallest and next-smallest macromolecular reactants, respectively – monomer and dimer for a self-associating solute; ligand (S) and acceptor (A) for an interaction between dissimilar solutes. This program incorporates evaluation of the meniscus concentration, $J(r_a)$, by the MSTAR approach [14, 15] and hence applies to the analysis of Rayleigh distributions obtained by the low-speed [16] as well as meniscus-depletion [17] methods of sedimentation equilibrium. Although the value of the meniscus concentration, $J(r_a)$, evaluated by such means may differ from the actual value by as much as 15% [18], the subsequent prediction (step h) of the loading concentration (J_0) provides an indication of the reliability of $J(r_a)$ in instances where an experimental value of J_0 has been determined from a synthetic boundary run on the loaded mixture. Furthermore, an iterative procedure has been incorporated at that stage to allow adjustment of $J(r_a)$ until consistency between the predicted and experimental values of J_0 is achieved. Such determination of $J(r_a)$ then becomes

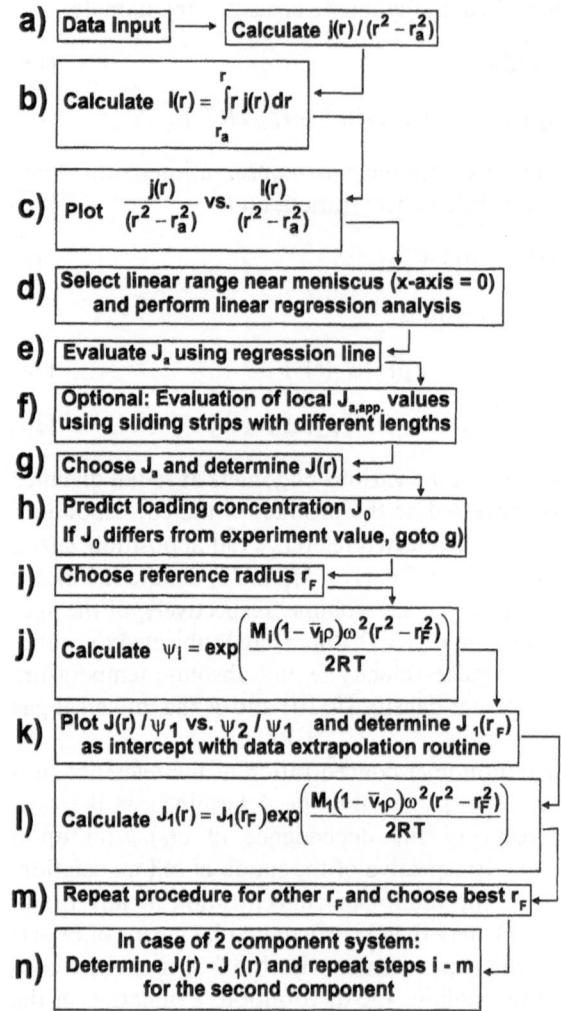

Fig. 1 Flow chart of the psi program for evaluating the concentration of the smallest species contributing to the Rayleigh sedimentation equilibrium distribution for an interacting system

the equivalent of the mass conservation procedure devised by Richards and Schachman [19].

Because the initial part of the program for analysis of the absorption scan of a sedimentation equilibrium distribution is simpler, steps a–h of Fig. 1 are replaced by steps a–c of Fig. 2, after which steps i–n becomes a common pathway.

Values of total concentration, $J(r)$ or $A(r)$, as well as that of the smallest solute, $J_1(r)$ or $A_1(r)$, and (where relevant) next smallest reactant, $J_2(r)$ or $A_2(r)$, are stored as ASCII files ready for export to any suitable nonlinear regression software for analysis of their interdependence. In addition, the data files for diagrams illustrating the evaluation of $J_1(r_F)$ or $A_1(r_F)$, etc., are also saved in ASCII format to allow their export to plotting software for purposes of figure production. This emphasizes that the

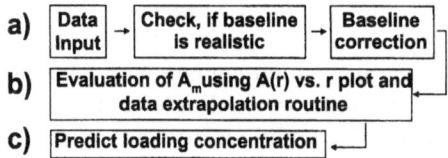

Fig. 2 Adaptation of the initial steps of the flow chart (Fig. 1) for application of the psi procedure to absorption scans of the sedimentation equilibrium distribution for an interacting system: the original flow chart is then re-entered at step i

software is designed as an open platform for the evaluation of sedimentation equilibrium distributions – just like the MSTAR program [15].

Application to simulated sedimentation distributions

In order to illustrate the use of the psi program we have simulated sedimentation equilibrium distributions for a range of interacting systems, and then tested the ability of the program to return the magnitudes of parameters used in the simulations. Inasmuch as the psi procedures have already been illustrated by application to sedimentation equilibrium distributions reflecting self-association [8] and interaction between dissimilar reactants [11, 13], the use of simulated data is considered to provide a more stringent test of the program's capabilities. The power and simplicity of this psi approach is first illustrated by its application to simulated Rayleigh distributions for a self-associating solute with characteristics akin to those of α-chymotrypsin under slightly acid conditions (pH 3.9, I = 0.2).

Characterization of solute self-association

The monomer distribution, $J_1(r)$, in a high-speed sedimentation equilibrium experiment on α-chymotrypsin ($M_1 = 25\,000$ g/mol) was simulated via Eq. (1) on the basis of $J_1(r_F) = 1.000$ fringe at a reference radius (r_F) of 7.10 cm within a liquid column with radial limits of 6.90 cm (r_a) and 7.15 cm (r_b) in an experiment conducted at 20 °C and 35 000 rpm. The buoyancy term was based on a buffer density (ρ) of 1.005 g/ml and partial specific volume $(\bar{v}_1)$ of 0.736 ml/g, the value for the enzyme. Corresponding values of $J_2(r)$ were then calculated from the expression $c_2(r) = X_2[c_1(r)]^2$ with the dimerization constant taken as 3 l/g [20] and a value of 3.0 for the number of fringes generated by a 1 mg/ml protein solution in the Beckman XL-I ultracentrifuge. The total distribution, $J(r) = J_1(r) + J_2(r)$, is shown as the solid line in Fig. 3a, whereas the evaluation of $J_1(r_F)$ by the psi procedure (Eq. 3b) is

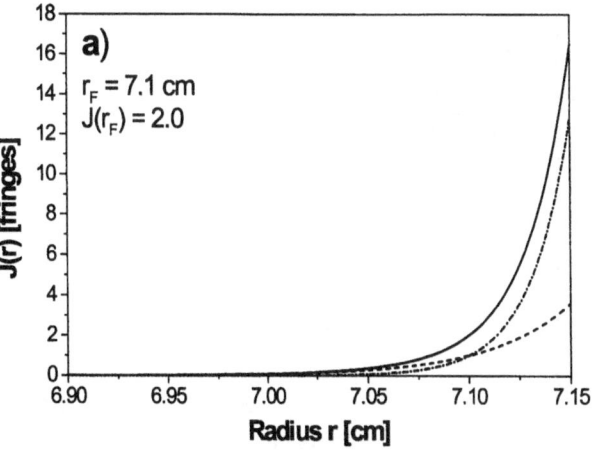

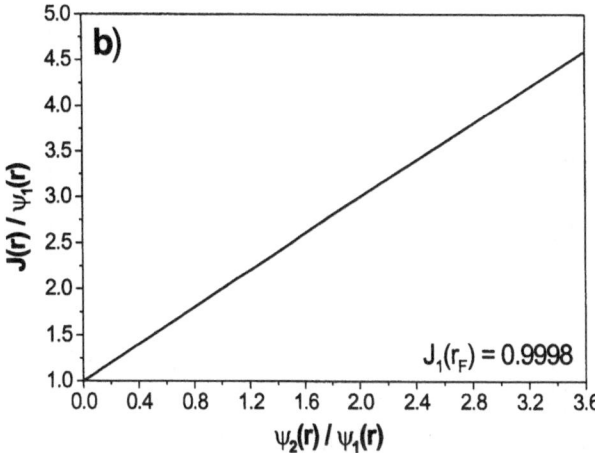

Fig. 3 Application of the psi program to the Rayleigh distribution in a simulated high-speed experiment (35 000 rpm) on α-chymotrypsin. (a) The simulated total concentration distribution (——), together with the derived distributions for monomer (- - -) and dimer (-·-··-··-). (b) Plot of the total distribution data in terms of Eq. (3b) to obtain $J_1(r_F)$ and hence the monomer distribution [Eq. (1)]

summarized in Fig. 3b. Clearly, the ordinate intercept of 0.9998 fringes returns the input value of 1.000 fringes for $J_1(r_F)$. Because the linearity of Fig. 3b signifies the absence of oligomers larger than dimer [Eq. (4)], $J_2(r)$ may be determined as $J(r) - J_1(r)$. The calculated radial dependences of the monomer and dimer concentrations (Fig. 3a) duplicate the simulated distributions.

In order to illustrate the use of the psi program with Rayleigh data from a low-speed experiment, the speed of the simulated run was decreased to 15 000 rpm and the value of $J_1(r_F)$ increased to 2.000 fringes. Figure 4a presents the simulated total distribution (——), which was also duplicated by iterative application of the MSTAR program to the "experimental" fringe data, $j(r) = J(r) - J(r_a)(\cdots)$ to determine $J(r_a)$ and hence $J(r)$ throughout the distribution. From the subsequent analysis of the total

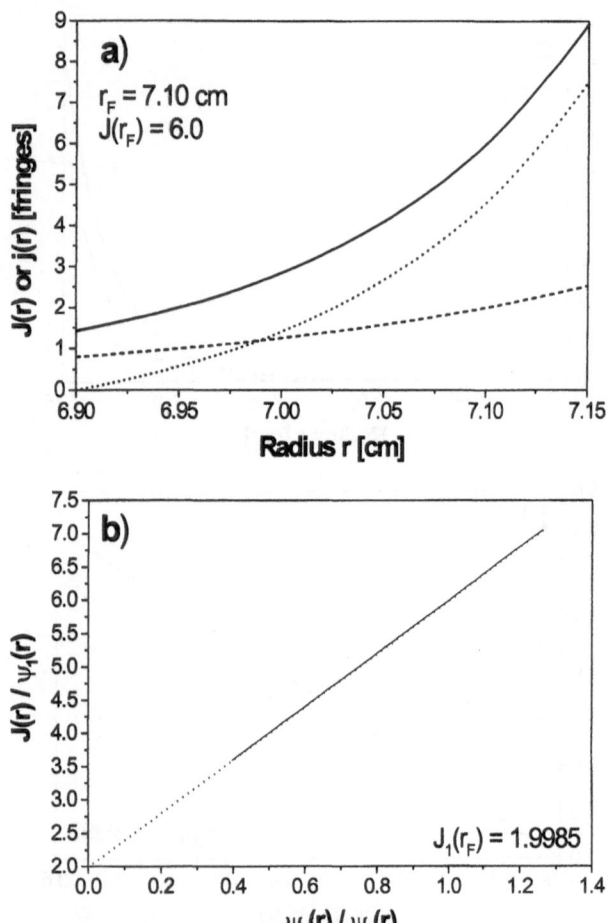

Fig. 4 Use of the psi program to analyze the Rayleigh distribution in a simulated low-speed sedimentation experiment (15 000 rpm) on α-chymotrypsin. (a) The simulated total concentration distribution $J(r)$ (——), together with the observable record $j(r)$ ($\cdots$) and the distribution for monomer (- - -). (b) Analysis of the total distribution data in terms of Eq. (3b) to obtain $J_1(r_F)$ and hence the calculated monomer distribution [Eq. (1)]. The dotted line shows the extrapolation to the intercept ($\cdots$)

distribution in terms of Eq. (3b), it is evident that the input value of 2.000 fringes for $J_1(r_F)$ is again being returned by the psi program (Fig. 4b). The inferred radial dependence of $J_1(r)$ (- - -) therefore duplicates the simulated monomer distribution.

Interactions between dissimilar reactants

For the simulation and analysis of distributions reflecting heterogeneous association, we have returned to a model 1:1 interaction used previously [11] in relation to the interaction of trimethylamine dehydrogenase (A) and electron-transferring flavoprotein (S) – a system examined by the absorption optical system of the Beckman XL-A

ultracentrifuge. Simulations were based on an absorbance of 0.150 for both $A_S(r_F)$ and $A_A(r_F)$ at a reference radial position (r_F) of 7.05 cm in a liquid column with 6.90 and 7.10 cm as the respective inner (r_a) and outer (r_b) extremities. The molar masses of A and S were taken as 150 000 and 50 000 g/mol, respectively, in simulated experiments conducted at 4 °C and 15 000 rpm, a value of 0.25 for the buoyancy term, $(1 - \bar{v}_i\rho)$, having been used for both reactants. The magnitude of $A_{AS}(r)$ was calculated as $KA_A(r)A_S(r)\varepsilon_{AS}/[\varepsilon_A\varepsilon_S]$ on the basis of an equilibrium constant of 10^5 M^{-1} and the molar absorption coefficients at 280 nm (ε_i) reported in ref. [11]. In practice, the experimental system represented the situation in which only the distribution in terms of total absorbance could be measured, and accordingly we commence illustration of the use of the psi program with that most unfavorable case [Eqs. (7) and (8)].

Psi analysis in terms of total absorbance

Analysis of the simulated total absorbance (——, Fig. 5a) in terms of Eq. (7b) with $A(r_F) = 0.320$ is shown in Fig. 5b, where the subscripts 1 and 2 refer to S and A (smallest and next-smallest species), respectively. The inference [Eq. (7b)] that $A_S(r_F)$, the absorbance due to free ligand at the reference radial position, should emanate from such analysis is borne out by the identity of the ordinate intercept and the input value of 0.150 for $A_S(r_F) \equiv A_1(r_F)$.

The next stage of the psi procedure for analyzing this least-favorable situation entails the generation of a revised total absorbance distribution in which acceptor (A) is the smallest species – a distribution obtained by subtracting $A_S(r)$(- - -, Fig. 5a) from $A(r)$ to obtain $A'(r)$ ($\cdots$). Analysis of the revised distribution in terms of Eq. (8b) with the subscripts 2 and 3 denoting A and AS ($M_{AS} = 200\,000$ g/mol) respectively is presented in Fig. 5c. Although this plot should be strictly linear, there is obvious departure from such behavior in the low range of $[\psi_{AS}(r)/\psi_A(r)] \equiv [\psi_2(r)/\psi_1(r)]$. On the grounds that neither A nor AS is contributing to $A'(r)$ in the region near the meniscus (Fig. 5a), this part (reflecting the presence of free S alone) should be disregarded for the evaluation of $A_A(r_F)$. The dotted vertical line denotes the lower limit of $[\psi_{AS}(r)/\psi_A(r)]$ used to obtain an ordinate intercept of 0.152 for $A_A(r_F)$, which is certainly an acceptable estimate of the input value (0.150). Because the revised absorption distribution $A'(r)$ incorporates the error associated with the evaluation of $A_S(r)$, the estimate of $A_A(r_F)$ includes not only the uncertainty associated with its evaluation from $A'(r)$ but also that inherent in the subtraction of $A_S(r)$ from $A(r)$ to obtain the revised distribution being analyzed. The consequent estimate of K from Eq. (9) is thus subject to

Progr Colloid Polym Sci (1997) 107:36–42
© Steinkopff Verlag 1997

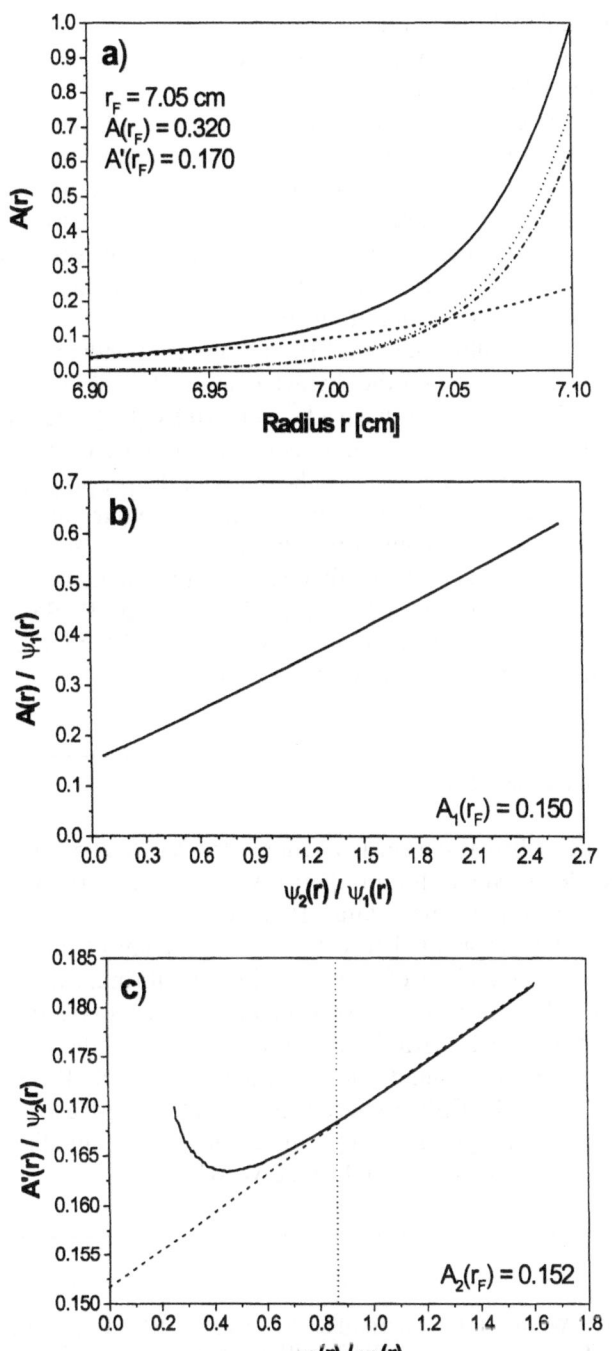

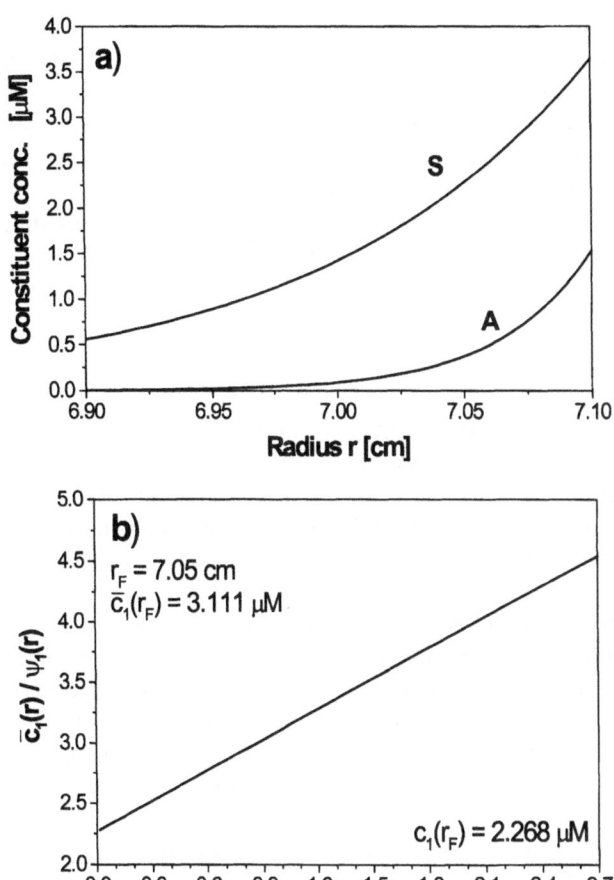

Fig. 6 Use of the psi program to analyze the sedimentation equilibrium distribution for an acceptor-ligand interaction in instances where the separate constituent distributions, $\bar{C}_S(r)$ and $\bar{C}_A(r)$, are available: (a) simulated distributions for total ligand and total acceptor; (b) Determination of $C_S(r_F)$ by means of Eq. (5b)

Fig. 5 Application of the psi program to the total absorbance distribution in a simulated sedimentation equilibrium experiment (15000 rpm) on an acceptor–ligand system undergoing 1:1 interaction ($M_A = 150\,000$ g/mol, $M_S = 50\,000$ g/mol). (a) The simulated total distribution (——), together with the distribution determined for free ligand (- - -), the revised total distribution (···) and the determined distribution for free acceptor (-·-·-·-). (b) Evaluation of $A_S(r_F)$ by means of Eq. (7b). (c) Determination of $A_A(r_F)$ by means of Eq. (8b). The dashed line shows the extrapolation to the intercept (- - -)

greater uncertainty than a value deduced from sedimentation equilibrium experiments in which separate distributions are available for the acceptor and ligand constituents.

Separate distributions available for the acceptor and ligand constituents

To illustrate the evaluation of the free ligand distribution in instances where separate concentration distributions, $\bar{C}_A(r)$ and $\bar{C}_S(r)$, may be determined from a sedimentation equilibrium experiment, we use the same simulated data set, but consider that the values of $\bar{C}_S(r)$ and $\bar{C}_A(r)$ were obtained by combining the $A(r)$ values at different wavelengths with the relevant absorption coefficients. Those separate distributions are shown as the solid lines in Fig. 6a. In applying the psi program to the distribution for the S constituent it must be remembered that subscript

2 refers to the second smallest species contributing to the distribution – the complex AS [Eq. (5)]. The ordinate intercept of the requisite psi plot (Fig. 6b) signifies a value of 2.268 μM for $C_S(r_F)$, which is essentially the input value (2.269 μM). On the grounds that $C_S(r)$ may therefore be calculated satisfactorily throughout the distribution, the psi program is terminated at this stage, and the binding function determined from Eq. (6).

Discussion

Two major features emerge from these illustrative applications of the psi analysis: (i) the simplicity of the analysis to obtain the free concentration (thermodynamic activity) of smallest reactant throughout the sedimentation equilibrium distribution; and (ii) the independence of that free concentration distribution upon any specific model of the interaction. Thus, in relation to the characterization of solute self-association, the psi plot (Figs. 3b and 4b) yields the value of $c_1(r_F)$ irrespective of the number of oligomeric species that coexist in equilibrium with monomer. Similarly, the evaluation of $C_S(r_F)$ (Figs. 5b and 6b) and $C_A(r_F)$ (Fig. 5c) is independent of the reaction stoichiometry between acceptor and ligand. An obvious advantage of the psi analysis [8, 11, 13] and this programmed version thereof is their ability to provide an unequivocal $[c(r), c_1(r)]$ data set for consideration in terms of possible self-association models: the corresponding data set for heterogeneous association is either $[\bar{C}_A(r), \bar{C}_S(r), C_S(r)]$ (Fig. 6) or $[\{A(r) - A_A(r)\} - A_S(r), A_A(r), A_S(r)]$ (Fig. 5). Although models of the interaction certainly have to then be introduced for quantitative interpretation of such information, the experimental data sets themselves are model-independent. In that regard the psi approach has definite advantages over its counterparts based on simulation of the

experimental distribution(s) [1–4], for which the magnitudes of the free concentrations become dependent upon the model being considered.

As noted elsewhere [8, 13], the psi analysis also has the potential to accomodate allowance for effects of thermodynamic nonideality superimposed on the interaction. Under those circumstances the parameter evaluated by psi analysis is the thermodynamic activity, $z_i(r)$, of the smallest species that is determined. The procedure for characterizing nonideal self-association from the $[c(r), c_1(r)]$ data set has already been illustrated [8, 13], and a corresponding procedure is imminent for the characterization of a 1 : 1 interaction between reactants from a $[\bar{C}_A(r), \bar{C}_S(r), C_S(r)]$ data set.

We hope that this emphasis on the simplicity, model independence and general applicability of the psi program may convince a wider audience of the advantages of this approach to the quantitative characterization of macromolecular interactions. To that end the program described in this communication has been placed on the RASMB server for the benefit of others who may wish to take advantage of its application free of charge.

Program availability

The described program is available for XL-A and XL-I data files as well as for another ASCII file input format of Rayleigh interference data. It is combined with the MSTAR analysis [15] to extend the spectrum of model independent analysis of sedimentation equilibrium experiments. The program can be downloaded using anonymous FTP from BBRI.HARVARD.EDU. After login, the program can be found in /RASMB/SPIN/MS_DOS/ MSTAR-COELFEN. In the event of problems with downloading the program, a copy can be obtained by emailing to COELFEN@MPIKG-TELTOW.MPG.DE.

References

1. Johnson ML, Correia JJ, Yphantis DA, Halvorson HR (1981) Biophys J 36:575
2. Johnson ML, Straume M (1994) In: Schuster TM, Laue TM (eds) Modern Analytical Ultracentrifugation. Birkhäuser, Boston, p 37
3. Laue TM, Senear DF, Eaton S, Ross AJB (1993) Biochemistry 32:2469
4. Kim T, Tsuykiyama T, Lewis MS, Wu C (1994) Protein Sci 3:1040
5. Milthorpe BK, Jeffrey PD, Nichol LW (1975) Biophys Chem 3:169
6. Nichol LW, Jeffrey PD, Milthorpe BK (1976) Biophys Chem 4:259
7. Jeffrey PD, Nichol LW, Teasdale RD (1979) Biophys Chem 10:379
8. Wills PR, Jacobsen MP, Winzor DJ (1996) Biopolymers 38:119
9. Wills PR, Winzor DJ (1992) In: Harding SE, Rowe AJ, Horton JC (eds) Analytical Ultracentrifugation in Biochemistry and Polymer Science. Roy Soc Chem. Cambridge, p 311
10. Wills PR, Comper WD, Winzor DJ (1993) Arch Biochem Biophys 300:206
11. Wilson EK, Scrutton NS, Cölfen H, Harding SE, Jacobsen MP, Winzor DJ (1997) Eur J Biochem 243:393
12. Klotz IM (1946) Arch Biochem 9:109
13. Wills PR, Jacobsen MP, Winzor DJ (1997) Progr Colloid Polym Sci (this issue)
14. Harding SE, Horton JC, Morgan PJ (1992) In: Harding SE, Rowe AJ, Horton JC (eds) Analytical Ultracentrifugation in Biochemistry and Polymer Science. Roy Soc Chem, Cambridge, p 275
15. Cölfen H, Harding SE (1997) Eur Biophys J 25:333
16. Van Holde KE, Baldwin RL (1958) J Phys Chem 62:734
17. Yphantis DA (1964) Biochemistry 3:297
18. Creeth JM, Harding SE (1982) J Biochem Biophys Methods 7:25
19. Richards EG, Schachman HK (1959) J Phys Chem 63:1578
20. Tellam R, de Jersey J, Winzor DJ (1979) Biochemistry 18:5316

Progr Colloid Polym Sci (1997) 107:43–57
© Steinkopff Verlag 1997

H. Durchschlag
P. Zipper

Calculation of hydrodynamic parameters of biopolymers from scattering data using whole-body approaches

Dr. H. Durchschlag (✉)
Institute of Biophysics and Physical
Biochemistry
University of Regensburg
Universitätsstraße 31
93040 Regensburg, Germany

P. Zipper
Institute of Physical Chemistry
University of Graz
8010 Graz, Austria

Abstract A calculation procedure is presented which relates solution scattering and hydrodynamic parameters. Biopolymers are modeled by whole-body approaches, approximating their overall shape by spheres or prolate/oblate ellipsoids of revolution. Molar masses, partial specific volumes, radii of gyration, volumes and surface-to-volume ratios are used for predicting sedimentation and diffusion coefficients and intrinsic viscosities of a variety of biopolymers, in addition to the derivation of several further parameters such as frictional coefficients, Simha factors, Stokes and viscosity radii. The establishment of a comprehensive set of structural and hydrodynamic data including several correlations allows the examination of observed and predicted parameters. In this context also the validity of some empirical relations was tested. A variety of roughly globular biopolymers (simple and conjugated proteins, ribonucleic acids) of different molar mass and shape have been examined. The comparisons comprise both the native states of the biopolymers under analysis and structural alterations in response to changes in environment or state of ligation. Far-reaching conformity between experimental values and anticipated parameters was achieved. Detailed error propagation calculations allow a close scrutiny of the accuracy of the parameters to be predicted.

Key words Biopolymers – analytical ultracentrifugation – viscometry – scattering – parameter predictions – modeling – whole-body approaches

Introduction

High-resolution crystallography and NMR spectroscopy provide the most detailed insight into the structure of biopolymers regarding gross structure as well as structural details. For the analysis of the overall structure in solution, the solution scattering techniques (small-angle X-ray and neutron scattering, light scattering) are also highly effective, while the hydrodynamic techniques (analytical ultracentrifugation, size-exclusion chromatography, viscometry and densimetry) generally give less structural details but are more convenient to use. Among the physical parameters obtainable from the structural and hydrodynamic analyses are: radii of gyration, volumes, surface areas, axial ratios, frictional coefficients, Stokes radii, sedimentation and diffusion coefficients, intrinsic viscosities, molar masses, partial specific volumes, and hydration, in addition to atomic coordinates obtainable from crystallographic work. In order to understand the information provided by these techniques and to allow a quantitative comparison of data, we are striving for correlations

Abbreviations *Methods*: AUC, analytical ultracentrifugation; SAS, small-angle scattering; SAXS, small-angle X-ray scattering; *Models*: OE, oblate ellipsoid; PE, prolate ellipsoid; S, sphere

between all the parameters mentioned. This prompted us to apply several approaches to establish possible relations between scattering/diffraction and hydrodynamic properties.

Hydrodynamic and solution scattering properties of biopolymers may be predicted by means of whole-body or multibody approaches, approximating macromolecular structures by simple geometric bodies or assemblies of such bodies, respectively (e.g., refs. [1–5]). In the first case, roughly globular biopolymers can be modeled as prolate or oblate ellipsoids of revolution or spheres. In general, axial ratios are obtained by exploiting the values for radius of gyration and hydrated volume or surface-to-volume ratio. Combining these axial ratios with other parameters (volume, molar mass, partial specific volume), several hydrodynamic parameters may be estimated [2, 4, 6–10]: frictional ratios, sedimentation and diffusion coefficients, Simha factors, intrinsic viscosities, radii of equivalent spheres (such as Stokes and viscosity radii).

On the basis of scattering or diffraction data, hydrodynamic properties of several globular proteins, a few viruses and small ribonucleic acids have been predicted successfully by means of whole-body approaches [2, 6–8]. Recently, we applied these approaches also to various nonstandard proteins (such as conjugated proteins, anisometric and multisubunit proteins) and high-molecular-weight ribonucleic acids, and to the anticipation of subtle structural changes [4, 9, 10]. Above all, we extended the repertoire of parameters to be predicted and the methods involved considerably. Reliable predictions based on crystal data were only achieved, if appropriate hydrational contributions were considered in the models. We also tested and improved several empirical equations relating scattering and hydrodynamic data.

In this study, we present a set of further examples, demonstrating the anticipation and derivation of several hydrodynamic and structural properties of several classes of biopolymers. The performance of error propagation calculations, taking into account typical errors for all individual input parameters, allows the estimation of the accuracy of the predicted hydrodynamic parameters. Such error estimations are crucial for possible statements on conformational changes.

Theory

Correlations between hydrodynamic and scattering data

For homogeneous particles of roughly globular shape, modeling the solution conformation of biopolymers is based on a whole-body approach, using prolate ($p > 1$) or oblate ($p < 1$) ellipsoids of revolution or spheres and the

theoretical formalism outlined by Kumosinski and Pessen [2, 6, 7] and us [4, 9, 10]. In general, axial ratios p are determined by combining the radius of gyration R_G with either the hydrated volume V or the surface-to-volume ratio S/V. In the case of inhomogeneous particles, we have to use special calculation procedures, in order to meet the requirements of the particle peculiarities.

The theoretical background and the basic equations for calculations and predictions of hydrodynamic data have been outlined in ref. [9]. In this paper, therefore, we restrict ourselves mainly to equations and conclusions not mentioned there.

Sedimentation and diffusion coefficients, s and D, are predicted on the basis of a modified Svedberg equation:

$$s = \frac{M(1 - \bar{v}\rho)}{(f/f_0)6\pi\eta N R_0} , \tag{1}$$

$$D = \frac{kT}{(f/f_0)6\pi\eta R_0} , \tag{2}$$

where M and $\bar{v}$ are the anhydrous molar mass and anhydrous partial specific volume of the macrosolute, respectively; M is obtained from the sequence or composition or, alternatively, from the small-angle scattering (SAS) experiment, and $\bar{v}$ by densimetry or by a calculation approach (cf. ref. [11]). The quantities f/f_0 and R_0 denote the frictional ratio and a fictive radius of the hydrated particle, respectively. As usual, N symbolizes Avogadro's number, k the Boltzmann constant, T the temperature, and ρ and η are the density and viscosity of water at 20 °C, respectively.

The fictive radius R_0 represents the radius of a sphere with a volume identical to the hydrated particle volume V:

$$R_0 = \left(\frac{3V}{4\pi}\right)^{1/3} , \tag{3}$$

where the hydrated volume V is obtainable directly from a SAXS study by means of a Porod analysis in terms of the correlation volume, without any assumption of a geometric model [12, 13]. If a direct analysis of the hydrated volume is not available, the volume of a geometric model considering the hydrodynamic properties of the particle adequately may be used instead.

By contrast, as is well known, combination of M and $\bar{v}$ yields only the "dry" volume of the particle. Use of V_{dry} instead of V in Eq. (3), neglecting the particle hydration, only yields the radius R_{dry} of an unhydrated particle:

$$V_{dry} = M \cdot \bar{v}/N , \tag{4}$$

$$R_{dry} = \left(\frac{3V_{dry}}{4\pi}\right)^{1/3} . \tag{5}$$

The frictional ratio f/f_0 of the hydrated particle is calculated according to Perrin's formulae for prolate ($p > 1$) or oblate ($p < 1$) ellipsoids of revolution, where the axial ratio $p = a/b$ is defined as the ratio of the semiaxis a of revolution to the equatorial semiaxis b of the ellipsoid (cf. refs. [2, 9]).

Usually, the axial ratios p of the hydrated particle can be obtained by combining R_G with $V (p \neq 1)$ (cf. refs. [2, 9]). It should be emphasized that for the estimation of axial ratios, the hydrated volume has to be used. Applying the dry volume would result in a dramatic overestimation of particle anisometry. Indeed, neglect of particle hydration has led to many misinterpretations of hydrodynamic results in the past.

If the surface-to-volume ratio S/V of the hydrated particle is additionally known, two further rather complex equations may be used for the estimation of p ($p >$ or < 1) (cf. refs. [2, 9]). Alternatively, the value for S/V may be roughly estimated from V and S of a smooth model [2].

The respective relations between p, R_G and V or S/V are based on the geometric relations between V and R_G, on the one hand, and the ellipsoid dimensions a and b, on the other:

$$V = \tfrac{4}{3}\pi a b^2 , \tag{6}$$

$$R_G = [(a^2 + 2b^2)/5]^{1/2} , \tag{7}$$

and rather complicated expressions for S or S/V in terms of a and b [14].

In special cases another manner of acting has to be adopted for the calculation of p: For strictly spherical particles (e.g., some viruses) p has to be taken as 1. If only the particle size is known (e.g., from SAS studies), p can be calculated from the given particle dimensions [4].

The so-called Stokes (hydrodynamic) radius R_D is given by the product of R_0 and f/f_0:

$$R_D = R_0(f/f_0) . \tag{8}$$

Since the Stokes radius is usually obtained from experimental diffusion coefficients D (using the Stokes–Einstein relation by combining Eqs. (2) and (8)), the subscript D [4, 10] seems to be more appropriate than subscript 0 [2, 6, 7, 9].

The intrinsic viscosity $[\eta]$ of the hydrated particle is obtained from V and M, combining the respective quotient with tabulated values for the Simha factor v derived for various p values [15]:

$$[\eta] = \frac{VN}{M} v . \tag{9}$$

Combining R_0 and v allows the prediction of the viscosity radius R_η, a quantity defined similar to R_D.

However, R_η may also be obtained from experimental values of $[\eta]$ [16]:

$$R_\eta = R_0 \left(\frac{v}{2.5} \right)^{1/3} , \tag{10}$$

$$R_\eta = \left(\frac{3}{4\pi} \frac{M[\eta]}{2.5 N} \right)^{1/3} . \tag{11}$$

On the basis of the definitions for the above-mentioned radii (R_G, R_0, R_D, R_η) theoretical expressions for various radii ratios have been deduced [4]. An analysis of these expressions revealed that the resulting quantities are distinctly different for spheres and prolate and oblate ellipsoids of revolution of varying axial ratio p. For example, R_G/R_0 shows the strongest dependence on p, and R_D/R_0 the weakest; R_η/R_0 resembles R_D/R_0 (thus, $R_\eta/R_D \simeq 1$), especially for oblate ellipsoids.

Use of the experimental values for both scattering and hydrodynamic parameters (R_G, and s, D, $[\eta]$), allows the calculation of experimental radii ratios [4]:

$$\left(\frac{R_G}{R_D} \right)_s = \frac{R_G \, s \, 6\pi\eta N}{M(1 - \bar{v}\rho)} , \tag{12}$$

$$\left(\frac{R_G}{R_D} \right)_D = \frac{R_G \, D \, 6\pi\eta}{kT} , \tag{13}$$

$$\left(\frac{R_G}{R_\eta} \right)_\eta = R_G \left(\frac{10\pi N}{3M[\eta]} \right)^{1/3} . \tag{14}$$

Correlations between hydrodynamic and diffraction data

Exploiting the information from crystallographic data (atomic coordinates) allows the calculation of the radius of gyration R_G (cf. refs. [5, 13, 17]). As may be anticipated, use of the anhydrous volume (sum of atomic volumes) leads to erroneous predictions of hydrodynamic data. Instead, a hydrated volume V has to be generated, assuming a hydration of $\delta_1 = 0.35$ g/g or adopting an experimentally determined hydration (e.g., obtained from SAXS or NMR studies) [9, 10]. Again, the axial ratio p may then be calculated from R_G and V.

Empirical relations between hydrodynamic and scattering/diffraction data

On the basis of a set of experimental values for structural and hydrodynamic parameters as well as averages of structural parameters derived from crystallographic data, attempts have been made to establish purely empirical correlations, especially with respect to application to globular proteins.

The sedimentation coefficient s may be estimated by combining M and R_D [18], and applying corrections for the partial specific volume in the case of nonstandard (e.g., conjugated) proteins [4]:

$$s_{M, RD, corr.} = \frac{M}{4.3 \times 10^{13} R_D} \frac{1 - \bar{v}\rho}{1 - 0.735\,\rho}, \tag{15}$$

where s is given in s, M in kg mol^{-1} and R_D in nm. If only viscosity data are available, R_D can be replaced with R_η.

For estimating the diffusion coefficient D_{RG} of proteins from R_G, Tyn and Gusek [19] proposed an empirical correlation based on the Stokes–Einstein equation:

$$D_{RG} = \frac{1.69 \times 10^{-6}}{R_G} \tag{16}$$

for D_{RG} given in $\text{cm}^2\,\text{s}^{-1}$ and R_G in nm.

For monomeric proteins, Teller [20] established empirical relations between packing volume V_p and accessible surface area A_s, on the one hand, and molar mass M, on the other:

$$V_p = (1.273 \pm 0.006)\,M\,, \tag{17}$$

$$A_s = (11.116 \pm 0.161)\,M^{2/3}\,, \tag{18}$$

where V_p, A_s, and M are given in nm^3, nm^2, and kg mol^{-1}, respectively.

The radii, R_p and R_s (both in nm), may be deduced from V_p and A_s, when treating the proteins as spheres:

$$R_p = \left(\frac{3V_p}{4\pi}\right)^{1/3} = \left(\frac{3 \times 1.273}{4\pi}\right)^{1/3} M^{1/3} = 0.6723\,M^{1/3}\,, \tag{19}$$

$$R_s = \left(\frac{A_s}{4\pi}\right)^{1/2} = \left(\frac{11.116}{4\pi}\right)^{1/2} M^{1/3} = 0.9405\,M^{1/3}\,, \tag{20}$$

where R_s is $1.4\,R_p$, due to the roughness of the protein surface.

From SAXS studies, similar quantities (V/M, $S/M^{2/3}$, $R_0/M^{1/3}$, and $R_G/M^{1/3}$) as outlined in Eqs. (17–20) may be derived. Therefore, a comparison of the factors in the corresponding relations allows conclusions regarding differences in particle hydration and surface roughness (rugosity), as determined by solution and crystallographic methods, respectively.

Results and discussion

In previous papers [4, 9, 10] we have already considered several globular and fibrous proteins of different subunit composition, both simple proteins as well as glyco- and nucleoproteins, and also ribonucleic acids for our parameter calculations and predictions. In addition, we have tested the possibility to predict changes of hydrodynamic parameters due to ligand binding. For the current calculations we chose several further particles of different molecular properties with respect to M, $\bar{v}$, R_G, V, S/V and shape, in order to prove the applicability of our calculation approaches on a broader basis. Input parameters were adopted from the literature: scattering and hydrodynamic data mainly from compilations in [2, 8, 19]; accurate values of molar masses may be taken from protein sequence databanks (e.g., ref. [21] or SWISS-PROT), and partial specific volumes from refs. [22, 23].

Documentation of data

In Table 1 experimental and calculated structural and hydrodynamic parameters of the selected biopolymers are listed, together with a few details concerning source and shape of the polymer and the calculation approach which has been used. The data of each polymer comprise three types of information:

(i) The first line contains the available input parameters: M, $\bar{v}$, R_G, V and S/V, and additionally s, D and $[\eta]$ for comparison with the predicted values. From these data, all the parameters in the lines beneath are derived.

(ii) Quantities which may be derived directly from the input parameters are summarized in a data block starting in line 2; empty fields in this area are caused by missing input data. To mention a few possibilities of interesting comparisons: The calculated radii (R_0, R_p, R_s, R_D, R_η), volumes (V_{dry}, V_p) and the surface-to-volume ratio (A_s/V_p) may be compared with the corresponding input parameters R_G, V, and S/V, and the parameter combinations $R_G(S/V)$, $(S/V)M^{1/3}$, $S/M^{2/3}$, V/M, and $R_G/M^{1/3}$ with the factors established on the basis of crystal data (Eqs. (17–20)). The behavior of the radii ratios, R_G/R_0, $(R_G/R_D)_s$, $(R_G/R_D)_D$, and $(R_G/R_\eta)_\eta$, may be of interest in context with the axial ratios to be calculated. First hints concerning approximate shapes may be inferred from the Simha factor v and the axial ratio p derived therefrom. The validity of empirical relations may be concluded from the values for D_{RG}, $s_{M, RD, corr.}$, $s_{M, R_\eta, corr.}$, $(F_{corr.})_D$ and $(F_{corr.})_\eta$.

(iii) The parameter predictions using the previously mentioned approaches for shape modeling are marked in the table by the calculation procedure applied (S, PE, OE, indicating spheres, prolate or oblate ellipsoidal models, respectively). The results comprise the parameters p, f/f_0, s, D, R_D, v, $[\eta]$ and R_η, which may be compared with the data in (i) and (ii). For ellipsoidal models, the results derived from R_G and V are given in the first line of this data block, and those obtained from R_G and S/V in the second.

Table 1 Structural and hydrodynamic parameters of biopolymers from SAS and hydrodynamic analyses, together with various correlations of data, and predictions of hydrodynamic parameters from SAS data. Molecules are arranged according to increasing molar mass

Biopolymer (source)	1	2	3	4	5	6	7	8	9	10	11	12
{Description of polymer, shape}	M (kg mol⁻¹)	$\bar{v}$ (cm³ g⁻¹)	R_G (nm)	V (nm³)	S/V (nm⁻¹)	$s \times 10^{13}$ (s)	$D \times 10^7$ (cm² s⁻¹)	$[\eta]$ (cm³ g⁻¹)	V_p (nm³)	A_s (nm²)	R_p (nm)	R_s (nm)
	S (nm²)	$R_G(S/V)$	R_o (nm)	R_G/R_o	$(S/V)M^{1/3}$ (nm⁻¹ kg¹ᐟ³ mol⁻¹ᐟ³)	$S/M^{2/3}$ (nm² kg⁻²ᐟ³ mol²ᐟ³)	V/M (nm³ kg⁻¹ mol)	$R_G/M^{1/3}$ (nm kg⁻¹ᐟ³ mol¹ᐟ³)	$D_{RG} \times 10^7$ (cm² s⁻¹)	$R_G{\cdot}D \times 10^7$ (nm cm² s⁻¹)	$(R_G{\cdot}s)/M \times 10^{13}$ (nm s kg⁻¹ mol)	$(R_G/R_D)_s$
	A_s/V_p (nm⁻¹)	R_G/R_p	R_G/R_s	V/V_p	S/A_s	V_{dry} (nm³)	V/V_{dry}	$R_o/M^{1/3}$ (nm kg⁻¹ᐟ³ mol¹ᐟ³)	p	p	$s_{M,R_\eta,corr.} \times 10^{13}$ (s)	$(F_{corr})\eta_3$ (kg mol⁻¹ nm⁻¹ s⁻¹)
	R_D (nm)	$(R_G/R_D)_b$	$R_D/M^{1/3}$ (nm kg⁻¹ᐟ³ mol¹ᐟ³)	$s_{M,R_D,corr.} \times 10^{13}$ (s)	$(F_{corr})_D \times 10^{13}$ (kg mol⁻¹ nm⁻¹ s⁻¹)	R_η (nm)	$(R_G/R_\eta)\eta$	$R_\eta/M^{1/3}$ (nm kg⁻¹ᐟ³ mol¹ᐟ³)	R_G/R_D	$s_{M,R_D,corr.} \times 10^{13}$ (s)	$s_{M,R_\eta,corr.} \times 10^{13}$ (s)	
[Calculation approach]^a)	p	f/f_o	$s \times 10^{13}$ (s)	$D \times 10^7$ (cm² s⁻¹)	R_D (nm)	ν	$[\eta]$ (cm³ g⁻¹)	R_η (nm)	R_G/R_D	$s_{M,R_D,corr.} \times 10^{13}$ (s)	$s_{M,R_\eta,corr.} \times 10^{13}$ (s)	$(R_G/R_D)_s$
Ribonuclease (bovine pancreas) {globular protein}	13.69 [21]	0.702 [23]	1.48 [24]	22.0 [24]	2.9 [24]	1.78 [25]	10.68 [26]	3.30 [27]	17.43	63.61	1.61	2.25
	63.8	4.29	1.74	0.851	6.94	11.15	1.61	0.619	11.42	15.8	0.192	0.73
	3.65	0.92	0.66	1.26	1.00	16.0	1.38	0.727	3.41	2.68	1.86	4.48
	2.01	0.74	0.84	1.78	4.31	1.93	0.77	0.81	0.822	1.99	1.98	
[PE]	1.87	1.036	2.00	11.90	1.80	2.83	2.74	1.81	0.733	1.77	1.71	
	3.69	1.16	1.78	10.62	2.02	4.34	4.20	2.09				
α-Lactalbumin (bovine milk) {globular protein}	14.18 [21]	0.704 [22]	1.45 [24]	25.1 [24]	2.4 [24]	1.92 [28]	10.57 [29]	3.01 [19]	18.05	65.12	1.63	2.28
	60.2	3.48	1.82	0.798	5.81	10.28	1.77	0.599	11.66	15.3	0.196	0.75
	3.61	0.89	0.64	1.39	0.93	16.6	1.51	0.750	2.82	1.87	1.95	4.36
	2.03	0.72	0.84	1.82	4.07	1.89	0.77	0.78	0.789	2.00	2.00	
[PE]	1.43	1.012	2.02	11.66	1.84	2.60	2.77	1.84	0.726	1.84	1.81	
	2.81	1.099	1.86	10.73	2.00	3.52	3.76	2.04				
Lysozyme (chicken egg white) {globular protein}	14.31 [21]	0.702 [30]	1.43 [24]	24.2 [24]	2.5 [24]	1.91 [31]	11.2 [31]	3.0 [19]	18.22	65.52	1.63	2.28
	60.5	3.58	1.79	0.797	6.07	10.26	1.69	0.589	11.82	16.0	0.191	0.73
	3.60	0.88	0.63	1.33	0.92	16.7	1.45	0.739	2.95	2.06	1.97	4.44
	1.91	0.75	0.79	1.95	4.40	1.90	0.75	0.78	0.788	2.06	2.06	
[PE]	1.42	1.011	2.08	11.81	1.81	2.59	2.64	1.82	0.720	1.88	1.84	
	2.92	1.107	1.90	10.79	1.99	3.62	3.68	2.03				
α-Chymotrypsin (bovine pancreas) {globular protein}	22.0 [32]	0.736 [33]	1.80 [32]	37.2^b)	1.57 [32]	2.4 [33]	10.2 [33]		28.01	87.28	1.88	2.64
	58.4	2.83	2.07	0.869	4.40	7.44	1.69	0.642	9.39	18.4	0.196	0.84
	3.12	0.96	0.68	1.33	0.67	26.9	1.38	0.739	0.833	2.36	2.34	
	2.10	0.86	0.75	2.43	4.35				0.831	2.35	2.34	
[PE]	2.00	1.044	2.37	9.91	2.16	2.91	2.96	2.18				
	2.02	1.045	2.37	9.90	2.16	2.92	2.98	2.18				
tRNA^phe (yeast) {nucleic acid}	25.0 [34]	0.54 [34]	2.31 [34]	36.6 [34]	2.66 [34]	11.40	7.8 [35]		31.83	95.04	1.97	2.75
	97.5	6.15	2.06	1.12	7.79	22.4	1.46	0.790	7.32	18.0		
	2.99	1.18	0.84	1.15	1.03		1.63	0.704	0.90	3.92	3.88	
	2.75	0.84	0.94	3.66	2.57	5.00	4.40	2.59	0.88	3.82	3.78	
[OE]	0.184	1.247	3.94	8.34	2.63	5.38	4.75	2.66				
	0.166	1.278	3.85	8.14								

Table 1 (continued)

Each data cell is stacked (top-to-bottom) according to the stacked column headings. The first four stacked labels correspond to the main calculation; the last two correspond to the [PE] calculation.

Biopolymer (source) {Description of polymer, shape} [Calculation approach][a]	M (kg mol⁻¹) / S (nm²) / A_s/V_p (nm⁻¹) / R_D (nm) / p / R_D (nm)	v̄ (cm³ g⁻¹) / $R_G(S/V)$ / R_G/R_p / $(R_G/R_D)_D$ / f/f_o / $(R_G/R_p)_D$	R_G (nm) / R_o (nm) / R_G/R_s / $R_D/M^{1/3}$ (nm kg⁻¹ᐟ³ mol¹ᐟ³) / $s\times10^{13}$ (s) / $R_D/M^{1/3}$	V (nm³) / R_G/R_o / V/V_p / $s_{M,RD,corr.}\times10^{13}$ (s) / $D\times10^7$ (cm² s⁻¹) / $s_{M,RD,corr.}\times10^{13}$	S/V (nm⁻¹) / $(S/V)M^{1/3}$ (nm⁻¹ kg¹ᐟ³ mol⁻¹ᐟ³) / S/A_s / $(F_{corr.})_D\times10^{13}$ (kg mol⁻¹ nm⁻¹ s⁻¹) / R_D (nm)	$s\times10^{13}$ (s) / $S/M^{2/3}$ (nm² kg⁻²ᐟ³ mol²ᐟ³) / V_{dry} (nm³) / R_η (nm) / v	$D\times10^7$ (cm² s⁻¹) / V/M (nm³ kg⁻¹ mol) / V/V_{dry} / [η] (cm³ g⁻¹) / R_η (nm)	[η] (cm³ g⁻¹) / $R_G/M^{1/3}$ / $R_\sigma/M^{1/3}$ / $R_p/M^{1/3}$ (nm kg⁻¹ᐟ³ mol¹ᐟ³) / R_η (nm) / R_G/R_D	V_p (nm³) / $D_{RG}\times10^7$ (cm² s⁻¹) / v / R_G/R_D	A_s (nm²) / $R_G\cdot D\times10^7$ (nm cm² s⁻¹) / p / $s_{M,RD,corr.}\times10^{13}$ (s)	R_p (nm) / $(R_G\cdot s)/M\times10^{13}$ (nm s kg⁻¹ mol) / $s_{M,R\eta,corr.}\times10^{13}$ (s)	R_s (nm) / $(R_G/R_D)_s$ / $(F_{corr.})\eta\times10^{13}$ (kg mol⁻¹ nm⁻¹ s⁻¹)
Chymotrypsinogen A (bovine pancreas) {globular protein} [PE]	25.67[c] / 60.5 / 2.96 / 2.26 / 2.00 / 2.12	0.736 [33] / 2.90 / 0.91 / 0.80 / 1.044 / 1.051	1.81 [32] / 2.08 / 0.65 / 0.76 / 2.75 / 2.73	37.8[b] / 0.869 / 1.16 / 2.64 / 9.86 / 9.79	1.60 [32] / 4.72 / 0.63 / 4.39 / 2.17 / 2.19	2.58 [36] / 6.95 / 31.4 / 2.17 / 2.91 / 2.99	9.5 [37] / 1.47 / 1.21 / 0.84 / 2.58 / 2.65	2.5 [2] / 0.614 / 0.706 / 0.73 / 2.19 / 2.21	32.67 / 9.34 / 2.82 / 0.833 / 0.827	96.72 / 17.2 / 1.86 / 2.74 / 2.72	1.98 / 0.182 / 2.74 / 2.72 / 2.69	2.77 / 0.78 / 4.57
Riboflavin-binding protein, pH 3.7 (chicken egg white) {globular protein, apoprotein} [PE]	32.5 [38] / 135.0 / 2.74 / 1.63 / 3.58	0.720 [38] / 4.18 / 0.96 / 1.021 / 1.153	2.06 [38] / 2.51 / 0.69 / 3.13 / 2.77	66.5 [38] / 0.820 / 1.61 / 8.35 / 7.39	2.03 [38] / 6.48 / 1.19 / 2.57 / 2.90	2.70 [38] / 13.26 / 38.9 / 2.69 / 4.23	2.05 / 1.71 / 3.31 / 5.21	0.646 / 0.788 / 2.58 / 2.99	41.37 / 8.20 / 0.802 / 0.711	113.2 / 3.11 / 2.75	2.15 / 0.171 / 3.10 / 2.67	3.00 / 0.69
Riboflavin-binding protein, pH 7.0 (chicken egg white) {globular protein, holoprotein} [PE]	32.5 [38][d] / 118.4 / 2.74 / 1.76 / 3.62	0.720 [38] / 4.22 / 0.92 / 1.029 / 1.156	1.98 [38] / 2.37 / 0.66 / 3.30 / 2.94	55.6 [38] / 0.836 / 1.34 / 8.80 / 7.83	2.13 [38] / 6.80 / 1.05 / 2.44 / 2.74	2.92 [38] / 11.63 / 38.9 / 2.76 / 4.27	1.71 / 1.43 / 2.84 / 4.39	0.620 / 0.742 / 2.45 / 2.83	41.37 / 8.54 / 0.813 / 0.724	113.2 / 3.28 / 2.92	2.15 / 0.178 / 3.26 / 2.82	3.00 / 0.72
Pepsin (not specified) {globular protein} [PE]	34.2 [39] / 142.7 / 2.69 / 2.46 / 2.00 / 4.76	0.725 [40] / 5.33 / 0.94 / 0.83 / 1.044 / 1.234	2.05 [41] / 2.36 / 0.67 / 0.76 / 3.38 / 2.86	54.9[b] / 0.869 / 1.26 / 3.35 / 8.71 / 7.37	2.60 [41] / 8.44 / 1.22 / 4.51 / 2.46 / 2.91	3.20 [42] / 13.54 / 41.2 / 2.63 / 2.91 / 5.52	8.71 [42] / 1.61 / 1.33 / 0.78 / 2.81 / 5.33	3.35 [42] / 0.632 / 0.726 / 0.81 / 2.48 / 3.07	43.54 / 8.24 / 3.47 / 0.833 / 0.705	117.1 / 17.9 / 2.75 / 3.35 / 2.84	2.18 / 0.192 / 3.14 / 3.33 / 2.69	3.05 / 0.79 / 4.22
β-Lactoglobulin A, dimer (bovine milk) {globular protein} [PE]	36.73 [21] / 100.0 / 2.63 / 2.74 / 2.13 / 2.93	0.750 [30] / 3.58 / 0.97 / 0.79 / 1.052 / 1.108	2.16 [43] / 2.43 / 0.69 / 0.82 / 3.17 / 3.01	60.3 [2] / 0.888 / 1.29 / 2.94 / 8.37 / 7.95	1.66 [2] / 5.51 / 0.81 / 4.41 / 2.56 / 2.69	2.87 [44] / 9.05 / 45.7 / 2.71 / 2.99 / 3.62	7.82 [45] / 1.64 / 1.32 / 0.80 / 2.96 / 3.58	3.4 [19] / 0.650 / 0.732 / 0.81 / 2.58 / 2.75	46.76 / 7.82 / 3.44 / 0.844 / 0.802	122.8 / 16.9 / 2.71 / 3.15 / 2.99	2.23 / 0.169 / 2.98 / 3.12 / 2.93	3.13 / 0.76 / 4.46

5S rRNA (E. coli) {nucleic acid}

44.0 [46]	0.54 [46]	3.27 [46]	56.0 [46]	2.88 [46]	5.3 [46]	6.2 [46]	0.926	56.01	138.5	2.37	3.32
161.0	9.40	2.37	1.38	10.15	12.92	1.27	0.672	5.17	20.3	0.394	0.97
2.47	1.38	0.98	1.00	1.16	39.5	1.42	3.53	0.938	5.08	5.02	
3.46	0.95	0.98	5.12	4.16	8.22	6.30	3.46	0.957	5.18	5.12	

[OE]

0.097	1.468	5.12	6.15	3.48	7.76	5.94					
0.104	1.440	5.22	6.27	3.42							

5S rRNA (rat liver) {nucleic acid}

45.0 [46]	0.54 [46]	3.31 [46]	58.0 [46]	2.92 [46]	4.9 [46]	5.9 [46]	0.931	57.29	140.6	2.39	3.35
169.5	9.67	2.40	1.38	10.39	13.40	1.29	0.675	5.11	19.5	0.360	0.89
2.45	1.38	0.99	1.01	1.21	40.4	1.44	3.57	0.938	5.14	5.07	
3.63	0.91	1.02	4.99	4.38	8.23	6.39	3.53	0.949	5.19	5.13	

[OE]

0.097	1.469	5.17	6.08	3.53	7.95	6.17					
0.101	1.452	5.23	6.14	3.49							

Albumin (bovine serum) {globular protein}

66.30 [21]	0.735 [22]	3.06 [47]	142.0 [47]	1.46 [47]	4.5 [48]	5.9 [48]	4.1 [48]	84.4	182.1	2.72	3.81
207.3	4.47	3.24	0.945	5.91	12.65	2.14	0.756	5.52	18.1	0.208	0.89
2.16	1.12	0.80	1.68	1.14	80.9	1.75	0.800	3.18	2.39	4.40	4.20
3.63	0.84	0.90	4.25	4.06	3.51	0.87	0.87	0.878	4.42	4.36	

[PE]

2.49	1.077	4.45	6.15	3.49	3.26	4.21	3.54	0.805	4.06	3.91	
3.88	1.174	4.09	5.64	3.80	4.53	5.85	3.95				

Histone core complex (calf thymus) {globular protein}

110 [49]	0.73 [49]	3.48 [49]	188.6 [49]	1.426 [8]	6.6 [49]	5.4 [49]					4.51
268.9	4.96	3.56	0.978	6.83	11.71	1.71					0.88
1.82	1.08	0.77	1.35	1.05	133.3	1.41					
3.97	0.88	0.83	6.57	4.28	3.73	3.85					

[OE]

0.287	1.135	6.50	5.31	4.04	4.50	4.65	4.07	0.862	6.46	6.41	
0.213	1.206	6.12	5.00	4.29			4.33	0.811	6.07	6.02	

Ceruloplasmin (human) {globular glycoprotein}

122.2[c]	0.714 [50]	3.45 [50]	262 [50]	0.816 [50]	7.2 [8]	5.30 [8]	0.695	155.6	273.7	3.34	4.67
213.9	2.82	3.97	0.869	4.05	8.69	2.14	0.800	4.90	18.3	0.203	0.80
1.76	1.03	0.74	1.68	0.78	144.9	1.81	4.17	0.833	7.40	7.34	
4.04	0.85	0.81	7.58	4.53	2.91	3.75	4.18	0.832	7.39	7.34	

[PE]

2.00	1.044	7.45	5.17	4.14	2.91	3.76					
2.01	1.045	7.44	5.17	4.15							

7S seed globulin (Phaseolus vulgaris) {globular protein}

137.5[d,e]	0.729 [51]	4.05 [51]	300 [51]	1.22 [51]	7.1 [8]	4.53 [51]	0.785	175.1	296.2	3.47	4.85
365	4.93	4.15	0.975	6.28	13.7	2.18	0.805	4.17	18.3	0.209	0.87
1.69	1.17	0.83	1.71	1.23	166.5	1.80	4.74	0.861	6.95	6.91	
4.73	0.86	0.92	6.91	4.19	3.71	4.87	5.04	0.810	6.54	6.49	

[OE]

0.290	1.133	7.00	4.56	4.70	4.48	5.88					
0.215	1.204	6.59	4.29	5.00							

Tryptophan synthase (E. coli) {globular protein}

143.15[c]	0.755 [52]	4.01 [53]	270 [53]		6.4 [52]	4.83 [52]	0.767	182.2	304.2	3.52	4.92
	1.14	4.01	1.00			1.89	0.766	4.21	19.4	0.179	0.83
1.67	0.90	0.82	1.48		179.5	1.50	4.65	0.869	6.67	6.62	
4.44	1.151	0.85	6.94	4.66	3.90	4.43					

[OE]

0.266		6.72	4.64	4.62							

Lactate dehydrogenase (dogfish) {globular protein}

145.17[c]	0.741 [2]	3.47 [2]	253.3 [2]	0.893[b]	7.54 [54]	5.05 [55][f]	3.8 [19]	184.8	307.0	3.53	4.94
226.2	3.10	3.93	0.884	4.69	8.19	1.74	0.660	4.87	17.5	0.180	0.79
1.66	0.98	0.70	1.37	0.74	178.6	1.42	0.747	3.62	0.30	7.43	4.24
4.24	0.82	0.81	7.78	4.44	4.44	0.78	0.84	0.828	7.88	7.84	

[OE]

0.414	1.068	7.93	5.11	4.19	3.08	3.24	4.21	0.826	7.86	7.82	
0.409	1.070	7.91	5.10	4.20	3.10	3.26	4.22				

Table 1 (continued)

The table is a dense multi-parameter compendium in which each biopolymer/calculation-approach entry lists a stacked set of quantities. The column-group headers (each a stack of sub-labels with units) are:

Group	Stacked parameters (units)
1	M (kg mol⁻¹) / S (nm²) / A_s/V_p (nm⁻¹) / R_D (nm) / p
2	$\bar v$ (cm³ g⁻¹) / $R_G(S/V)$ / R_G/R_p / $(R_G/R_D)_D$ / f/f_o
3	R_G (nm) / R_o (nm) / R_G/R_s / $R_D/M^{1/3}$ (nm kg$^{-1/3}$ mol$^{1/3}$) / $s\times10^{13}$ (s)
4	V (nm³) / R_G/R_o / V/V_p / $s_{M,RD,corr.}\times10^{13}$ (s) / $D\times10^{7}$ (cm²s⁻¹)
5	S/V (nm⁻¹) / $(S/V)M^{1/3}$ (nm⁻¹ kg$^{1/3}$ mol$^{1/3}$) / S/A_s / $(F_{corr})_D\times10^{-13}$ (kg mol⁻¹ nm⁻¹ s⁻¹) / R_D (nm)
6	$s\times10^{13}$ (s) / $S/M^{2/3}$ (nm² kg$^{-2/3}$ mol$^{2/3}$) / V_{dry} (nm³) / R_η (nm) / ν
7	$[\eta]$ (cm³ g⁻¹) / V/M (nm³ kg⁻¹ mol) / V/V_{dry} / $(R_G/R_\eta)_\eta$
8	$R_G/M^{1/3}$ (nm kg$^{-1/3}$ mol$^{1/3}$) / $R_o/M^{1/3}$ / $R_\eta/M^{1/3}$ / R_η (nm)
9	V_p (nm³) / $D_{RG}\times10^{7}$ (cm²s⁻¹) / ν / R_G/R_D
10	A_s (nm²) / $R_G\!\cdot\!D\times10^{7}$ (nm cm² s⁻¹) / p / $s_{M,RD,corr.}\times10^{13}$ (s)
11	R_p (nm) / $(R_G s)/M\times10^{13}$ (nm s kg⁻¹ mol) / $s_{M,R_\eta,corr.}\times10^{13}$ (s)
12	R_s (nm) / $(R_G/R_D)_s$ / $(F_{corr})\eta_3\times10^{13}$ (kg mol⁻¹ nm⁻¹ s⁻¹)

[Calculation approach]ᵃ

β-Lactoglobulin A, octamer (bovine milk) {globular protein}

146.94 [21] | 0.750 [30]ᵍ | 3.44 [43] | 3.72 | 0.652 | 187.1 | 309.5 | 3.55 | 4.96
268.8 | 4.30 | 0.69 | 0.704 | 4.91 | 7.91 | 0.173 | 0.78
1.65 | 0.97 | 7.97 | 4.10 | 0.844 | 7.47 | 7.87
0.348 | 1.096 | 7.52 | 4.35 | 0.797 | 7.41
0.255 | 1.161

215.0 [2] | 0.926 | 1.15 | 5.26 | 4.96
1.25 [2] | 6.60 | 0.87 | 4.07 | 4.32
7.38 [44] | 9.65 | 183.0 | 3.35 | 4.01
1.46 | 1.17 | 2.95 | 3.53

[OE]

Pyruvate kinase (brewer's yeast) {globular protein, apoenzyme}

219.5ᶜ⁾ᵉ⁾ | 0.754 [56] | 4.35 [56] | 4.59 | 0.721 | 279.4 | 404.5 | 4.06 | 5.67
356.9 | 3.82 | 0.77 | 0.762 | 3.89 | 18.3 | 0.172 | 0.79
1.45 | 1.07 | 9.35 | 5.14 | 0.852 | 9.28 | 9.23
5.10 | 0.85 | 9.22 | 5.21 | 0.840 | 9.16 | 9.10
0.321 | 1.112
0.298 | 1.127

406.0 [56] | 0.947 | 1.45 | 4.20 | 4.14
0.879ᵇ | 5.30 | 0.88 | 4.59 | 5.11
8.7 [57] | 9.81 | 274.8 | 3.50 | 5.18
1.85 | 1.48 | 3.90 | 3.65
4.2 [57] | | | 4.07

[OE]

347.1 | 4.59 | 4.59 | 0.703 | 281.1 | 406.1 | 4.06 | 5.68
1.44 | 1.05 | 0.75 | 0.760 | 3.98 | 9.47 | 0.170 | 0.78
0.349 | 1.096 | 9.54 | 5.06 | 0.844 | 9.42
0.320 | 1.113 | 9.40 | 5.14 | 0.832 | 9.28
0.925 | 1.44 | 4.26 | 4.19
5.17 | 0.85 | 5.04 | 5.11
8.81 [57] | 9.50 | 276.5 | 3.35 | 3.51
1.84 | 1.47 | 3.71 | 3.88

Pyruvate kinase + fructose diphosphate (brewer's yeast) {globular protein, holoenzyme}

220.8ᶜ⁾ᵉ⁾ʰ⁾ | 0.754 [56]ᵈ | 4.25 [56] | | | | | |
347.1 | 3.63 | ...

Catalase (bovine liver) {globular protein}

230.34ᶜ⁾ | 0.730 [30] | 3.98 [58] | 4.65 | 0.649 | 293.2 | 417.7 | 4.12 | 5.77
315.8 | 2.99 | 0.69 | 0.758 | 4.25 | 16.3 | 0.195 | 0.82
1.42 | 0.97 | 11.4 | 0.85 | 3.55 | 2.85 | 10.4 | 3.98
5.23 | 0.76 | 11.2 | 4.85 | 0.825 | 11.3 | 11.2
1.91 | 1.038 | | 4.97 | 0.809 | 11.1 | 11.0
2.24 | 1.060

420.0ᵇ | 0.857 | 1.43 | 4.44 | 4.35
0.752ᵇ | 4.61 | 0.76 | 4.82 | 4.92
11.3 [59] | 8.40 | 279.2 | 5.22 | 3.97
4.1 [59] | 1.82 | 1.50 | 0.76 | 3.13
| 3.99 | | | 2.85 | 3.07 | 3.37

[PE]

Pyruvate decarboxylase + TPP + Mg²⁺ (brewer's yeast) {globular protein, holoenzyme}

248.4ᶜ⁾ᵉ⁾ʰ⁾ | 0.751 [60] | 4.38 [61] | 4.64 | 0.697 | 316.2 | 439.3 | 4.23 | 5.91
391 | 4.09 | 0.74 | 0.738 | 3.86 | 24.3 | 0.175 | 0.80
1.39 | 1.04 | 0.61 | 5.18 | 0.851 | 10.5 | 10.5
3.86 | 1.13 | 10.6 | 5.36 | 0.823 | 10.2 | 10.1
0.325 | 1.109 | 10.3
0.272 | 1.146

419 [61] | 0.944 | 1.33 | 14.1 | 4.16 | 4.03
0.933 [61] | 5.87 | 0.89 | 6.08 | 5.15 | 5.32
9.95 [62] | 9.89 | 309.8 | 3.47 | 3.85
5.55 [62] | 1.69 | 1.35 | 3.53 | 3.91

[OE]

Protein / method	Col 1	Col 2	Col 3	Col 4	Col 5	Col 6	Col 7	Col 8	Col 9	Col 10	Col 11	Col 12
α-Globulin (sesame) {globular protein}	270 [63]; 442	0.730 [63]; 4.83	4.1 [63]; 4.47	375 [63]; 0.917	1.18 [63]; 7.62	12.8 [63]; 10.6; 327.3	3.95 [63]; 1.39; 1.15	0.634; 0.692	343.7; 4.12	464.4; 16.2	4.35; 0.194	6.08; 0.81
[OE]	1.35; 5.43; 0.361; 0.220	0.94; 0.76; 1.090; 1.197	0.67; 0.84; 13.2; 12.0	1.09; 11.8; 4.39; 4.00	0.95; 3.96; 4.88; 5.36	3.29; 4.41	2.75; 3.69	4.90; 5.40	0.841; 0.765	13.1; 11.9	13.0; 11.8	
11S seed globulin (rape seed) {globular protein}	300 [64]; 410	0.729 [65]; 3.74	4.1 [64]; 4.75	450 [64]; 0.862	0.911 [64]; 6.10	12.7 [64]; 9.15; 363.2	3.78 [65]; 1.50; 1.24	0.612; 0.710	381.9; 4.12	498.2; 15.5	4.50; 0.174	6.30; 0.72
[OE]	1.30; 5.67; 0.458; 0.308	0.91; 0.72; 1.053; 1.120	0.65; 0.85; 14.3; 13.5	1.18; 12.6; 4.28; 4.02	0.82; 4.26; 5.01; 5.33	2.95; 3.58	2.67; 3.24	5.02; 5.36	0.819; 0.770	14.2; 13.4	14.2; 13.3	
Legumin (Vicia faba) {globular protein}	312.48[c]; 426	0.729 [66]; 2.77	4.45 [66]; 5.47	685 [66]; 0.814	0.622 [66]; 4.22	13.0 [8]; 9.25; 378.3	3.38 [66]; 2.19; 1.81	0.656; 0.806	397.8; 3.80	511.9; 15.0	4.56; 0.185	6.38; 0.77
[OE]	1.29; 6.34; 0.603; 0.504	0.98; 0.70; 1.022; 1.041	0.70; 0.93; 13.4; 13.1	1.72; 11.7; 3.83; 3.76	0.83; 3.88; 5.59; 5.69	2.69; 2.85	3.55; 3.76	5.60; 5.71	0.796; 0.782	13.3; 13.1	13.3; 13.0	
Leucine aminopeptidase (bovine) {globular protein}	317.22[c]; 553	0.751 [23]; 5.07	4.45 [67]; 4.87	485 [67]; 0.913	1.14 [67]; 7.77	12.6 [8]; 11.9; 395.6	3.75 [8]; 1.53; 1.23	0.652; 0.715	403.8; 3.80	517.0; 16.7	4.59; 0.177	6.41; 0.80
[OE]	1.28; 5.71; 0.366; 0.208	0.97; 0.78; 1.088; 1.213	0.69; 0.84; 13.2; 11.8	1.20; 12.1; 4.04; 3.63	1.07; 4.14; 5.30; 5.91	3.27; 4.58	3.01; 4.22	5.33; 5.97	0.840; 0.753	13.1; 11.7	13.0; 11.6	
11S seed globulin (sunflower) {globular protein}	321.63[c]; 356	0.730 [68]; 3.43	3.95 [64]; 4.61	410 [64]; 0.857	0.868 [64]; 5.95	11.8 [64]; 7.58; 389.9	3.78 [8]; 1.27; 1.05	0.577; 0.673	409.4; 4.28	521.8; 14.9	4.61; 0.145	6.44; 0.61
[OE]	1.27; 5.67; 0.470; 0.348	0.86; 0.70; 1.050; 1.097	0.61; 0.83; 15.9; 15.2	1.00; 13.4; 4.43; 4.24	0.68; 4.90; 4.84; 5.05	2.92; 3.35	2.24; 2.57	4.85; 5.08	0.817; 0.782	15.8; 15.1	15.7; 15.0	
Glutamate dehydrogenase (bovine liver) {globular protein}	333.37[c]; 433	0.749 [69]; 3.05	4.70 [70]; 5.42	668 [70]; 0.867	0.648[b]; 4.49	11.4 [69]; 9.00; 414.6; 5.53	3.5 [19]; 2.00; 1.61; 0.85	3.2 [69]; 0.678; 0.782; 0.798	424.4; 3.60; 2.65	534.4; 16.5; 1.55	4.66; 0.161; 1.33	6.52; 0.72; 5.01
[PE]	1.26; 6.12; 1.98; 2.30	1.01; 0.77; 1.043; 1.064	0.72; 0.88; 13.1; 12.8	1.57; 12.0; 3.79; 3.71	0.81; 4.53; 5.65; 5.77	2.90; 3.12	3.49; 3.76	5.70; 5.84	0.831; 0.815	13.0; 12.7	12.9; 12.6	
Immunoglobulin IgM$_{ser}$ (bovine) {glycoprotein, special shape}	950 [71]; 2300	0.724 [71]; 14.7	11.5 [71]; 7.55	1800 [71]; 1.52	1.28 [71]; 12.6	17.7 [71]; 23.8; 1142	1.73 [71]; 1.89; 1.58	1.170; 0.768	1209; 1.47	1074; 19.9	6.61; 0.214	9.25; 0.88
[OE]	0.89; 12.4; 0.072; 0.066	1.74; 0.93; 1.601; 1.644	1.24; 1.26; 19.2; 18.7	1.49; 18.6; 1.77; 1.73	2.14; 4.51; 12.1; 12.4	10.7; 11.6	12.2; 13.2	12.3; 12.6	0.952; 0.927	19.0; 18.5	18.8; 18.3	
Immunoglobulin IgM$_{ser}$ 58°C (bovine) {glycoprotein, special shape}	950 [71]; 1550	0.741[i]; 8.62	8.9 [71]; 7.26	1600 [71]; 1.23	0.97 [71]; 9.52	19.6 [71]; 16.0; 1169	1.68; 1.37	0.905; 0.738	1209; 1.90	1074; 22.3	6.61; 0.184	9.25; 0.80
[OE]	0.89; 0.139; 0.115	1.35; 1.335; 1.405	0.96; 22.4; 21.3	1.32; 2.21; 2.10	1.44; 9.69; 10.2	6.16; 7.19	6.25; 7.29	9.80; 10.3	0.919; 0.873	21.2	22.0; 20.9	

H. Durchschlag and P. Zipper
Modeling by whole-body approaches

Table 1 (continued)

Legend of parameters (column/row labels)

M (kg mol⁻¹)	S (nm²)	A_s/V_p (nm⁻¹)	R_D (nm)	p	f/f_o
$\bar{v}$ (cm³ g⁻¹)	$R_G(S/V)$	R_G/R_p	$(R_G/R_D)_D$	$D\times10^7$ (cm² s⁻¹)	
R_G (nm)	R_o (nm)	R_G/R_s	$R_D/M^{1/3}$ (nm kg⁻¹ᐟ³ mol¹ᐟ³)	$s\times10^{13}$ (s)	
V (nm³)	R_G/R_o	V/V_p	$s_{\mathrm{M,RD,corr.}}\times10^{13}$ (s)	$D\times10^7$ (cm² s⁻¹)	
S/V (nm⁻¹)	$(S/V)M^{1/3}$ (nm⁻¹ kg¹ᐟ³ mol¹ᐟ³)	S/A_s	$(F_{\mathrm{corr.}})_D\times10^{-13}$ (kg mol⁻¹ nm⁻¹ s⁻¹)	R_D (nm)	
$s\times10^{13}$ (s)	$S/M^{2/3}$ (nm² kg⁻²ᐟ³ mol²ᐟ³)	V_{dry} (nm³)	R_η (nm)	v	
$D\times10^7$ (cm² s⁻¹)	V/M (nm³ kg⁻¹ mol)	V/V_{dry}	$(R_G/R_\eta)_\eta$	$[\eta]$ (cm³ g⁻¹)	
$[\eta]$ (cm³ g⁻¹)	$R_G/M^{1/3}$ (nm kg⁻¹ᐟ³ mol¹ᐟ³)	$R_s/M^{1/3}$ (nm kg⁻¹ᐟ³ mol¹ᐟ³)	$R_\eta/M^{1/3}$ (nm kg⁻¹ᐟ³ mol¹ᐟ³)	R_η (nm)	
V_p (nm³)	$D_{RG}\times10^7$ (cm² s⁻¹)	v	R_G/R_D		
A_s (nm²)	$R_G\cdot D\times10^7$ (nm cm² s⁻¹)	p	$s_{\mathrm{M,RD,corr.}}\times10^{13}$ (s)		
R_p (nm)	$(R_G\cdot s)/M\times10^{13}$ (nm s kg⁻¹ mol)	$s_{\mathrm{M,R\eta,corr.}}\times10^{13}$ (s)			
R_s (nm)	$(R_G/R_D)_s$				

Data

Parameter	Bacteriophage R17 (E. coli) {nucleoprotein, spherical}	Brome grass mosaic virus, pH 6 {nucleoprotein}	Brome grass mosaic virus, pH 6 [S]	Brome grass mosaic virus, pH 7 {nucleoprotein}	Brome grass mosaic virus, pH 7 [S]	Brome grass mosaic virus, pH 7.3 {nucleoprotein}	Brome grass mosaic virus, pH 7.3 [S]
M (kg mol⁻¹)	3700 [72]	4700 [74]		4700 [74]		4700 [74]	
$\bar{v}$ (cm³ g⁻¹)	0.673 [72]	0.692[j]		0.692[j]		0.692[j]	
R_G (nm)	10.44 [72]	10.6 [75]		12.8 [75]		15.2	
$s\times10^{13}$ (s)	80 [73]	86.2 [76]		86.2 [76][l]		86.2 [76][l]	
$D\times10^7$ (cm² s⁻¹)		1.55 [76]		1.55 [76][l]		1.55 [76][l]	
V (nm³)	9630 [72]	9200[k]		11000[k]		14700[m]	
A_s/V_p (nm⁻¹)		0.521		0.521		0.521	
R_D (nm)		13.8	13.0	13.8	13.8		15.2
p	0.565	1.000		1.14			
f/f_o	1.00	1.000	1.000	0.93	1.000		1.000
R_o (nm)	13.20	13.0	13.0	13.8	13.8	15.2	15.2
R_G/R_o	0.72	0.815	0.815	0.81	0.928		0.842
R_G/R_s	0.72	0.67	0.83	0.83	0.83	0.83	0.83
R_G/R_p		0.94	1.000	1.14			1.000
V_p (nm³)		4710		5983		5983	
A_s (nm²)		2659		3119		3119	
R_p (nm)		10.40		11.3		11.3	
R_s (nm)	80.9	14.55	15.8	15.8	15.8	15.8	15.2
$(R_G/R_D)_s$		0.78		0.72		0.86	
$[\eta]$ (cm³ g⁻¹)		2.60 / 2.33 / 3.92	2.95	2.34 / 2.04 / 3.52	3.52	3.13 / 2.72 / 4.71	4.71
$R_G/M^{1/3}$		0.675		0.633		0.764	
$R_s/M^{1/3}$		0.853		0.776		0.824	
V/M (nm³ kg⁻¹ mol)	2.04	1.96	1.55	2.34		3.13	
V/V_{dry}	1.62	1.70		2.04		2.72	
V/V_p		1.54	1.65	1.84	1.55	2.46	1.41
$R_G\cdot D\times10^7$ (nm cm² s⁻¹)				16.4		19.8	
$D_{RG}\times10^7$ (cm² s⁻¹)		1.62		1.32			
$(R_G\cdot s)/M\times10^{13}$		0.226		0.194		0.235	
$s_{\mathrm{M,RD,corr.}}\times10^{13}$ (s)		91.8	98.3	91.8	92.6	91.8	84.1
S (nm²)	4135			5401		5401	
$S/M^{2/3}$	2.50			2.50		2.50	
$(F_{\mathrm{corr.}})_D\times10^{-13}$			4.58		4.58		4.58
$s_{\mathrm{M,R\eta,corr.}}\times10^{13}$ (s)		80.3	97.6	92.0	92.0	83.5	83.5

Turnip yellow mosaic virus

{nucleoprotein, spherical virus}	5530 [74]	0.666 [77]	11.1 [74]	12250^n)		113.8 [78]	1.42 [78]				
	0.494	0.93	14.30	0.776			2.22	0.628	7040	3476	11.9	16.6
	15.1	0.74	0.67	1.74			2.00	0.809	1.52	15.8	0.228	0.78
	1.000	1.000	114.0	107.3	4.05		3.33	14.3	0.776	113.2	113.2	
[S]				1.50	14.3						

Wild cucumber mosaic virus

{nucleoprotein, spherical virus}	5820 [79]	0.655^i)	11.2 [80]	11494°)		119 [79]					
	0.485	0.93	14.00	0.800			1.97	0.623	7409	3597	12.1	16.9
	1.000	1.000	0.66	1.55			1.82	0.778	1.51		0.229	0.75
			126.5	1.53			2.97	14.0	0.800	125.7	125.7	
[S]					14.0						

Tomato bushy stunt virus

{nucleoprotein, spherical virus}	8700 [81]	0.700 [81]	13.2 [81]	24429^p)		132 [82]	1.15 [82]	3.44 [82]			
	0.425	0.95	18.00	0.733			2.81	0.642	11075	4702	13.8	19.3
	18.6	0.71	0.68	2.21			2.42	0.875	1.28	15.2	0.200	0.76
	1.000	1.000	0.91	122.8	4.00		0.79	0.817	2.03	127.1	136.2	4.44
[S]			128.0	1.19	18.0		4.23	18.0	0.733		127.1	

a) Calculation approaches: OE, PE, oblate or prolate ellipsoid of revolution, respectively (the axial ratio p was calculated from R_G and V (first line) or, if possible, from R_G and S/V (second line); S, solid sphere (the particle is either a sphere or a hollow sphere or may be approximated for hydrodynamic modeling by a sphere)
b) Secondary parameter, derived by means of a model [2]
c) According to SWISS-PROT Protein Sequence Data Bank
d) Value for the apoform was used
e) Slightly different molar masses were averaged, if no clear-cut decision concerning isozymes or chains was possible
f) Experimental value refers to pig-heart enzyme
g) Value of the dimer was used
h) The ligands were taken into account
i) The temperature dependence of $\bar{v}$ [22] was taken into account
j) Calculated from the composition
k) Volume of a solid sphere, calculated from the outer diameter [75]
l) Experimental value refers to pH 6
m) Volume of a solid sphere, calculated from the outer diameter [83]
n) Volume of a solid sphere, calculated from the outer diameter [74]
o) Volume of a solid sphere, calculated from the outer diameter [80]
p) Volume of a solid sphere, calculated from the outer diameter [84]

Correlations between structural and hydrodynamic data

Applying the calculation procedures outlined above shows that a variety of useful data may be obtained. To facilitate the proceeding of our approach and understanding the information obtained, some features and summarizing conclusions are given in the following.

As may be anticipated, the directly calculated radii of globular, roughly spherical particles, R_0, R_D and R_η, always exceed the experimental value of R_G, while particles of special shape (e.g., immunoglobulins, ribonucleic acids, anisometric proteins) may show a behavior different from that. Thus, statements concerning particle anisometry may also be drawn from the radii ratios, preferably from the ratio R_G/R_0 (cf. ref. [4]). Compatibility of the experimental data for s, D and $[\eta]$ may be inferred from identical or similar values for the ratios $(R_G/R_D)_s$, $(R_G/R_D)_D$ and $(R_G/R_\eta)_\eta$.

Due to hydration contributions, the particle volume V of biopolymers, derived from SAXS experiments, is always greater than V_{dry}. A similar conclusion holds for the comparison of V with the packing volume V_p of proteins, the latter obtained on the basis of crystallographic data, whereas ribonucleic acids may deviate from this norm. An inspection of V/V_{dry} and V/V_p values reflects this in a quantitative way. On the other hand, the SAXS surface area S is often essentially lower than the crystallographic accessible surface A_s, as may be seen quantitatively from the set of S/A_s values. This also leads to deviations between experimentally observed ratios S/V, V/M, $S/M^{2/3}$ and $(S/V)M^{1/3}$, on the one hand, and the theoretical ratios A_s/V_p, V_p/M, $A_s/M^{2/3}$ and $(A_s/V_p)M^{1/3}$, on the other. Differences in hydration and surface roughness of the particles under consideration may be responsible for the aforementioned discrepancies between solution and crystal data [2, 4, 9].

Use of Eqs. (15) and (16) for estimations of s and D ($s_{M,RD,corr.}$ or alternatively $s_{M,R_\eta,corr.}$, and D_{RG}), indeed, proves that these empirical relations may be used efficaciously for rapid estimations, provided the particles to be analyzed do not exhibit extreme anisotropy [4].

Most important, predictions of hydrodynamic properties on the basis of scattering data by means of whole-body approaches yield in nearly all cases reasonable results for s and D, and, though less pronounced, also for $[\eta]$. When looking at the axial ratios p, one should be aware of the fact that they have been derived by use of simple geometric models, and that we are striving for parameter predictions and not for an exact shape analysis. In principle, the results of the two calculation approaches for p (R_G and V or R_G and S/V) are equivalent. For predictions of s and D values of small biopolymers ($M < 40$ kg/mol), however, the combination of R_G and surface-to-volume ratio, S/V, seems to

be superior. This may be due to more realistic considerations of hydration plus surface roughness, when using this latter calculation approach. As has been shown in certain cases (e.g., with catalase), application of a realistic model and derivation of secondary parameters (cf. ref. [2]) may substitute the lack of sufficient experimental data to some extent. In the case of hollow bodies or inhomogeneous particles, the simple whole-body approaches have to be replaced with more sophisticated procedures [4]. For improving the accuracy of modeling or for predicting the solution behavior of more complex assemblies of macromolecules, multibody approaches have to be adopted (cf. ref. [5]).

Of course, a thorough comparison of observed and predicted hydrodynamic data suffers from quality differences in the experimental material. If some of the more pronounced discrepancies (e.g., s of glutamate dehydrogenase and immunoglobulins, D of pyruvate decarboxylase, s and D of some seed globulins, many viscosity data) are caused by true peculiarities of the molecules under analysis or are due to experimental artifacts or calculation deficits, remains to be established. However, both experimental determinations of viscosity data and their predictions seem to be of rather low reliability.

Accuracy of parameter predictions

The accuracy of parameter predictions and, above all, the anticipation of small structural alterations, which may occur in response to changes in environment (e.g., proteins at different temperatures) or state of ligation (e.g., apo- and holoforms of enzymes), require the critical assessment of possible errors. All input parameters are affected by experimental uncertainties. Under the assumption of typical error bars for $M(\pm 5\%)$, $\bar{v}$ ($\pm 1\%$), $R_G(\pm 1\%)$, $V(\pm 2\%)$, and $S(\pm 2\%)$ (cf. refs. [4, 9, 13], detailed error propagation calculations have been performed, considering a great number of permutations for the error bars of all variables. Some representative examples for models of different isometry (spheres, prolate and oblate ellipsoids of revolution with $p = 1.5$ or 0.7, respectively) are listed in Table 2. The mentioned examples comprise the propagation of errors (i) of the individual input parameters, (ii) of all input parameters afflicted with errors of the same sign, (iii) and of some selected examples of special interest leading to maximum deviations.

The results of these error calculations reveal that maximum errors can be found for predictions of p: except for very few cases, they do not exceed 5–10%. However, as discussed above, in the present context this quantity is of subordinate relevance. The more intriguing question aims at the possibility of predicting s and D accurately. Apart

Table 2 Critical assessment of errors for the prediction of hydrodynamic quantities from scattering data, illustrated for a globular model protein with average characteristics: molar mass $M = 100\,\text{kg/mol}$, partial specific volume $\bar{v} = 0.735\,\text{cm}^3/\text{g}$, hydration $\delta_1 = 0.35\text{g/g}$, spherical or prolate or oblate shape ($p = 1$ or 1.5 or 0.7), respectively. For error propagation calculations, typical error bars for the input parameters are assumed: $\pm 5\%$ for M, $\pm 1\%$ for $\bar{v}$ and R_G, and $\pm 2\%$ for V and S.[a] Hydrodynamic modeling was performed on the basis of spheres or prolate or oblate ellipsoidal models (S, PE, OE), the calculation approaches exploiting V (spheres) or R_G and V or S/V (ellipsoids)[b]

Deviation (in %)

Input parameters					Hydrodynamic parameters			
					Calculation from V			
M	$\bar{v}$	R_G	V	S	s	D	$[\eta]$	
Spheres								
+5					+5.0	0	−4.76	
−5					−5.0	0	+5.26	
	+1				−2.76	0	0	
	−1				+2.75	0	0	
		+1			0	0	0	
		−1			0	0	0	
			+2		−0.66	−0.66	+2.0	
			−2		+0.68	+0.68	−2.0	
				+2	0	0	0	
				−2	0	0	0	
+5	+1		+2		+1.44	−0.66	−2.86	
−5	−1		−2		−1.72	+0.68	+3.16	

Input parameters					Calculation from R_G and V				Calculation from R_G and S/V			
M	$\bar{v}$	R_G	V	S	p	s	D	$[\eta]$	p	s	D	$[\eta]$
Prolate ellipsoids of revolution												
+5					0	+5.0	0	−4.76	0	+5.0	0	−4.76
−5					0	−5.0	0	+5.26	0	−5.0	0	+5.26
	+1				0	−2.76	0	0	0	−2.76	0	0
	−1				0	+2.76	0	0	0	+2.76	0	0
		+1			+4.90	−0.37	−0.37	+1.33	+3.02	−0.22	−0.22	+0.82
		−1			−5.36	+0.37	+0.37	−1.28	−3.18	+0.23	+0.23	−0.78
			+2		−3.46	−0.41	−0.41	+1.14	−6.43	−0.22	−0.22	+0.45
			−2		+3.35	+0.43	+0.43	−1.11	+6.02	+0.22	+0.22	−0.40
				+2	0	0	0	0	+5.90	−0.44	−0.44	+1.60
				−2	0	0	0	0	−6.57	+0.45	+0.45	−1.55
		+1	+2		+1.69	−0.78	−0.78	+2.48	−3.11	−0.44	−0.44	+1.22
		+1	−2		+8.06	+0.06	+0.06	+0.16	+8.85	−0.01	−0.01	+0.39
		−1	+2		−9.23	−0.04	−0.04	−0.15	−10.03	+0.01	+0.01	−0.31
		−1	−2		−1.71	+0.80	+0.80	−2.42	+3.08	+0.45	+0.45	−1.18
		+1		+2	+4.90	−0.37	−0.37	+1.33	+8.73	−0.67	−0.67	+2.40
		+1		−2	+4.90	−0.37	−0.37	+1.33	−3.24	+0.23	+0.23	−0.79
		−1		+2	−5.36	+0.37	+0.37	−1.28	+2.96	−0.22	−0.22	+0.81
		−1		−2	−5.36	+0.37	+0.37	−1.28	−10.18	+0.68	+0.68	−2.30
+5	+1	+1	+2	+2	+1.69	+1.31	−0.78	−2.40	+3.02	+1.21	−0.88	−2.06
−5	−1	−1	−2	−2	−1.71	−1.60	+0.80	+2.72	−3.18	−1.50	+0.90	+2.35
Oblate ellipsoids of revolution												
+5					0	+5.00	0	−4.76	0	+5.00	0	−4.76
−5					0	−5.00	0	+5.26	0	−5.00	0	+5.26
	+1				0	−2.76	0	0	0	−2.76	0	0
	−1				0	+2.75	0	0	0	+2.75	0	0
		+1			−6.55	−0.46	−0.46	+1.49	−3.45	−0.23	−0.23	+0.72
		−1			+8.50	+0.45	+0.45	−1.42	+3.98	+0.23	+0.23	−0.72
			+2		+5.28	−0.36	−0.36	+1.04	+8.53	−0.21	−0.21	+0.55
			−2		−4.58	+0.37	+0.37	−1.04	−6.61	+0.21	+0.21	−0.53
				+2	0	0	0	0	−6.49	−0.45	−0.45	+1.47
				−2	0	0	0	0	+8.73	+0.46	+0.46	−1.45
		+1	+2		−2.36	−0.81	−0.81	+2.49	+3.90	−0.44	−0.44	+1.28

Table 2 (continued)

M	$\bar{v}$	R_G	V	S	Calculation from R_G and V				Calculation from R_G and S/V			
					p	s	D	$[\eta]$	p	s	D	$[\eta]$
		+1	−2		−10.33	−0.10	−0.10	+0.46	−9.39	−0.02	−0.02	+0.20
		−1	+2		+16.11	+0.07	+0.07	−0.45	+14.33	+0.02	+0.02	−0.27
		−1	−2		+2.53	+0.83	+0.83	−2.46	−3.51	+0.44	+0.44	−1.28
		+1		+2	−6.55	−0.46	−0.46	+1.49	−9.29	−0.68	−0.68	+2.21
		+1		−2	−6.55	−0.46	−0.46	+1.49	+4.07	+0.23	+0.23	−0.74
		−1		+2	+8.50	+0.45	+0.45	−1.42	−3.38	−0.22	−0.22	+0.71
		−1		−2	+8.50	+0.45	+0.45	−1.42	+14.59	+0.69	+0.69	−2.25
+5	+1	+1	+2	+2	−2.36	+1.28	−0.81	−2.39	−3.45	+1.21	−0.88	−2.15
−5	−1	−1	−2	−2	+2.53	−1.58	+0.83	+2.67	+3.98	−1.50	+0.91	+2.41

[a] The given limits of error stem from typical experiments cited in the literature (cf. refs. [4, 9, 13])
[b] The table comprises the error propagations of the individual input parameters and some selected combinations of interest

from a few exceptional cases, both sedimentation and diffusion coefficients may be anticipated very accurately, the accuracy for D being highest. In the majority of the cases of practical interest, the errors in s and D do not exceed $\pm 1\%$, i.e., they are less than the extent of the changes due to conformational alterations (cf. ref. [17]). In this context, two further points have to be stressed: under comparable experimental conditions, some of the experimental errors cancel out, and, if the molar mass is known from sequence data, this parameter does not affect the resultant hydrodynamic quantities.

The correct anticipation of sedimentation parameters has also been shown recently for the case of minute substrate-induced conformational changes of the enzymes citrate synthase and malate synthase [4, 9, 10], and may also be visualized from an inspection of the changes of s and D values for riboflavin-binding protein, pyruvate kinase, and immunoglobulin IgM_{ser}. In conclusion, even subtle changes of hydrodynamic parameters can be predicted correctly, although slight errors in the input parameters can never be avoided.

Conclusions

Taken together, the results of this study and the previous papers [4, 9, 10] clearly demonstrate that the structure of most globular proteins (either simple or conjugated) and even some ribonucleic acids of low molecular mass may be modeled by simple whole-body approaches, provided the molecules are relatively compact, homogeneous and isometric. Particles exhibiting mass inhomogeneities (i.e., hollow particles, multisubunit proteins, some conjugated proteins, high-molecular mass ribonucleic acids), structural anisometry and/or unusual binding of water, and liganded biopolymers require more complex calculation procedures.

A crucial point to be addressed in connection with the interpretation of hydrodynamic data is the necessary consideration of the hydrodynamically effective hydration. As is well known, many erroneous, extreme axial ratios derived in the past from hydrodynamic experiments stem from the total neglect of hydration contributions. By contrast, the present approach circumvents this problem by considering the hydration properly either by using appropriate values for the experimentally accessible *hydrated* SAXS volume V or by models exhibiting appropriate amounts of hydration.

The approaches obtained may be used for several purposes: correlations between scattering and hydrodynamic parameters, prediction of hydrodynamic parameters from scattering data, separation of the contributions of shape and hydration to frictional ratios, insight into structural details such as shape changes or surface roughness, verification of observable quantities and predicted models of biopolymers, compatibility tests of structural and hydrodynamic data, etc.

References

1. Harding SE (1989) In: Harding SE, Rowe AJ (eds) Dynamic Properties of Biomolecular Assemblies. Royal Society of Chemistry, Cambridge UK, p 32

2. Pessen H, Kumosinski, TF (1993) In: Baianu IC, Pessen H, Kumosinski TF (eds) Physical Chemistry of Food Processes, Vol 2: Advanced Techniques, Structures, and Applications. Van Nostrand Reinhold, New York, p 274

3. García de la Torre J, Navarro S, Lopez Martinez MC, Diaz FG, Lopez Cascalez JJ (1994) Biophys J 67:530

Progr Colloid Polym Sci (1997) 107:43–57
© Steinkopff Verlag 1997

4. Durchschlag H, Zipper P (1997) J Appl Cryst, in press
5. Zipper P, Durchschlag H (1997) Progr Coll Polym Sci 107:58–71
6. Kumosinski TF, Pessen H (1982) Arch Biochem Biophys 219:89
7. Kumosinski TF, Pessen H (1985) Meth Enzymol 117:154
8. Müller JJ, Damaschun H, Damaschun G, Gast K, Plietz P, Zirwer D (1984) Studia Biophysica 102:171
9. Durchschlag H, Zipper P, Purr G, Jaenicke R (1996) Colloid Polym Sci 274:117
10. Durchschlag H, Zipper P (1996) J Mol Struct 383:223
11. Durchschlag H, Zipper P (1997) J Appl Cryst, in press
12. Glatter O, Kratky O (eds) (1982) Small Angle X-ray Scattering. Academic Press, London
13. Durchschlag H (1993) In: Baianu IC, Pessen H, Kumosinski TF (eds) Physical Chemistry of Food Processes, Vol 2: Advanced Techniques, Structures, and Applications. Van Nostrand Reinhold, New York, p 18
14. Mittelbach P, Porod G (1965) Kolloid Z 202:40
15. Scheraga HA (1955) J Chem Phys 23:1526
16. le Maire M, Viel A, Møller, JV (1989) Anal Biochem 177:50
17. Durchschlag H, Zipper P, Wilfing R, Purr G (1991) J Appl Cryst 24:822
18. Potschka M (1988) J Chromatogr 441: 239
19. Tyn MT, Gusek TW (1990) Biotechnol Bioeng 35:327
20. Teller DC (1976) Nature (London) 260: 729
21. Dayhoff MO (ed) (1972) Atlas of Protein Sequence and Structure, Vol 5; (1973) Suppl 1; (1976) Suppl 2; (1978) Suppl 3. Natl Biomed Res Found, Washington, DC
22. Durchschlag H (1986) In: Hinz H-J (ed) Thermodynamic Data for Biochemistry and Biotechnology. Springer, Berlin p 45
23. Durchschlag H (1996) In: Hinz H-J (ed) Landolt-Börnstein New Series VII. Springer, Berlin, submitted
24. Pessen H, Kumosinski TF, Timasheff SN (1971) J Agric Food Chem 19:698
25. Yphantis DA (1959) J Phys Chem 63: 1742
26. Creeth JM (1958) J Phys Chem 62:66
27. Buzzell JG, Tanford C (1956) J Phys Chem 60:1204
28. Kronman MJ, Andreotti RE (1964) Biochemistry 3:1145
29. Polson A (1939) Kolloid Z 88:51
30. Lee JC, Timasheff SN (1974) Biochemistry 13:257
31. Sophianopoulos AJ, Rhodes CK, Holcomb DN, Van Holde KE (1962) J Biol Chem 237:1107
32. Krigbaum WR, Godwin RW (1968) Biochemistry 7: 3126
33. Schwert GW, Kaufman S (1951) J Biol Chem 190:807
34. Müller JJ, Damaschun G, Wilhelm P, Welfle H, Pilz I (1982) Int J Biol Macromol 4:289
35. Müller JJ, Zirwer D, Damaschun G, Welfle H, Gast P, Plietz P (1983) Studia Biophysica 96:103
36. Wilcox PE, Kraut J, Wade RD, Neurath H (1957) Biochim Biophys Acta 24:72
37. Schwert GW (1951) J Biol Chem 190: 790
38. Kumosinski TF, Pessen H, Farrell HM Jr (1982) Arch Biochem Biophys 214: 714
39. Rajagopalan TG, Moore S, Stein WH (1966) J Biol Chem 241:4940
40. McMeekin TL, Wilensky M, Groves ML (1962) Biochem Biophys Res Commun 7:151
41. Vazina AA, Lednev VV, Lemagikhin BK (1966) Biochimija (Moscow) 31:720
42. Edelhoch H (1957) J Am Chem Soc 79: 6100
43. Witz J, Timasheff SN, Luzzati V (1964) J Am Chem Soc 86:168
44. Kumosinski TF, Timasheff SN (1966) J Am Chem Soc 88:5635
45. Cecil R, Ogston AG (1949) Biochem J 44:33
46. Müller JJ, Zalkova TN, Zirwer D, Misselwitz R, Gast K, Serdyuk IN, Welfle H, Damaschun G (1986) Eur Biophys J 13:301
47. Luzzati V, Witz J, Nicolaieff A (1961) J Mol Biol 3:379
48. Peters T Jr (1985) Adv Prot Chem 37: 161
49. Damaschun H, Zalenskaya IA, Damaschun G, Vorob'ev VI, Misselwitz R, Zirwer D (1983) Studia Biophysica 97:105
50. Damaschun G, Damaschun H, Dembo AT, Kayushina RL, Kröber R, Moshkov KA, Müller JJ, Neifakh SA, Rolbin JA, Shavlovsky MM, Zirwer D (1978) Studia Biophysica 71:53
51. Plietz P, Damaschun G, Zirwer D, Gast K, Schlesier B (1983) Int J Biol Macromol 5:356
52. Lane AN, Kirschner K (1983) Eur J Biochem 129:675
53. Wilhelm P, Pilz I, Lane AN, Kirschner K (1982) Eur J Biochem 129:51
54. Pesce A, Fondy TP, Stolzenbach F, Castillo F, Kaplan NO (1967) J Biol Chem 242:2151
55. Jaenicke R, Gregori E, Laepple M (1979) Biophys Struct Mechanism 6:57
56. Müller K, Kratky O, Röschlau P, Hess B (1972) Hoppe-Seyler's Z Physiol Chem 353:803
57. Bischofberger H, Hess B, Röschlau P (1971) Hoppe-Seyler's Z Physiol Chem 352:1139
58. Malmon AG (1957) Biochim Biophys Acta 26:233
59. Sumner JB, Gralén N (1938) J Biol Chem 125:33
60. Pilz I, Ullrich J (1973) Eur J Biochem 34: 256
61. Müller JJ, Damaschun G, Hübner G (1979) Acta biol med germ 38:1
62. Ullrich J, Kempfle M (1969) FEBS Letters 4:273
63. Plietz P, Damaschun G, Zirwer D, Gast K, Schwenke KD, Prakash V (1986) J Biol Chem 261:12686
64. Plietz P, Damaschun G, Müller JJ, Schwenke K-D (1983) Eur J Biochem 130:315
65. Schwenke KD, Schultz M, Linow K-J, Gast K, Zirwer D (1980) Int J Peptide Protein Res 16:12
66. Plietz P, Zirwer D, Schlesier B, Gast K, Damaschun G (1984) Biochim Biophys Acta 784:140
67. Damaschun G, Damaschun H, Hanson H, Müller JJ, Pürschel H-V (1973) Studia Biophysica 35:59
68. Plietz P, Damaschun H, Zirwer D, Gast K, Schwenke KD, Paehtz W, Damaschun G (1978) FEBS Letters 91:227
69. Reisler E, Pouyet J, Eisenberg H (1970) Biochemistry 9:3095
70. Pilz I, Sund H (1971) Eur J Biochem 20: 561
71. Müller JJ, Kayushina RL (1987) Studia Biophysica 120:15
72. Zipper P, Kratky O, Herrmann R, Hohn T (1971) Eur J Biochem 18:1
73. Gesteland RF, Boedtker H (1964) J Mol Biol 8:496
74. Jacrot B, Chauvin C, Witz J (1977) Nature (London) 266:417
75. Anderegg JW (1967) In: Brumberger H (ed) Small-Angle X-ray Scattering. Gordon & Breach, New York, p 243
76. Bockstahler LE, Kaesberg P (1962) Biophys J 2:1
77. Markham R (1951) Faraday Soc Discuss 11:221
78. Harding SE, Johnson P (1985) Biochem J 231:549
79. Hadidi AF, Fraenkel-Conrat H (1974) In: King RC (ed) Handbook of Genetics, Vol 2. Plenum Press, New York, p 381
80. Anderegg JW, Geil PH, Beeman WW, Kaesberg P (1961) Biophys J 1:657
81. Chauvin C, Witz J, Jacrot B (1978) J Mol Biol 124:641
82. Tanford C (1961) Physical Chemistry of Macromolecules. Wiley, New York
83. Jacrot B, Pfeiffer P, Witz J (1976) Phil Trans R Soc Lond B 276:109
84. Krüse J, Krüse KM, Witz J, Chauvin C, Jacrot B, Tardieu A (1982) J Mol Biol 162:393

Progr Colloid Polym Sci (1997) 107:58–71
© Steinkopff Verlag 1997

P. Zipper
H. Durchschlag

Calculation of hydrodynamic parameters of proteins from crystallographic data using multibody approaches

Prof. Dr. P. Zipper (✉)
Institute of Physical Chemistry
University of Graz
Heinrichstraße 28
8010 Graz, Austria

H. Durchschlag
Institute of Biophysics and Physical
Biochemistry
University of Regensburg
93040 Regensburg, Germany

Abstract Bead models, composed of overlapping spheres of different or identical radii, were generated from the crystal structures of selected proteins (citrate synthase, lactate dehydrogenase, and glyceraldehyde-3-phosphate dehydrogenase). Starting from the atomic coordinates taken from the Brookhaven Protein Data Bank, different methods ("running mean" or "cubic grid") were used for reducing the initially large number of coordinates to a practicable size. The structure of the multibody models was checked by comparing the pair distance distribution function and the radius of gyration of the reduced models with the corresponding quantities of an initial model. For models consisting of beads of unequal size both reduction methods may be used. The concept of equal beads was found to be compatible with the "running mean" method, whereas the use of the "cubic grid" method for generating models composed of equal spheres led to quite distorted structures. The program HYDRO (by García de la Torre et al. (1994) Biophys J 67:530–531) was applied for predictions of the sedimentation coefficient s, diffusion coefficient D, and intrinsic viscosity $[\eta]$. Reliable and numerically stable results were only obtained after the program had been modified in order to improve the treatment of overlapping unequal spheres. The hydrodynamic parameters thus predicted were compared with experimental values as well as with the results from whole-body approaches. In general, a good conformity of predicted and observed values for s and D, including the extent of changes due to conformational alterations, was obtained by multibody approaches if appropriate refinements of the models with respect to radius of gyration and volume were applied. By contrast, the values predicted for $[\eta]$ usually exceeded the experimental results considerably.

Key words Proteins – analytical ultracentrifugation – viscometry – scattering – crystal structure – parameter predictions – modeling – multibody approaches

Introduction

Hydrodynamic and solution scattering properties of biopolymers may be modeled by whole-body or multibody approaches [1–9]. While the first approach approximates macromolecular conformations by spheres, ellipsoids of revolution, triaxial ellipsoids or other simple geometric structures, the second procedure utilizes structures built up from many spheres ("beads") to fit the properties of

Progr Colloid Polym Sci (1997) 107: 58-71
© Steinkopff Verlag 1997

complex and inhomogeneous macromolecular assemblies more correctly or to simulate subtle conformational changes. It is obvious that both approaches have their merits. For the application of multibody approaches, however, essentially enhanced computational efforts are required, unless realistic simplifications (e.g., concerning number and size of beads) are used. While even rather complicated bead models have been employed in solution scattering studies for many years (Debye modeling; cf. refs. [3–5, 10, 11]), hydrodynamic modeling of molecules of arbitrary shape became feasible only recently due to the improvements of the theory and the development of appropriate computer programs by García de la Torre and others (cf. refs. [2, 7, 12] and references therein).

The 3-D structures of a large number of proteins and nucleic acids have been solved to high resolution and their atomic coordinates are usually deposited in the Brookhaven Protein Data Bank (PDB). Therefore, it is tempting to use this information for constructing hydrodynamic and/or scattering models. Modeling of proteins may be achieved by performing several steps: use of the atomic coordinates stored in the PDB files, appropriate transformation and superposition of the coordinate elements in the case of multisubunit proteins, model building (e.g., by approximating a certain number ($n = 1–20$) of amino acid residues by spheres; their radii may be deduced from the volumes which may be calculated as sum of the constituent atomic volumes). Problems arise from consideration of the appropriate hydration contributions. If small-angle scattering data are available, application of the experimentally determined radius of gyration or amount of hydration may help to mimic the solution parameters exactly (e.g., by normalizing the calculated radius of gyration to the experimentally observed value). Evaluation of small-angle scattering curves or distance distribution functions may also be used as additional tools for probing the validity of hydrodynamic models under consideration.

In this study, the hydrodynamic and solution structural parameters of selected proteins (apo- and holoform of citrate synthase, lactate dehydrogenase, and glyceraldehyde-3-phosphate dehydrogenase) are considered as representative examples. The values predicted from crystal data by means of various multibody approaches are compared with experimental values from ultracentrifugal, viscometric and small-angle scattering experiments, as well as with the results from whole-body approaches (cf. ref. [9]).

Methods

Atomic coordinates of protein structures were taken from the Brookhaven Protein Data Bank. Two different approaches were developed in order to reduce the initially very large number of coordinates to a practicable size. The first step in both approaches is a transformation of the atomic coordinates to the coordinates of the centers of gravity of the amino acid residues, a procedure that has also been applied in the calculation of small-angle X-ray scattering (SAXS) curves from crystal data (cf. ref. [5]). The resulting "initial model" of the protein structure contains not only the spatial coordinates x, y, z of all amino acid residues but also the calculated volume v of each residue as additional parameter. All further reductions of coordinates by means of appropriate averaging (methods A and B, see below) refer to this initial model.

Reduction method A

In this approach the number of coordinates is reduced by a kind of "running mean". Proceeding along the peptide chain, the center of gravity of every N amino acid residues in sequential order is calculated and taken as new set of coordinates, together with the total volume of the N constituent residues. The eligible reduction factor N is held constant (except for the very last portion of the chain where deviations may become unavoidable). Finally, spheres of appropriate radii r (derived from the calculated volumes) are placed at the calculated positions x, y, z. Alternatively, an average radius can be calculated and assigned to all spheres.

Reduction method B

In this approach the initial model is placed into a cubic grid of eligible cell dimension C. Searching through all occupied cells, the center of gravity of all amino acid residues located in a given cell is calculated and taken as new set of coordinates, together with the total volume of the respective residues. Finally, spheres of appropriate radii r (derived from the calculated volumes) are placed at the calculated positions x, y, z. Alternatively, also in this approach an average radius may be assigned to all spheres. On the whole, this reduction method is similar to the algorithm presented recently [13].

In the approach of method A, the symmetry relations of oligomeric proteins are not affected by the reduction procedure. This may not be the case if reduction method B is applied. To maintain the symmetry relations also in this approach, it was necessary to place the origin of the cubic grid at the symmetry center (center of gravity) of the initial model.

Control of reduction

Reducing the number of coordinate elements that describe a protein structure necessarily leads to differences in the surface structure and the internal mass distribution of the "reduced model" if compared to the "initial model". To guarantee that the reduction process does not lead to intolerable distortions of the initial structure, visual control of the reduced models is helpful but may not be sufficient. For visual controls, we used drawings prepared by the molecular graphics program RASMOL (by R. Sayle). As an additional control we applied the comparison of the pair distance distribution function (PDDF) $p(r)$ of the reduced model with that of the initial model. The $p(r)$ function plays an important role in SAXS (cf. ref. [5]). For our purpose the PDDF was calculated indirectly by means of a Fourier transform of the SAXS curve derived from the model, but a direct calculation from the spatial arrangement and the radii of the spheres is also possible (cf. refs. [4, 14]).

Hydrodynamic calculations

Hydrodynamic parameters of the bead models were calculated by means of García de la Torre's program HYDRO [7] which was downloaded from the website of the Biophysical Society and installed on an AXP/VMS mainframe system. In general, the calculations were carried out as recommended with the program control parameter IND = 1 (i.e., using, if possible, a modified Burgers–Oseen tensor). Some additional calculations were also performed with the parameter IND = 0 (i.e., always using an unmodified Burgers–Oseen tensor). Because the results obtained with the original program turned out to be affected by some insufficiencies, we finally used a slightly modified version of the program as described in the section Results and Discussion.

Refinement of models

The models established on the basis of the coordinates from the Protein Data Bank did usually not meet the experimentally determined (SAXS) values for the radius of gyration, R_G, and the particle volume, V. This fact necessitated the application of additional transformations of the model data, either by rescaling the x, y, z coordinates of the spheres by a scaling factor F_1 (to improve the fit of the radius of gyration), or by changing the size of the spheres by a scaling factor F_2 (to improve the fit of the particle volume), or by a combination of both measures.

Results and discussion

Citrate synthase

Construction of bead models

Of the three enzymes selected for this study, citrate synthase from pig heart is the most extensively studied example. It was chosen because the apoform and the holoform of the dimeric enzyme have different crystal structures [15, 16], they have been characterized by both SAXS [5, 17] and hydrodynamic techniques (cf. ref. [17] and references therein). Moreover, citrate synthase turned out to be very useful for testing the different methods for the construction of bead models, and for comparisons with whole-body approaches [17–19].

Starting from the atomic coordinates of the open form (apoform) of the enzyme, as deposited in the file PDB1CTS of the Brookhaven Protein Data Bank, the first step of data reduction, common to all reduction approaches, resulted in the establishment of the "initial model" shown in Fig. 1. This model consists of 874 beads, each bead representing a single amino acid residue. This number exceeds by far the limit of 300 beads (imposed by our implementation of the program HYDRO) and had to be reduced further.

Application of the "running mean" approach (method A), using reduction factors N ranging from 5 to 20, led to the reduced models presented in Figs. 2a–e. It should

Fig. 1 Drawing of the "initial model" for the open form of citrate synthase. The model consists of 874 spheres. Each sphere represents a single amino acid residue, the calculated volume of which determines the size of the sphere

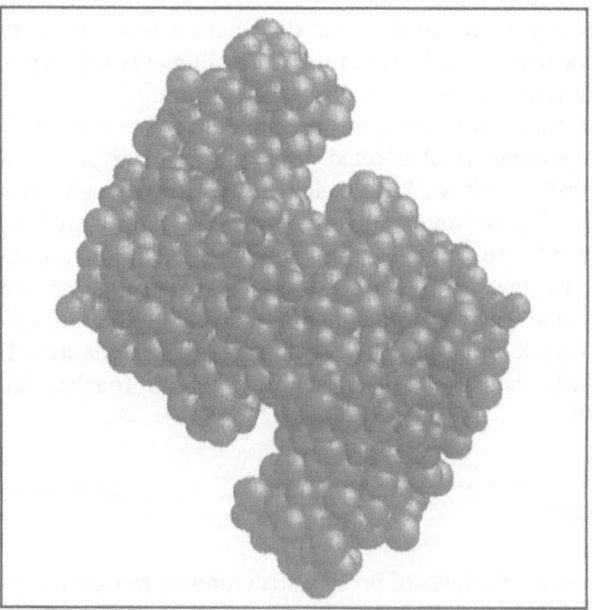

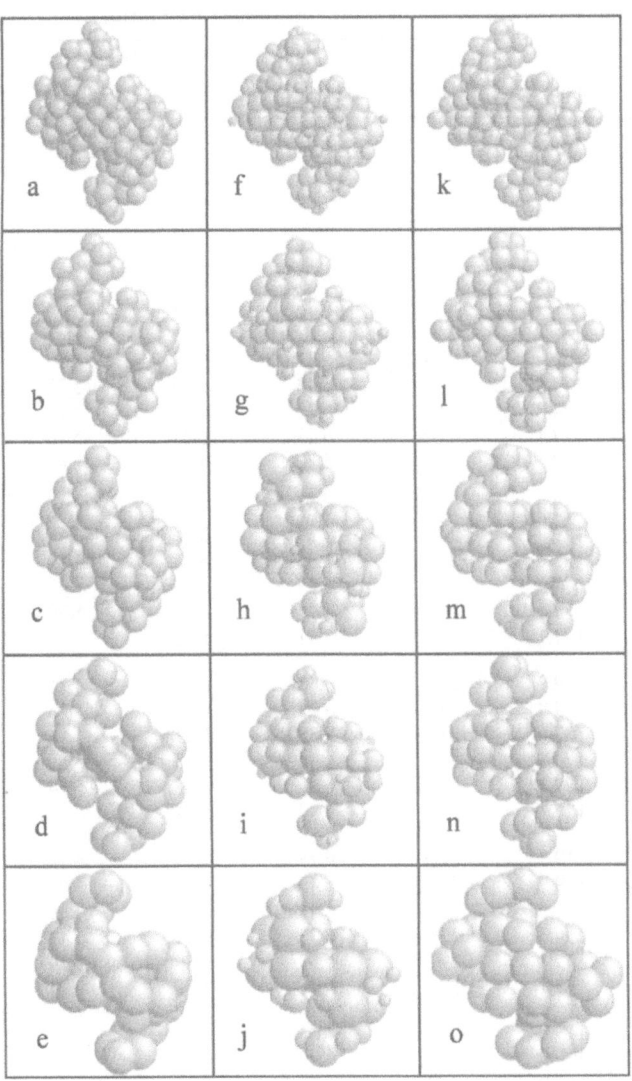

Fig. 2 Drawings of "reduced models" for the open form of citrate synthase. The models in the left lane (a–e) were generated from the initial model by applying reduction method A, those in the central lane (f–j) by applying reduction method B. Replacement of the unequal spheres in models f–j with equal spheres leads to the models shown in the right lane (k–o). From top to bottom, the reduction factor N equals 5, 7, 10, 15, 20, and the cell dimension C amounts to 1, 1.25, 1.5, 1.75, 2 nm, respectively. For further details of the models see Table 1

be noted that these models consist of unequal spheres, their radii, however, do not differ very much because each sphere (except the two spheres at the carboxyl termini) represents the same number of amino acid residues. It is obvious that the characteristic shape of the molecule is maintained rather well even if the number of beads of the initial model is reduced by a factor of 20 (Fig. 2e). Replacing the unequal spheres with equal spheres of average size

only leads to marginal changes in the visual appearance of the models (not shown).

The models presented in Figs. 2f–j were constructed by means of the reduction method B, using centered cubic grids with cell dimensions C ranging from 1 to 2 nm. A striking feature of these models is the broad size spectrum of spheres, ranging from very small spheres on the surface to large spheres in the core. The width of the size distribution increases with the cell dimension C. The models shown in the third lane of Fig. 2 (denoted k–o) were obtained by replacing the unequal spheres of the models in the second lane with equal spheres of average size, leaving the spatial coordinates of the spheres unchanged. Evidently, the replacement of the small spheres on the surface with larger ones enhances the overall size of the particles and changes the internal mass distribution.

Objective measures to judge the validity of the reduced models are the comparison of the calculated radii of gyration, R_G, of the models (cf. Table 1, the rows $F_1 = F_2 = 1$) with the value calculated for the initial model (Table 1, separate row), and the comparison of the pair distance distribution functions $p(r)$ in Fig. 3. We recall that the radius of gyration and the $p(r)$ function of a particle are closely related and R_G can also be calculated from $p(r)$ [4].[1] The $p(r)$ functions presented in Figs. 3a and b show that both reduction methods, when generating models composed of unequal spheres, only cause marginal alterations of the PDDF if compared to that of the initial model, except for the extreme degrees of reduction, $N = 20$ and $C = 2$ nm, respectively. As follows from the numbers of spheres constituting the models (cf. Table 1), the method B model with $C = 2$ nm corresponds to a reduction factor of about 18. Perhaps the higher reduction factor of $N = 20$ for the method A model explains the poorer fit of the PDDF of this model if compared to the fit for the method B model. In accordance with the findings from the $p(r)$ functions, an inspection of the R_G values of the reduced models in Table 1 (rows $F_1 = F_2 = 1$) reveals that at high degrees of reduction the radii of gyration of the models constructed according to method B deviate from the R_G of the initial model to a lesser extent than the R_G values of the models generated by method A.

The situation changes considerably if the unequal spheres are replaced with equal spheres of average size. As follows from Fig. 3c, in the case of models generated by reduction method A this replacement does not cause any pronounced deterioration of the fit of the PDDF. Similarly, the replacement affects the R_G values of method

---

[1]In this context it has to be noted that here the R_G values and $p(r)$ functions are calculated for mass distributions, whereas in SAXS the corresponding quantities refer to electron density distributions

Table 1 Experimental and calculated structural and hydrodynamic parameters of citrate synthase from pig heart (open form)

Database	N	Number of spheres	F_1	F_2	V [nm³]	R_G [nm]	$s \times 10^{13}$ [s]	$D \times 10^7$ [cm²/s]	$[\eta]$ [cm³/g]
Experimental[a] [17,19]					174.4	2.91	6.2	5.8	3.95
Initial model (from PDB1CTS)		874			119.9	2.80			
Reduced models, method A with unequal spheres	5	176	1	1	119.9	2.79	6.57	6.39	4.37
			1.041	1	119.9	2.91	6.35	6.17	4.64
			1	1.041	135.2	2.80	6.52	6.35	4.66
			1.041	1.041	135.2	2.91	6.31	6.14	4.93
			1	1.133	174.3	2.80	6.43	6.26	5.37
			1.039	1.133	174.3	2.91	6.23	6.06	5.64
			1	1.139	177.1	2.80	6.43	6.25	5.42
			1.139	1.139	177.1	3.18	5.76	5.61	6.46
	7	126	1	1	119.9	2.79	6.61	6.43	4.32
			1.042	1	119.9	2.91	6.39	6.22	4.59
			1.042	1.042	135.6	2.91	6.35	6.17	4.89
			1.039	1.133	174.3	2.91	6.27	6.10	5.60
			1	1.139	177.1	2.80	6.46	6.28	5.39
			1.139	1.139	177.1	3.18	5.81	5.65	6.39
	10	88	1	1	119.9	2.78	6.68	6.50	4.25
			1.047	1	119.9	2.91	6.44	6.26	4.53
			1.047	1.047	137.6	2.91	6.38	6.21	4.88
			1.043	1.133	174.3	2.91	6.30	6.13	5.55
			1	1.139	177.1	2.80	6.50	6.33	5.34
			1.139	1.139	177.1	3.17	5.87	5.71	6.28
	15	60	1	1	119.9	2.75	6.78	6.59	4.14
			1.057	1	119.9	2.90	6.49	6.31	4.45
			1.057	1.057	141.5	2.91	6.41	6.24	4.88
			1.052	1.133	174.3	2.91	6.33	6.16	5.49
			1	1.139	177.1	2.77	6.57	6.39	5.24
			1.139	1.139	177.1	3.14	5.95	5.79	6.11
	20	44	1	1	119.9	2.73	7.04	6.84	3.89
			1.068	1	119.9	2.90	6.70	6.52	4.21
			1.068	1.068	146.0	2.91	6.59	6.41	4.74
			1.063	1.133	174.3	2.91	6.50	6.32	5.27
			1	1.139	177.1	2.75	6.78	6.59	5.01
			1.139	1.139	177.1	3.10	6.18	6.01	5.75
Reduced models, method A with equal spheres	5	176	1	1	119.9	2.78	6.52	6.34	4.42
			1.048	1	119.9	2.91	6.27	6.10	4.74
			1.048	1.048	138.0	2.91	6.22	6.05	5.08
			1.045	1.133	174.3	2.91	6.16	5.99	5.74
			1	1.139	177.1	2.79	6.38	6.20	5.48
			1.139	1.139	177.1	3.16	5.73	5.57	6.53
	7	126	1	1	119.9	2.77	6.59	6.41	4.35
			1.051	1	119.9	2.91	6.32	6.15	4.68
			1.051	1.051	139.1	2.91	6.27	6.10	5.05
			1.048	1.133	174.3	2.91	6.20	6.03	5.69
			1	1.139	177.1	2.78	6.43	6.25	5.42
			1.139	1.139	177.1	3.15	5.78	5.63	6.42
	10	88	1	1	119.9	2.76	6.68	6.50	4.25
			1.057	1	119.9	2.91	6.39	6.21	4.59
			1.057	1.057	141.5	2.91	6.32	6.15	5.02
			1.052	1.133	174.3	2.91	6.25	6.08	5.61
			1	1.139	177.1	2.77	6.50	6.32	5.34
			1.139	1.139	177.1	3.14	5.86	5.70	6.28

Table 1 (continued)

Database	N or C[nm]	Number of spheres	F_1	F_2	V [nm^3]	R_G [nm]	$s \times 10^{13}$ [s]	$D \times 10^7$ [cm^2/s]	$[\eta]$ [cm^3/g]
	15	60	1	1	119.9	2.75	6.70	6.51	4.21
			1.057	1	119.9	2.90	6.41	6.24	4.53
			1.057	1.057	141.5	2.91	6.33	6.16	4.97
			1.053	1.133	174.3	2.91	6.25	6.08	5.59
			1	1.139	177.1	2.77	6.49	6.32	5.32
			1.139	1.139	177.1	3.14	5.88	5.72	6.21
	20	44	1	1	119.9	2.70	7.05	6.86	3.88
			1.078	1	119.9	2.90	6.67	6.48	4.24
			1.078	1.078	150.1	2.91	6.54	6.36	4.86
			1.074	1.133	174.3	2.91	6.47	6.29	5.32
			1	1.139	177.1	2.73	6.79	6.61	5.00
			1.139	1.139	177.1	3.08	6.19	6.02	5.73
Reduced models, method B with unequal spheres	1.0	203	1	1	119.9	2.82	6.42	6.25	4.55
			1.032	1	119.9	2.91	6.25	6.08	4.78
			1	1.032	131.7	2.82	6.39	6.22	4.77
			1.032	1.032	131.7	2.91	6.22	6.05	5.00
			1	1.133	174.3	2.83	6.31	6.14	5.54
			1.03	1.133	174.3	2.91	6.15	5.98	5.76
			1	1.139	177.1	2.83	6.30	6.13	5.59
			1.139	1.139	177.1	3.21	5.64	5.48	6.72
	1.25	126	1	1	119.9	2.81	6.48	6.30	4.48
			1.034	1	119.9	2.91	6.29	6.12	4.72
			1.034	1.034	132.5	2.91	6.26	6.09	4.96
			1.03	1.133	174.3	2.91	6.20	6.03	5.70
			1	1.139	177.1	2.83	6.34	6.17	5.54
			1.139	1.139	177.1	3.21	5.69	5.53	6.62
	1.5	85	1	1	119.9	2.80	6.55	6.37	4.40
			1.039	1	119.9	2.91	6.35	6.17	4.66
			1.039	1.039	134.4	2.91	6.31	6.14	4.94
			1.034	1.133	174.3	2.91	6.24	6.07	5.66
			1	1.139	177.1	2.82	6.40	6.23	5.48
			1.139	1.139	177.1	3.19	5.75	5.60	6.51
	1.75	67	1	1	119.9	2.80	6.57	6.39	4.37
			1.041	1	119.9	2.90	6.36	6.19	4.63
			1.041	1.041	135.2	2.91	6.31	6.14	4.93
			1.034	1.133	174.3	2.91	6.25	6.08	5.63
			1	1.139	177.1	2.82	6.42	6.24	5.45
			1.139	1.139	177.1	3.18	5.77	5.61	6.46
	2.0	49	1	1	119.9	2.79	6.61	6.43	4.32
			1.044	1	119.9	2.90	6.39	6.21	4.58
			1.044	1.044	136.4	2.91	6.33	6.16	4.91
			1.036	1.133	174.3	2.91	6.26	6.09	5.61
			1	1.139	177.1	2.82	6.42	6.25	5.43
			1.139	1.139	177.1	3.17	5.80	5.64	6.38
Reduced models, method B with equal spheres	1.0	203	1	1	119.9	2.97	6.23	6.06	4.80
			0.98	1	119.9	2.91	6.34	6.17	4.65
			1	1.133	174.3	2.98	6.11	5.94	5.82
			0.977	1.133	174.3	2.91	6.23	6.06	5.64
	1.25	126	1	1	119.9	3.02	6.25	6.08	4.76
			0.961	1	119.9	2.91	6.46	6.28	4.49
			1	1.133	174.3	3.03	6.11	5.95	5.80
			0.958	1.133	174.3	2.91	6.33	6.15	5.50
	1.5	85	1	1	119.9	3.12	6.30	6.13	4.72
			0.932	1	119.9	2.91	6.67	6.49	4.27
			1	1.133	174.3	3.13	6.15	5.98	5.78
			0.927	1.133	174.3	2.91	6.52	6.35	5.28

Table 1 (continued)

Database	C [nm]	Number of spheres	F_1	F_2	V [nm³]	R_G [nm]	$s \times 10^{13}$ [s]	$D \times 10^7$ [cm²/s]	$[\eta]$ [cm³/g]
	1.75	67	1	1	119.9	3.17	6.21	6.04	4.83
			0.916	1	119.9	2.91	6.66	6.48	4.27
			1	1.133	174.3	3.18	6.05	5.88	5.92
			0.910	1.133	174.3	2.91	6.50	6.32	5.29
	2.0	49	1	1	119.9	3.17	6.26	6.09	4.74
			0.915	1	119.9	2.91	6.70	6.52	4.21
			1	1.133	174.3	3.19	6.08	5.91	5.85
			0.908	1.133	174.3	2.91	6.52	6.34	5.25
Whole-body models (prolate ellipsoids)	From SAXS data [19]				174.4	2.91	6.30[b] 6.35[c]	6.00[b] 6.04[c]	2.98[b] 2.90[c]
	From crystal data, without hydration				119.9	2.80	6.91	6.58	2.30
	From crystal data, hydration 0.35 g/g				177.1	2.80	6.37	6.07	2.86

[a] $M = 97\,938$ g/mol, $\bar{v} = 0.740$ cm³/g [17]
[b] Calculated from R_G and V
[c] Calculated from R_G and surface-to-volume ratio S/V

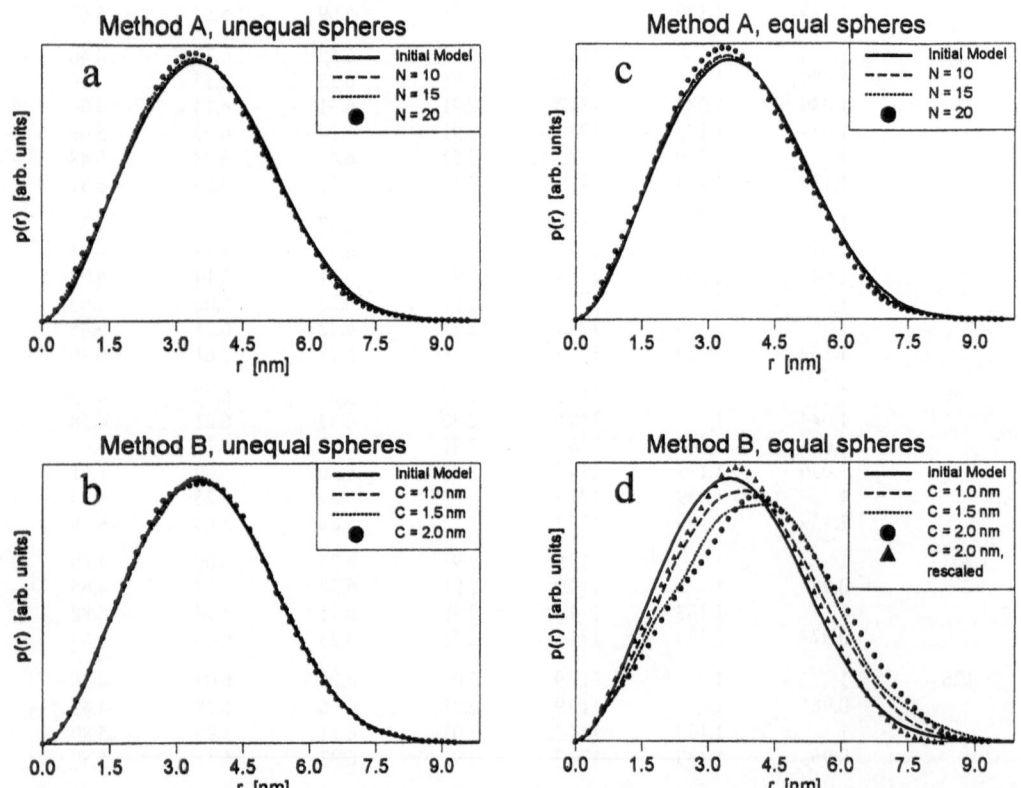

Fig. 3 Pair distance distribution functions $p(r)$ calculated from reduced models for the open form of citrate synthase and $p(r)$ function of the initial model. Plots a and b refer to reduced models composed of unequal spheres (cf. left and central lanes in Fig. 2), plots c and d to models built up from equal spheres (for d cf. right lane in Fig. 2). For further details see the text

A models only slightly (cf. Table 1). On the contrary, in the case of method B models the mere replacement of unequal spheres with equal spheres causes dramatic alterations of the PDDF (Fig. 3d) and, accordingly, a pronounced increase of the R_G values (Table 1). To compensate for the increase of R_G, the spatial coordinates of the spheres may be rescaled, i.e., transformed by a factor which is chosen to yield the $R_G = 2.80$ nm of the initial model. As expected, the rescaling also improves the fit of the PDDF (one example is shown in Fig. 3d), but nevertheless the fit remains poorer than for the corresponding method A models at comparable degrees of reduction. These findings allow the statement that, considering the conservation of the structure as criterion, the combination of reduction method B with spheres of equal size is dangerous and should be used only with caution, while reduction method A and spheres of equal size may be combined without causing serious problems.

In an analogous way as for the open form of citrate synthase, models were also generated for one of the closed forms (holoforms) of the enzyme. The initial model was based on the coordinates deposited in the Protein Data Bank file PDB4CTS.

Probing the program HYDRO

We started our calculations of hydrodynamic parameters (sedimentation coefficient s, diffusion coefficient D, and intrinsic viscosity $[\eta]$) from the various reduced models of citrate synthase with the original version of the program HYDRO, with the control parameter IND = 1 as recommended by the program's author. However, when we calculated the hydrodynamic parameters for models built from unequal spheres, we observed a low numerical stability of the results. Sometimes even very slight alterations of a model (by slightly changing one of the scaling factors) led to drastic changes of calculated parameters. Some results for s and D were completely outside the range of expectation. Moreover, we often encountered considerable deviations of the centers of diffusion and of viscosity from the center of gravity, in some cases even very large deviations (up to 12 nm) were observed. We ascribed these problems to the overlapping of unequal spheres because we never observed comparable phenomena with the models built up from equal spheres. Alternative calculations performed with the program control parameter IND = 0 (forcing the permanent use of the unmodified Burgers–Oseen tensor) did not improve the results. Indeed, these calculations usually led to larger deviations of the diffusion and/or viscosity center, and even to large deviations in the case of models built from equal spheres.

Eventually, an inspection of the source code of the program HYDRO revealed an insufficiency in handling the case of overlapping unequal spheres (a founded theory for this case is still lacking). As stated by García de la Torre [2], the expression for the interaction tensor, modified for unequal spheres (cf. Eq. (13) in ref. [12]), is not valid if the spheres overlap. Nevertheless, in the program (at least in the downloaded version) just this expression is used for overlapping spheres, too. Byron [13] claims that the program reverts to an unmodified Burgers–Oseen tensor in the event of an overlapping bead pair, but this is obviously not the case. Therefore, we modified the program in such a way that, whenever a pair of overlapping unequal spheres is encountered, the program treats them (and only in this event) as if they were equal, thus the expression by Rotne and Prager [20] (cf. Eq. (13) in ref. [2]) can be applied. In this context the radii of the two spheres are averaged so that the total volume of the pair remains constant, but it turned out that other kinds of averages lead to almost the same results (however, larger discrepancies are obtained if instead of an averaged radius the true radius of one of the two spheres is used in the Rotne–Prager expression). Calculations using this modified version of the program led to numerically stable results and deviations of the diffusion or viscosity center from the center of gravity were always found to be less than 0.1 nm. On the other side, an alternative modification of the program, implementing a temporary reversion to the unmodified Burgers–Oseen tensor in the case of overlap, turned out to be less successful (e.g., deviations of diffusion or viscosity center still occurred) and was discarded.

Calculation of hydrodynamic parameters

The values of s, D, and $[\eta]$, as obtained from the various reduced models of the open form of citrate synthase by means of the modified program HYDRO, are compiled in Table 1, together with the model parameters (N or C, number of spheres, scaling factors F_1 and F_2, volume V, R_G). The experimental values for V and R_G (from SAXS) and for s, D, and $[\eta]$ are also given in the table.

The rows with $F_1 = F_2 = 1$ refer to the unrefined reduced models. Comparison of the calculated s and D values of these models with the experimental ones shows that the calculated values are always higher. The smallest discrepancies are encountered for the method B models with equal spheres. The calculated $[\eta]$ values of the unrefined models are, with two exceptions, also higher than the experimental ones. The exceptions are the method A models, with equal or unequal spheres, for the highest degree of reduction ($N = 20$). In all groups of models, except method B models with equal spheres, there is a clear tendency of the s and D values to increase and of $[\eta]$ to decrease if the degree of reduction increases. For models generated by method A, the differences between models with equal and with unequal spheres are only marginal.

The calculated R_G values of the unrefined models are significantly lower than the experimental R_G value, except again for the method B models with equal spheres where the calculated values exceed the experimental one (the fact that these models are a special case has already been discussed). All unrefined models possess the same volume which is identical to the calculated sum of atomic volumes. Its value is considerably lower than the experimental volume of $174.4 \, \text{nm}^3$ of the hydrated enzyme particle as derived from SAXS and agrees well with the anhydrous volume of $120.3 \, \text{nm}^3$ that can be calculated from the molar mass and the experimental partial specific volume (cf. refs. [17, 19]).

The discrepancies observed between calculated and experimental hydrodynamic parameters, and V and R_G values, respectively, suggested the application of some refinement of the models. For this purpose we used the scaling factors F_1 and/or F_2. F_1 acts only on the spatial coordinates and F_2 only on the radii of the spheres. Thus, it is possible to adjust only the R_G value of the model (through F_1) or only the volume (through F_2) or radius of gyration and volume simultaneously. Detailed examples of refinements are given in Table 1 for the models with $N = 5$ (method A) and $C = 1 \, \text{nm}$ (method B), and less detailed examples for the other models. The presented examples include the adjustment of R_G to the experimental value by uniform expansion ($F_1 = F_2 > 1$), the uniform expansion ($F_1 = F_2 = 1.139$) to account for an assumed hydration of $0.35 \, \text{g/g}$ [17, 18], and the non-uniform expansion ($F_1 > 1$, $F_2 = 1.133$) to adjust both R_G and V to the experimental values. The models according to method B with equal spheres had to be refined in a different way ($F_1 < 1$) because of the initially too high R_G values.

With all models according to method A and also with the method B models with unequal spheres, the closest fit of the calculated s and D values to the experimental data was always achieved by the non-uniform expansion by which the experimental values for both R_G and V could be modeled. The uniform expansion to a volume corresponding to a hydration of $0.35 \, \text{g/g}$ generally yielded too low values for s and D, except for the method A models with $N = 20$. The fit of $[\eta]$ values got even worse by the refinements. The values calculated for the refined models are always higher than those of the unrefined models, except in the special case of method B models with equal spheres. Surprisingly, the values of s and D calculated for the latter models are not very different from the experimental parameters, even without any refinement of the models and

in spite of the disagreement of the R_G and V values with the experimental data. It has to be noted that for these models the adjustment of both R_G and V to the experimental values by non-uniform rescaling of radii and coordinates did not lead to any improvement in the fit of the hydrodynamic parameters.

At the end of Table 1 we present results obtained from whole-body models (prolate ellipsoids of revolution) of the open form of citrate synthase. These results, which were derived from SAXS data and from crystal data (actually from the "initial model", without and with hydration) agree fairly well with results for s and D from multibody models presented in Table 1. They are also in accord with the corresponding experimental results, provided appropriate hydration is assumed for the prediction from crystal data. By contrast, the intrinsic viscosity is obviously underestimated as compared to the experimental value, whereas the multibody models overestimate $[\eta]$.

In order to allow a better comparison of the multibody (MB) and whole-body (WB) approaches, for selected reduced models the values of R_G and V as given in Table 1 were taken as a basis for the whole-body calculations of s, D and $[\eta]$. The ratios between the tabulated values for s, D and $[\eta]$ and those predicted by the whole-body approach were calculated and finally averaged. As follows from the results listed in Table 2, the mean ratios for s and D values, $\langle s_{MB}/s_{WB} \rangle$ and $\langle D_{MB}/D_{WB} \rangle$, are nearly 1 and the standard deviation is 3%. These findings clearly show that the multibody as well as the whole-body approach lead to essentially consistent predictions for s and D. On the other hand, the mean ratio for the intrinsic viscosity, $\langle [\eta]_{MB}/[\eta]_{WB} \rangle$, is always much greater than 1, thus indicating a pronounced discrepancy between the $[\eta]$ values predicted by the two approaches. At present, we have no definite explanation of this discrepancy, but find it noteworthy that also for the intrinsic viscosity a mean ratio close to 1 may be obtained if no volume correction (cf. ref. [2]) is applied in the calculation of $[\eta]$ by means of the program HYDRO. In that case, however, the standard deviation of the ratio would increase up to 9%.

In Table 3 we present a compilation of results obtained from calculations of hydrodynamic parameters for the bead models of the closed form of citrate synthase. The calculated values for s and D of models with unequal spheres are slightly higher than those obtained for the open form (Table 1) and similar to the results from whole-body approaches (Table 3). A direct comparison with experimental parameters is only possible for the sedimentation coefficient since other experimental data are lacking. The comparison unveils that the experimentally observed difference in the s values of the two forms of citrate synthase is predicted quite correctly both by the multibody and the whole-body approaches.

Table 2 Comparison of hydrodynamic parameters predicted from multibody (MB) and whole-body (WB) models of citrate synthase from pig heart (open form)

Models			$\langle s_{MB}/s_{WB} \rangle$	$\langle D_{MB}/D_{WB} \rangle$	$\langle [\eta]_{MB}/[\eta]_{WB} \rangle$
Method A	$N = 5$		0.97 ± 0.03	0.99 ± 0.03	1.90 ± 0.01
	$N = 20$		1.02 ± 0.03	1.04 ± 0.03	1.76 ± 0.02
Method B	$C = 1.0$ nm		0.95 ± 0.03	0.97 ± 0.03	1.95 ± 0.02
	$C = 2.0$ nm		0.97 ± 0.03	0.99 ± 0.03	1.89 ± 0.01

Table 3 Experimental and calculated structural and hydrodynamic parameters of citrate synthase from pig heart (closed form)

Database	C [nm]	Number of spheres	F_1	F_2	V [nm^3]	R_G [nm]	$s \times 10^{13}$ [s]	$D \times 10^7$ [cm^2/s]	$[\eta]$ [cm^3/g]
Experimental [17, 19]					164.0	2.80	6.4		
Initial model (from PDB4CTS)		874			119.9	2.70			
Reduced models, method B with unequal spheres	1.0	194	1	1	119.9	2.72	6.54	6.36	4.40
			1.029	1	119.9	2.80	6.38	6.21	4.60
			1.029	1.029	130.6	2.80	6.36	6.18	4.79
			1.026	1.11	163.9	2.80	6.30	6.13	5.38
			1	1.139	177.1	2.73	6.41	6.24	5.44
			1.139	1.139	177.1	3.10	5.74	5.59	6.50
	1.5	88	1	1	119.9	2.71	6.62	6.44	4.31
			1.035	1	119.9	2.80	6.43	6.26	4.53
			1.035	1.035	132.9	2.80	6.40	6.22	4.78
			1.031	1.11	163.9	2.80	6.35	6.17	5.33
			1	1.139	177.1	2.73	6.48	6.30	5.36
			1.139	1.139	177.1	3.08	5.81	5.66	6.37
	2.0	50	1	1	119.9	2.69	6.69	6.51	4.22
			1.043	1	119.9	2.79	6.47	6.29	4.48
			1.043	1.043	136.0	2.80	6.42	6.24	4.79
			1.036	1.11	163.9	2.80	6.37	6.20	5.29
			1	1.139	177.1	2.72	6.51	6.34	5.32
			1.139	1.139	177.1	3.06	5.88	5.72	6.24
Reduced models, method B with equal spheres	1.0	194	1	1	119.9	2.88	6.36	6.19	4.61
			0.971	1	119.9	2.80	6.52	6.34	4.41
			1	1.11	163.9	2.89	6.25	6.08	5.43
			0.968	1.11	163.9	2.80	6.43	6.25	5.20
	1.5	88	1	1	119.9	2.98	6.29	6.12	4.69
			0.938	1	119.9	2.80	6.62	6.44	4.29
			1	1.11	163.9	2.99	6.16	5.99	5.56
			0.934	1.11	163.9	2.80	6.50	6.32	5.11
	2.0	50	1	1	119.9	3.16	6.17	6.01	4.85
			0.88	1	119.9	2.80	6.80	6.62	4.09
			1	1.11	163.9	3.18	6.02	5.86	5.76
			0.875	1.11	163.9	2.80	6.64	6.46	4.93
Whole-body models (prolate ellipsoids)	From SAXS data [19]				164.0	2.80	6.51[a]	6.16[a]	2.73[a]
							6.62[b]	6.28[b]	2.56[b]
	From crystal data, without hydration				119.9	2.70	7.01	6.67	2.19
	From crystal data, hydration 0.35 g/g				177.1	2.70	6.46	6.15	2.73

[a] Calculated from R_G and V
[b] Calculated from R_G and surface-to-volume ratio S/V

Lactate dehydrogenase

The atomic coordinates for the apoform of the tetrameric enzyme from dogfish were taken from the Protein Data Bank file PDB6LDH. Reduction of coordinates and calculation of hydrodynamic parameters were performed in an analogous way as described above for citrate synthase. Similarly, the R_G values and the volumes of the initial model and of the unrefined reduced models derived therefrom are definitely lower than the corresponding experimental (SAXS) parameters.

The results of calculations of hydrodynamic parameters are presented in Table 4. It can be seen that for this protein the scaling factors needed to adjust the radius of gyration (F_1) and the hydrated volume (F_2) are very similar. Therefore, the various refinements of the models do not lead to much differing hydrodynamic parameters. There are, however, some differences in the parameters s and D between models according to method A and method B: at comparable degrees of reduction, the latter models yield lower values. The s values calculated for the unrefined models are always essentially larger than the experimental value, while those of the various refined models are much closer to the experimental sedimentation coefficient. In the case of refined method B models, the agreement between predicted and observed values is very high. The values predicted for D on the basis of the refined models always agree very well with the experimental quantity. The calculated $[\eta]$ values are always larger than the experimental result, the smallest discrepancies are observed with the unrefined models. The comparison with the parameters predicted by whole-body approaches (Table 4) unveils a fair agreement of s and D, but not of $[\eta]$, with the data derived from SAXS. By contrast, the predictions based on the crystal data (without other refinements than considering hydration) are insufficient.

Glyceraldehyde-3-phosphate dehydrogenase

The atomic coordinates for the apoform of the tetrameric enzyme from lobster were taken from the Protein Data Bank file PDB4GPD. The calculated hydrodynamic data for various reduced models are compiled in Table 5. Since experimental SAXS and hydrodynamic results for the lobster enzyme are lacking, we have to revert to the known data for the enzyme from baker's yeast as reference data. This, of course, impedes the comparison of experimental and predicted data. Above all, it turned out that the R_G values of the initial model and of the unrefined reduced models exceed slightly the experimental radius of gyration (cf. Table 5). This necessitated the use of

a different strategy in the refinement of the models. The simultaneous adjustment of R_G and V yielded s and D values which are higher than the experimental hydrodynamic data for the yeast enzyme but similar to the data predicted by whole-body approaches, either from SAXS data for the yeast enzyme or the crystal data for the enzyme from lobster. The experimental value for $[\eta]$ is predicted neither by the multibody nor the whole-body approaches.

Conclusions

The comparison of the two different approaches for reducing the number of coordinates to a practicable size shows that in principle both approaches can be applied. One should be aware, however, that the use of a very large reduction factor N in the "running mean" method A may distort the structure of the initial model to a higher degree than the use of a large cell dimension in the "cubic grid" method B. A too high degree of reduction should be avoided with both methods (e.g., $N = 20$ or $C = 2$ nm) and a concomitant control, e.g., by means of the pair distance distribution function and the radius of gyration, is recommended. When method B is used for the data reduction, the size of the resulting spheres has to be taken into account properly. In the case of method B, replacement with spheres of equal size would lead to a distorted structure as can be concluded from the deviations found for the radius of gyration and the distance distribution function, and also from the different behavior of hydrodynamic parameters. On the other hand, in models generated by method A such restrictions do not apply or are less important.

The hydrodynamic parameters obtained from the multibody approaches by means of program HYDRO are sensitive to the appropriate choice of the model, especially with regard to radius of gyration and volume. A rescaling of the model may be useful. Though a general rule cannot be given because currently our database is too small, an adjustment of both the radius of gyration and the volume of the model to the experimental values (e.g., from SAXS) may be appropriate. It is obvious that the calculated values for sedimentation and diffusion coefficients meet the experimental values better than the calculated intrinsic viscosities. Possibly, each hydrodynamic parameter requires a special model (e.g., a special hydration) in order to be fitted properly, and all three parameters can never be fitted simultaneously by the same model. At the present stage, at least in the case of globular proteins, the results obtained by means of program HYDRO do not appear to be superior to the predictions based on whole-body approaches.

Table 4 Experimental and calculated structural and hydrodynamic parameters of lactate dehydrogenase from dogfish

Database	N or C[nm]	Number of spheres	F_1	F_2	V [nm³]	R_G [nm]	$s \times 10^{13}$ [s]	$D \times 10^7$ [cm²/s]	$[\eta]$ [cm³/g]
Experimental[a] [21]					253.3	3.47	7.54[b]	5.05[c]	3.8[d]
Initial model (from PDB6LDH)		1316			178.9	3.14			
Reduced models, method A with unequal spheres	10	132	1	1	178.9	3.11	8.55	5.64	4.34
			1.116	1.116	248.7	3.47	7.67	5.05	6.03
			1.116	1.123	253.4	3.47	7.66	5.04	6.09
			1.119	1.119	250.7	3.48	7.64	5.04	6.08
	15	88	1	1	178.9	3.08	8.71	5.74	4.19
			1.125	1.125	254.8	3.47	7.74	5.10	5.97
			1.125	1.123	253.4	3.47	7.74	5.10	5.95
			1.119	1.119	250.7	3.45	7.78	5.13	5.87
	20	68	1	1	178.9	3.05	8.79	5.79	4.12
			1.136	1.136	262.3	3.47	7.74	5.10	6.04
			1.137	1.123	253.4	3.47	7.75	5.11	5.93
			1.119	1.119	250.7	3.42	7.85	5.17	5.78
Reduced models, method A with equal spheres	10	132	1	1	178.9	3.08	8.58	5.65	4.32
			1.126	1.126	255.5	3.47	7.62	5.02	6.17
			1.126	1.123	253.4	3.47	7.62	5.02	6.14
			1.119	1.119	250.7	3.45	7.66	5.05	6.05
	15	88	1	1	178.9	3.06	8.74	5.76	4.17
			1.135	1.135	261.6	3.47	7.70	5.07	6.09
			1.135	1.123	253.4	3.47	7.71	5.08	5.99
			1.119	1.119	250.7	3.42	7.81	5.14	5.84
	20	68	1	1	178.9	3.08	8.73	5.75	4.16
			1.127	1.127	256.1	3.47	7.74	5.10	5.96
			1.127	1.123	253.4	3.47	7.75	5.11	5.92
			1.119	1.119	250.7	3.45	7.80	5.14	5.83
Reduced models, method B with unequal spheres	1.1	251	1	1	178.9	3.13	8.32	5.48	4.55
			1.108	1.108	243.4	3.47	7.51	4.95	6.19
			1.107	1.123	253.4	3.47	7.50	4.94	6.30
			1.119	1.119	250.7	3.51	7.43	4.90	6.37
	1.5	131	1	1	178.9	3.12	8.36	5.51	4.51
			1.112	1.112	246.1	3.47	7.52	4.95	6.20
			1.111	1.123	253.4	3.47	7.52	4.95	6.28
			1.119	1.119	250.7	3.49	7.47	4.92	6.32
	1.75	85	1	1	178.9	3.11	8.47	5.58	4.40
			1.115	1.115	248.1	3.47	7.59	5.00	6.10
			1.114	1.123	253.4	3.47	7.59	5.00	6.16
			1.119	1.119	250.7	3.48	7.57	4.98	6.17
	2.0	63	1	1	178.9	3.11	8.50	5.60	4.37
			1.117	1.117	249.4	3.47	7.61	5.01	6.09
			1.116	1.123	253.4	3.47	7.61	5.01	6.13
			1.119	1.119	250.7	3.48	7.59	5.00	6.12
Whole-body models (oblate ellipsoids)	From SAXS data [18]				253.3	3.47	7.93	5.11	3.24
	From crystal data, without hydration				178.9	3.14	8.82	5.69	2.35
	From crystal data, hydration 0.35 g/g				250.7	3.14	8.36	5.39	2.74

[a] $M = 145\,169$ g/mol (according to SWISS-PROT Protein Sequence Data Bank), $\bar{v} = 0.741$ cm³/g [21]
[b] Identical values for dogfish and pig-heart lactate dehydrogenase (cf. refs. [22, 23])
[c] Value for pig-heart lactate dehydrogenase [23]
[d] Taken from ref. [24]

Table 5 Experimental and calculated structural and hydrodynamic parameters of glyceraldehyde-3-phosphate dehydrogenase

Database	C [nm]	Number of spheres	F_1	F_2	V [nm^3]	R_G [nm]	$s \times 10^{13}$ [s]	$D \times 10^7$ [cm^2/s]	$[\eta]$ [cm^3/g]
Experimental (baker's yeast)[a] [19]					264.2	3.21	7.6	5.0	3.45
Initial model (lobster) (from **PDB4GPD**)		1332			175.1	3.23			
Reduced models, method B with unequal spheres	1.1	238	1	1	175.1	3.24	8.18	5.39	4.70
			0.992	1	175.1	3.21	8.23	5.43	4.64
			0.989	1.147	264.2	3.21	8.10	5.34	5.73
			1.138	1.138	258.1	3.68	7.18	4.74	6.93
	1.5	129	1	1	175.1	3.22	8.28	5.46	4.60
			0.996	1	175.1	3.21	8.31	5.48	4.57
			0.990	1.147	264.2	3.21	8.18	5.39	5.65
			1.138	1.138	258.1	3.67	7.28	4.80	6.77
	1.75	92	1	1	175.1	3.22	8.35	5.50	4.54
			0.998	1	175.1	3.21	8.36	5.51	4.52
			0.991	1.147	264.2	3.21	8.20	5.41	5.63
			1.138	1.138	258.1	3.66	7.33	4.83	6.69
	2.0	71	1	1	175.1	3.21	8.34	5.49	4.53
			0.992	1.147	264.2	3.21	8.17	5.39	5.64
			1.138	1.138	258.1	3.65	7.33	4.83	6.68
Whole-body models (oblate ellipsoids)	From SAXS data [19]				264.2	3.21	8.19	5.29	2.95
	From crystal data, without hydration				175.1	3.23	8.70	5.61	2.50
	From crystal data, hydration 0.35 g/g				258.1	3.23	8.20	5.29	2.95

[a] $M = 142\,868$ g/mol (according to SWISS-PROT Protein Sequence Data Bank), $\bar{v} = 0.737$ cm^3/g [19]

A crucial problem in the application of the multibody approach by means of the program HYDRO was an insufficiency in handling the case of overlapping unequal spheres. We believe that our way of handling this case is a practicable one and better than all other ways tested, though it is certainly not founded by a theory. Anyhow, our modified version of the program yields numerically stable results. It never led to obviously wrong results for s and D, which sometimes happened with the initially used original program version. Founded comparisons between various models or ways of refinement became possible only after the program had been modified as described. A striking argument for the feasibility of our program modification is the good accordance of the parameters calculated for the models of citrate synthase, which have been generated by method A with equal or unequal spheres.

A safer way to circumvent the problems of overlapping beads would be to avoid the use of the "cubic grid" method B, because this method implies models composed of overlapping spheres of often very different size. Instead of method B, one may apply the "running mean" method A, with a not too high reduction factor, in conjunction with equal spheres of average size for the calculation of hydrodynamic parameters. An even better alternative would be a change of the algorithm of the running mean procedure: instead of averaging a constant number of amino acid residues, a variable number of atoms along the chain may be averaged in such a way that the volume (and thus also the size) of the resulting spheres remains nearly constant. Finally, these may be replaced with equal ones without any significant alteration of the structure of the model. Whatever procedure is applied to generate models built up from equal spheres, also such models have to be regarded with caution since the Rotne–Prager interaction tensor for overlapping equal spheres [2, 20] does not account for the case that one bead overlaps two or even more beads [25].

Progr Colloid Polym Sci (1997) 107:58–71
© Steinkopff Verlag 1997

References

1. Harding SE (1989) In: Harding SE, Rowe AJ (eds) Dynamic Properties of Biomolecular Assemblies. Royal Society of Chemistry, Cambridge UK, pp 32–56
2. García del la Torre J (1989) In: Harding SE, Rowe AJ (eds) Dynamic Properties of Biomolecular Assemblies. Royal Society of Chemistry, Cambridge UK, pp 3–31
3. Glatter O (1972) Acta Phys Austriaca 36:307–315
4. Glatter O (1982) In: Glatter O, Kratky O (eds) Small Angle X-ray Scattering. Academic Press, London, pp 119–165
5. Durchschlag H, Zipper P, Wilfing R, Purr G (1991) J Appl Cryst 24:822–831
6. Müller JJ, Pankow H, Poppe B, Damaschun G (1992) J Appl Cryst 25:803–806
7. García de la Torre J, Navarro S, Lopez Martinez MC, Diaz FG, Lopez Cascalez JJ (1994) Biophys J 67:530–531
8. Beavil AJ, Young RJ, Sutton BJ, Perkins SJ (1995) Biochemistry 34:14449–14461
9. Durchschlag H, Zipper P (1997) Progr Colloid Polym Sci 107:43–57
10. Durchschlag H (1975) Biophys Struct Mechanism 1:169–188
11. Lehner D, Zipper P, Henriksson G, Pettersson G (1996) Biochim Biophys Acta 1293:161–169
12. García de la Torre J, Bloomfield VA (1977) Biopolymers 16:1747–1763
13. Byron O (1997) Biophys J 72:408–415
14. Glatter O (1980) Acta Phys Austriaca 52:243–256
15. Remington S, Wiegand G, Huber R (1982) J Mol Biol 158:111–152
16. Wiegand G, Remington S, Deisenhofer J, Huber R (1984) J Mol Biol 174:205–219
17. Durchschlag H, Zipper P, Purr G, Jaenicke R (1996) Colloid Polym Sci 274:117–137
18. Durchschlag H, Zipper P (1996) J Mol Struct 383:223–229
19. Durchschlag H, Zipper P (1997) J Appl Cryst (in press)
20. Rotne J, Prager S (1969) J Chem Phys 50:4831–4837
21. Pessen H, Kumosinski TF (1993) In: Baianu IC, Pessen H, Kumosinski TF (eds) Physical Chemistry of Food Processes, Vol 2: Advanced Techniques, Structures and Applications. Van Nostrand Reinhold, New York, pp 274–306
22. Pesce A, Fondy TP, Stolzenbach F, Castillo F, Kaplan NO (1967) J Biol Chem 242:2151–2167
23. Jaenicke R, Gregori E, Laepple M (1979) Biophys Struct Mechanism 6:57–65
24. Tyn MT, Gusek TW (1990) Biotechnol Bioeng 35:327–338
25. García de la Torre J (1997) Personal communication

Progr Colloid Polym Sci (1997) 107:72–76
© Steinkopff Verlag 1997

M. Pitschke
K. Post
D. Riesner

Analytical ultracentrifugation with fluorescence detection and biosafety containment and its application to the prion protein

M. Pitschke · K. Post ·
Prof. Dr. D. Riesner (✉)
Heinrich-Heine-Universität Düsseldorf
Institute für Physikalische Biologie
40225 Düsseldorf, Germany

Abstract In order to utilize the high sensitivity of the fluorescence detection system for sedimentation equilibrium in analytical ultracentrifugation, a new evaluation method was developed. In comparison to absorption recording the sensitivity could be raised 100–1000 fold as shown for a series of proteins (bovine serum albumin, immunoglobulin G and chymotrypsin). Sedimentation equilibrium runs can be analysed down to sample concentrations of $0.25 \, \text{ng}/\mu l$ with a minimal sample volume of $40 \, \mu l$. The accuracy of the molecular weights is comparable to that recorded by the absorption optics. Analysis of infectious prions in the analytical ultracentrifuge requires strict safety conditions. A biohazard safety containment with a vacuum system is constructed to prevent contamination of the laboratory. Decontamination of the cells is ensured by the use of self-manufactured titanium cells, which could be autoclaved at $134\,^{\circ}\text{C}$ in 1 N NaOH. Analysis of SDS-solubilized PrP(27–30) which is a N-terminal truncated fragment of the whole prion protein, showed by analytical ultracentrifugation relatively homogeneous fractions with molecular masses between 20.000 Da and 120.000 Da, depending on the conditions of solubilization.

Key words Molecular weight determination – analytical ultracentrifugation – fluorescence detection – prion protein – fluorescence labelling

Introduction

Molecular properties such as molecular weight or shape parameters can be determined by analytical ultracentrifugation. The detection limit in sedimentation equilibrium runs with absorption optics is about $50 \, \mu\text{g}/\text{ml}$ for proteins. With the fluorescence detection system, as described earlier [1] an increase in detection sensitivity was achieved, if sedimentation or density gradient runs were carried out [2]. So far the same advantage although highly desirable was not achieved in sedimentation equilibrium runs. It would enable us to study molecules, which can be prepared only in low concentrations. One attractive example would be the prion protein because of its tendency of self-aggregation. Prions are the agent of transmissible spongiforme encephalopathies such as Scrapie in sheep, bovine spongiform encephalopathy in cattle and Creutzfeld-Jakob disease in humans [3–6]. Prions are proteinaceous, infectious particles that are composed largely, if not entirely of an abnormal PrP isoform designated PrP^{SC}. No relevant nucleic acid, that might be responsible for infectivity is detectable [7]. Both PrP^{SC} and the cellular isoform PrP^{C} are encoded by a single copy gene of

Progr Colloid Polym Sci (1997) 107:72–76
© Steinkopff Verlag 1997

the host; PrP^{SC} is derived from PrP^{C} by a posttranslational process [8]. Limited proteolysis of PrP^{SC} produce a N-terminally truncated protein designated PrP 27–30 under conditions in which prion infectivity is retained. PrP 27–30 forms rod-shaped polymers with the properties of amyloid that are referred to as prion rods [9]. Disruption of the prion rods is possible with sonication in the presence of 0.2% [10]. After pelleting the non-disrupted rods at $100.000 \times g$ for 1 h, about 10–20% of the total prion material remains in the supernatant but has lost its infectivity. Sucrose gradient centrifugation of the solubilized prion particles show a peak with a sedimentation coefficient of about 7S. Electron microscopy indicates spherical particles with a diameter of 8–20 nm. However accurate molecular weights and state of aggregation could not be determined so far, but will be undertaken in this work with fluorescence recorded sedimentation equilibrium runs.

Materials and methods

Chemicals and proteins

All chemicals and buffers were of the highest purity commercially available. FLUOS (for fluorescence labeling) and bovine serum albumin (BSA) was from Boehringer Mannheim, chymotrypsin from Sigma and immunoglobulin G (IgG-goat anti-mouse) from dianova (Hamburg). PrP 27–30 was kindly provided by Dr. S.B. Prusiner, San Francisco. The material was solubilized with sonication in the presence of 0.2% SDS [10]. The sonications were carried out in a biosafety hood with two different water cooled cup horn sonicators (Branson or Braun-Melsungen)

Fluorescence-labeling of proteins

The model proteins BSA, IgG and chymotrypsin were fluorescence-labeled with FLUOS, a fluorescein derivative conjugated to succinimidylester which reacts with primary aminogroups of the protein under physiological conditions (phosphate buffered saline, PBS). The protocol for fluorescence labelling from the producers manual was adapted to lower protein concentrations. Labeling was carried out with 3×10^{-6} M protein and 2×10^{-3} M FLUOS in a volume of 0.1 ml in PBS followed by incubation for 1 h at 20 °C in the dark. Unreacted FLUOS was separated by size-exclusion chromatography with a Sephadex G-25 column (NAP-5, Pharmacia). To decrease the amount of unbound FLUOS this step was carried out twice.

Construction of titanium cells for analytical ultracentrifugation

Cell housing, screw ring, window holders and centerpiece were constructed of titanium by our mechanical workshop. The optical pathway of the titanium centerpieces is 4 mm.

Biosafety equipment

To prevent contamination of the laboratory in the case of a cell leakage or distortion, vacuum chamber of the model E centrifuge was modified for continuous subpressure even if opened. Two exhaust layer at the wall behind the chamber and at the bottom of the chamber combined with two powerful exhaust pumps and the original AUC vacuum pump results in a continuous air flow into the centrifuge after opening the chamber. The waste air is filtered through HEPA-filters with a separation degree of 99.999% at 0.03 μm particle size (Camfil, Germany).

Ultracentrifugation

Sedimentation equilibrium runs were carried out in an analytical ultracentrifuge Spinco model E. The optical systems used were either the absorption optics equipped with a high-intensity illumination system [11] or a laser-excited fluorescence detection system [1]. For excitation at 488 nm an argon-ion laser (Spectra Physics) is focussed to an area of 120 μm $\times$ 120 μm in the center of the cell and is moved radially for scanning. The fluorescence light around 518 nm is collected on a photomultiplier. Measurements with absorption detection was made at 280 nm. All runs were performed in a four-hole rotor AN-F Ti. Sedimentation equilibrium runs of the model proteins were carried out with charcoal-filled epon single sector centerpieces with an optical pathlength of 3 mm (fluorescence detection) or charcoal-filled epon double sector centerpieces with an optical pathlength of 12 mm (absorption detection) in aluminium cells. Centrifugations of prion proteins were performed in titanium cells with an optical pathlength of the centerpiece of 4 mm (cf. above).

The calculation of the molecular parameters are performed according to Yphantis [12] which permits measurement of the baseline and the actual concentration profile at the same rotor speed; for determination of molecular weights from flourescence detected runs corrections were carried out as described in the following section.

Partial specific volumes were taken from Durchschlag [13], and the partial specific volume of the prion protein was calculated due to the amino acid and carbohydrate composition [14].

Results and discussion

Correction of molecular weight determination by sedimentation equilibrium and fluorescence detection

In order to analyse molecular weights from sedimentation equilibrium runs recorded with the fluorescence detection system, the fluorescence profile over the radius of the cell has to be transformed into a concentration profile. Therefore the ratio between fluorescence intensity and concentration has to be either independent of the radius and concentration or the function has to be known. Consequently, in the present work, two corrections have been carried out:

i) radius dependent fluorescence emission and
ii) concentration dependent fluorescence quenching.

Correction of the radius dependent fluorescence emission is necessary, because the yield of fluorescence light detection changes with the radius. For example, near the bottom of the centerpiece the emitted fluorescence light is partially shadowed or reflected by the bottom of the centerpiece. Unfavorably the largest increase of the signal is detected at the bottom of the cell, so a discrimination of this point would result in significant loss of accuracy. Therefore, a scan that has to show theoretically the same fluorescence intensity at every point of the radius was used as a standard (Fig. 1). Fluorescence of the homogenous solution of the non-sedimenting dye decreases near the bottom of the centerpiece (solid line). If the decrease in

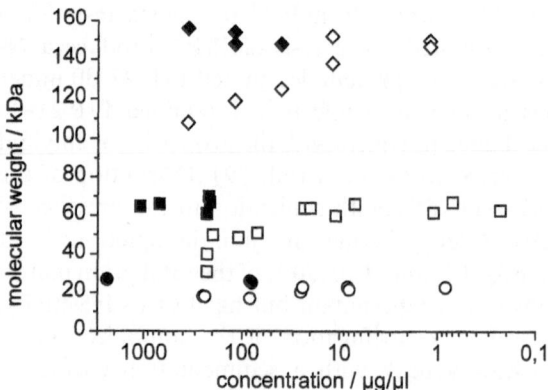

Fig. 2 Concentration dependence of apparent molecular weights. Detection with absorption optics (closed symbols) and with fluorescence optics (open symbols). Immunoglobulin G (IgG, diamonds), bovine serum albumin (BSA, squares) and chymotrypsin (circles)

yield was corrected for sedimentation–diffusion, equilibrium profile (dotted line) changed markedly due to the correction (interrupted line); with this correction, molecular weights could be evaluated from sedimentation–diffusion profiles with fluorescence recording.

The dependence of fluorescence emission upon sample concentration is not linear due to quenching effects. In equilibrium runs the higher fluorescence at the bottom of the centerpiece is quenched more than the fluorescence at the meniscus. To determine the concentration range where linearity in fluorescence is fulfilled, measurements of the good characterized model proteins (Fig. 2) IgG (diamonds), BSA (squares) and chymotrypsin (circles) were carried out with the fluorescence optics (open symbols) and compared to the results from the absorption optics (closed symbols). Results from absorption measurements showed no dependence upon the protein concentration. With decreasing sample concentration the dispersion of the measured molecular weights increases because of the unfavorable signal-to-noise ratio (Fig. 2 refer to BSA). Molecular weights, determined with fluorescence detection, showed however a strong dependence on the sample concentration. Higher concentrated samples, with a concentration comparable to concentrations in absorption measurements, resulted in significant lower molecular weights. With decreasing concentrations the deviation decreased and the resulting molecular weights are very close or identical to those determined with absorption detection at higher concentrations. An extrapolation to an infinitely small concentration is optimal. In practice a 10–30 fold dilution compared with concentrations for absorption detection is sufficient. The concentration of fluorophors should be less than 10^{-6} M. The extension to lower concentrations was 400 times for IgG and Chymotrypsin and 4000 times for BSA, respectively.

Fig. 1 Correction of the radius-dependent fluorescence emission. Radius dependent fluorescence emission of a homogenous solution (solid line), of the equilibrium scan before (interrupted line) and after correction (dotted line)

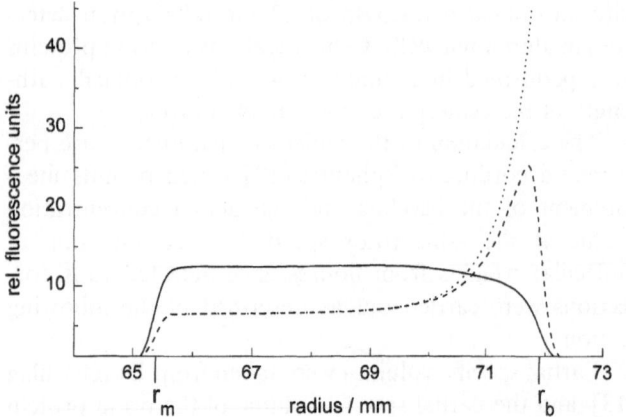

Progr Colloid Polym Sci (1997) 107:72–76
© Steinkopff Verlag 1997

Table 1 Comparison of molecular weight determined with absorption optics, fluorescence optics and from the amino acid composition

Sample	Molecular weight (kDa) determined with		Theoretical molecular weight (kDa)
	absorption	fluorescence	
BSA	66.4 ± 3.4	65.8 ± 4.4	67
Chymotrypsin	24.5 ± 1.7	22.6 ± 1.4	23.9
IgG	152 ± 4	148 ± 6	150

The determined molecular weights of the model either measured with absorption or fluorescence optics or calculated by their amino acid composition are listed in Table 1. They agree within the limits of error; the accuracy of the absorption optics is probably slightly higher.

Biosafety containment of the analytical ultracentrifuge

Ultracentrifugation studies on infectious prion material require particular biosafety conditions. For all runs self-constructed titanium cells were used, which can be autoclaved at 134 °C in 1 M NaOH, the most effective decontamination procedure. A modification for the rotor chamber of the analytical ultracentrifuge was constructed (scheme in Fig. 3A) to prevent environment contamination. Continuous air flow into the rotor chamber is guaranteed by two exhaust grids behind and one in the bottom of the rotor chamber connected with two exhaust pumps, which are self-starting when opening the chamber. An air flow rate of 0.5 to 1 m/s over the open rotor chamber from the front to the back is achieved. The mechanical vacuum pump of the centrifuge provides the

air flow to the bottom of the chamber. The waste air is cleaned with three HEPA-filters. In Fig. 3B the technical realization with the filter and the exhaust pumps at the back of the centrifuge is presented.

Molecular weight determination of solubilized prion-protein (PrP 27–30)

PrP 27–30 rods are solubilized by sonication in the presence of 0.2% SDS as described earlier [10]. The resulting molecular weights (Table 2) as detected with the absorption optics depends on the sonication energy we used for the solubilization. The first solubilization with a molecular weight distribution between 78 and 120 kDa which corresponds to trimer and pentamers, respectively, was carried out in cup horn sonicator (Branson) at 40 W for 5 min. The following sonications (2–4) were made in a water-cooled cup horn sonicator (Braun, Melsungen, FRG) with a sonication energy of about 70 W for 5 min. Consequently, the increasing sonication energy led to a higher degree of PrP 27–30-rod disruption with a decrease of the molecular weights. In the fourth solubilization the solubilized PrP

Table 2 Molecular weight of the solubilized prion protein determined with absorption and fluorescence detection

Sample		Absorption detection	Fluorescence detection
PrP (27–30) 0.2% SDS	1. Solubilization	78–120 kDa	N.D.
	2. Solubilization	23–40 kDa	22.4 kDa
	3. Solubilization	27–62 kDa	N.D.
	4. Solubilization	23 kDa	25 kDa

Fig. 3 Biosafety containment with subpressure system. (A) schematic drawing of the vacuum chamber of the model E.centrifuge; 1: rotor, 2: rotor chamber, 3: vacuum elbow to the diffusion on pump, 4: optical pathway for absorption detection, 5: optical pathway for fluorescence detection, 6: exhaust layer at the back of the rotor chamber, 7: air flow to the back and the bottom of the rotor chamber, 8: exhaust pumps for subpressure, 9: HEPA-filters, 10: exhaust layer at the bottom of the rotor chamber; (B) HEPA-filter and exhaust pumps at the back of the centrifuge

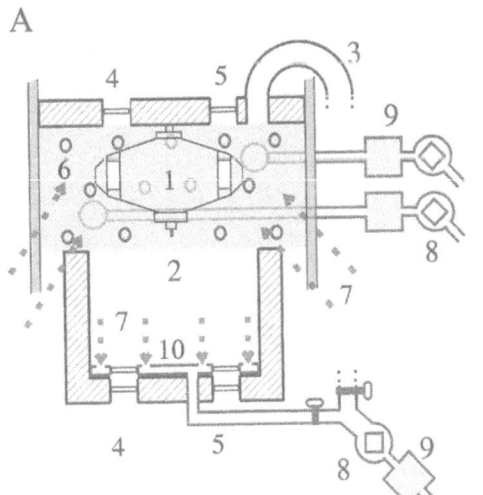

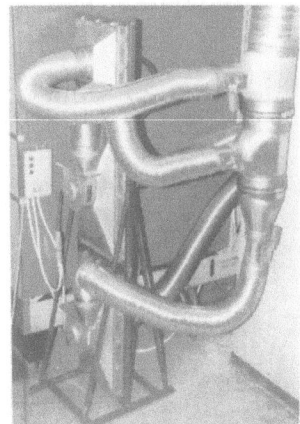

76

M. Pitschke et al.
AUC studies on the prion protein

27–30 material consisted of the monomer with a molecular weight of about 23 kDa which is in good accordance with the molecular weight of syrian golden hamster monomeric PrP 27–30 of 22 kDa, calculated from the amino acid composition, the carbohydrate content and the glycolipid anchor. With fluorescence detection we evaluated only the monomeric molecular weight for solubilized PrP 27–30. Loss of the oligomers is due to the labelling procedure. Possibly higher oligomers than monomers were broken, e.g., because of 10% DMSO in the labelling reaction, which is necessary for solubilization of the fluorescence label. Otherwise oligomeric PrP 27–30 is retarded by

the gelfiltration Sephadex G-25 column which is used for the separation of the unlabeled dye from the conjugates. Our results have shown that homogeneous monomeric PrP 27–30 can now be prepared, which is the appropriate material for further biophysical studies on PrP 27–30, for example in attempts to reaggregate prion rods.

Acknowledgements We are indebted to the employees of our mechanical workshop, especially to Mr. Beckman, to Mr. Seidel for the practical realization of the vacuum system for the analytical ultracentrifuge and Mr. Srejic for the construction of the analytical cells of titanium.

References

1. Rappold W (1986) Ph D thesis, Universität Düsseldorf
2. Schmidt B, Rappold W, Rosenbaum V, Fischer R, Riesner D (1990) Colloid Polym Sci 268:45–54
3. Gajdusek DC (1977) Science 197:943–960
4. Parry HB (1983) In: Oppenheimer DR (ed) Academic Press, New York, pp. 1–192
5. Pruisiner SB (1982) Science 216:136–144
6. Wells GAH, Wilesmith (1995) Brain Pathol. 5:91–103
7. Kellings K, Meyer N, Mierenda C, Prusiner SB, Riesner (1992) J Gen Virol 73:1025–1029
8. Prusiner SB (1991) Science 252:1515–1522
9. Prusiner SB, McKinley MP, Bowman KA, Bolton DC, Bendheim PE, Groth DF, Glenner GG (1983) Cell 35:349–358
10. Riesner D, Kellings K, Post K, Wille H, Serban H, Groth D, Baldwin MA, Prusiner SB (1996) J. Virol 70:1714–1722
11. Floßdorf J (1980) Makromol Chem 181:715–724
12. Yphantis DA (1964) Biochemistry 3:297
13. Durchschlag H (1986) In Hinz, HJ (ed). Springer, Berlin, pp. 45–128
14. Durchschlag H (1989) Colloid Polym Sci 267:1139–1150

Progr Colloid Polym Sci (1997) 107:77–81
© Steinkopff Verlag 1997

F. Dölle
D. Schubert

Dye-labelling as a means to study ternary protein complexes by analytical ultracentrifugation: the band 3/ankyrin/aldolase complex from erythrocyte membranes

F. Dölle · Prof. Dr. D. Schubert (✉)
Institut für Biophysik
der Johann Wolfgang Goethe-Universität
Haus 74
Theodor-Stern-Kai 7
60590 Frankfurt, Germany

Abstract Demonstration of the formation of ternary complexes of proteins, in the presence of all constituents and binary complexes, and analysis of their stoichiometries is a difficult task. For the band 3/ankyrin/aldolase complex from erythrocyte membranes in detergent solutions, we have solved this problem by sedimentation equilibrium analysis in the analytical ultracentrifuge. Labelling of the ankyrin with a dye (fluorescein isothiocyanate) and measuring the absorbance versus radius profiles at a wavelength where only the dye absorbs allowed us to focus on the ankyrin-containing complexes. So,

the different oligomers of uncomplexed band 3, the uncomplexed aldolase and the various binary complexes of band 3 and aldolase could be disregarded in the analysis. The ternary band 3/ankyrin/aldolase complex could be unambiguously detected. Its stoichiometry (band 3/ankyrin/aldolase tetramer) was found to vary between 4:1:1 and 4:1:4, depending on the abundance of the enzyme.

Key words Sedimentation equilibrium – ternary protein complexes – dye labelling – band 3/ankyrin/aldolase complex

Introduction

The formation and properties of stable ternary protein complexes can be studied by a variety of techniques, following separation of these complexes from binary ones and unreacted constituents. On the other hand, the corresponding analysis for unstable complexes, which are in an association equilibrium with their constituents and intermediates and are too short-lived to be purified, is much more complicated. Sedimentation equilibrium analysis in the analytical ultracentrifuge represents the sole or at least a far superior method available for the purpose. In cases where the molar mass of the ternary complexes is substantially higher than that of the binary ones, a straightforward analysis may be possible. In others, dye-labelling of one of the constituents of the complex and collecting the absorb-

ance versus radius data $A(r)$ at a wavelength where only the dye absorbs [1–3] may simplify the system to such a degree that an unambiguous analysis is feasible. An example of the latter will be given below. The ternary complex to be studied is assembled from two proteins of the human erythrocyte membrane: the intrinsic membrane protein band 3 (AE1) and the peripheral membrane protein ankyrin or band 2.1 (which represents the main linkage between the lipid bilayer and the membrane skeleton [4, 5]), and the cytoplasmic erythrocyte enzyme aldolase [4–6]. Both latter proteins bind to band 3: virtually 100% of ankyrin [4–6] and, probably, up to two-thirds of the cellular aldolase [7] are bound to the cytoplasmic domain of that protein, most probably in a dynamic equilibrium. In the present paper, we will demonstrate that the ternary complex between the three proteins is also formed by the purified proteins in solutions of a nonionic detergent, and that its stoichiometry varies between $4:1:1$ and $4:1:4$ (band 3/ankyrin/aldolase tetramer), depending on the abundancy of the enzyme.

Experimental

Proteins

Band 3 protein from human erythrocyte membranes was solubilized and purified, applying the nonionic detergent nonaethyleneglycol lauryl ether ("$C_{12}E_9$", Boehringer, Mannheim), as described earlier [8].

Ankyrin was isolated according to a modified version of a method described by Pinder et al. [9]. Details will be given elsewhere. The protein was labelled with a dye, fluorescein isothiocyanate (FITC) (Roth, Karlsruhe), as described by Mulzer et al. [2, 10]. Aldolase, as well as most of the reagents used, were purchased from Boehringer (Mannheim).

Analytical ultracentrifugation

A Beckman Optima XL-A ultracentrifuge, in connection with an An-60Ti rotor, was used. Sedimentation equilibrium experiments used Epon 6-channel centerpieces and were performed at rotor speeds between 6000 and 8000 rpm. Sample volume was 120–130 μl. In each experiment on the ternary band 3/ankyrin/aldolase system, the binary band 3/ankyrin system was run in one double sector as a control for the intactness of the two proteins. Sedimentation velocity runs were performed at 35000 rpm and used Epon double sector centerpieces. The absorbance versus radius profiles $A(r)$ were measured at 497 nm, an absorption maximum of FITC. Rotor temperature was 4 °C. Protein concentrations were 100–200 μg/ml (band 3), 100–300 μg/ml (ankyrin), and 150–600 μg/ml (aldolase), respectively. The buffers used contained 20 mM Hepes (pH 7.2), 20 mM NaCl, 2 mM EDTA, 0.5 mM dithiothreitol, 0.3% $C_{12}E_9$, 0.5 mM phenylmethylsulfonyl fluoride (or 0.5 mM Pefabloc SC).

The evaluations of the sedimentation equilibrium data assumed the absence of thermodynamic nonidealities and were of the type described earlier [11, 12]. The computer programs used were developed by Schuck [12, 13]; they included an estimation of the statistical accuracy of the results obtained. The fixed parameters in the equations applied, $m(1 - \bar{v}\rho_0)$ for the different proteins, were the following: 32 kDa for the band 3 protomer/detergent complex [14], 56 kDa for ankyrin [2], and 43 kDa for tetrameric aldolase [15].

Results

The binary systems: band 3/aldolase and band 3/ankyrin

The cytoplasmic domain of the band 3 protein of the human erythrocyte membrane represents a binding site for a variety of proteins of the membrane and cytoplasm, including ankyrin and aldolase [4–6]. The binding sites for the two ligand proteins do not overlap, so that both ligands can bind simultaneously [6]. A detection of the different types of hypothetical complexes is complicated by the self-association of band 3: At least in our hands [8, 16], the detergent-solubilized protein represents a mixture of monomers, dimers and tetramers in dynamic equilibrium (the same behavior is assumed to be present in the intact membrane [8, 17]). In earlier studies, we have applied dye-labelling of the ligand proteins to facilitate the evaluation [2, 3, 10, 15]. We have found that the actual situation is far less complex than could be anticipated: The band 3 tetramer represents the sole or at least predominant binding site for both ankyrin and aldolase; band 3 monomers or dimers contribute little or nothing to ligand binding [2, 10, 15]. Differences exist, however, with respect to the number of ligand binding sites: Whereas the band 3 tetramer can bind up to four aldolase tetramers depending on the abundancy of the enzyme [15], we have found in the present study that the stoichiometry of the band 3/ankyrin complex is $4:1$, regardless of the molar ratio of both proteins (which was in the range of 0.5–3.0).

The ternary system: band 3/ankyrin/aldolase

According to the results obtained with binary complexes described above, ternary complexes containing band 3, ankyrin, and aldolase are expected to have effective molar masses $m_{eff} = \sum m_i(1 - \bar{v}_i\rho_0)$ between 274 and 358 kDa, depending on the number of aldolase molecules bound. On the other hand, m_{eff} of the binary complexes will be lower by only approximately 20%, both at low and high abundancy of aldolase. Considering the large number of possible protein particles in the sample, this difference in m_{eff} of only 20% does not allow a reliable detection of the ternary complexes if all particles are contributing to $A(r)$. As in the case of the studies on the binary complexes, dye-labelling of one of the ligand molecules and collecting the $A(r)$-data at a wavelength where only the dye absorbs, apparently simplifies the analysis considerably. It is obvious that ankyrin is the best candidate for labelling: In this case, besides ternary complexes only the free ligand and one type of binary complex contribute to $A(r)$, and m_{eff} of the ternary complexes will exceed that of the binary band 3/ankyrin complex by up to almost a factor of two (Fig. 1).

Progr Colloid Polym Sci (1997) 107:77–81
© Steinkopff Verlag 1997

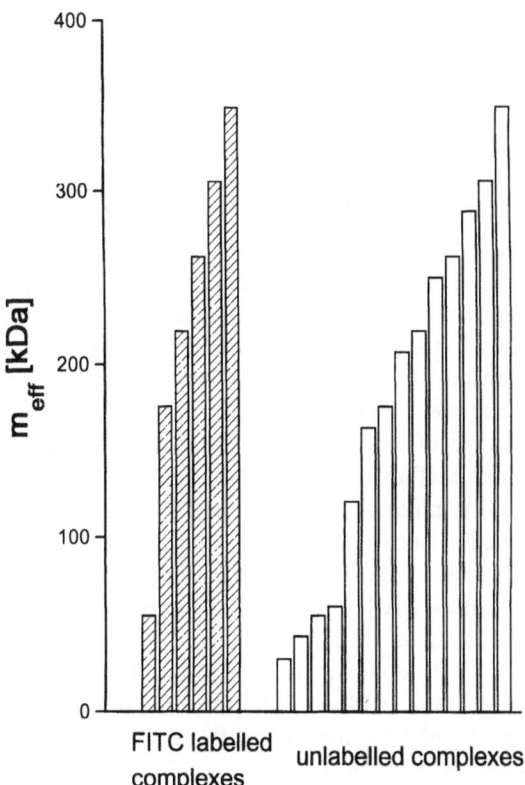

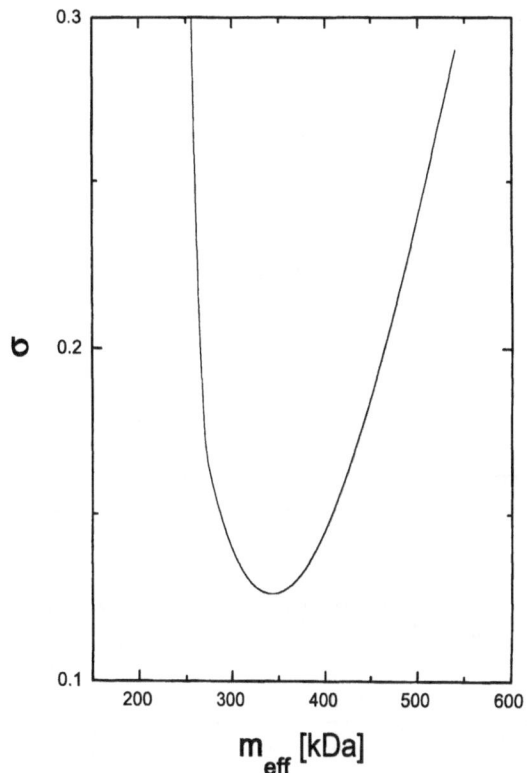

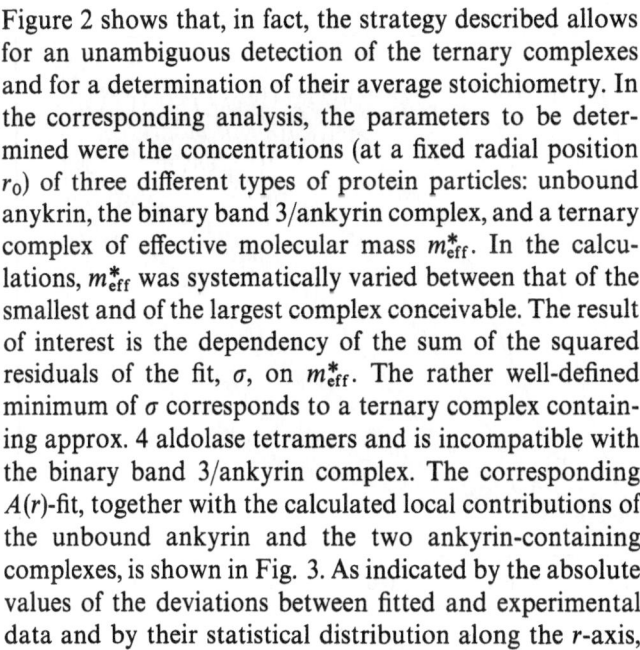

Fig. 1 A plot of the effective molecular masses, $m_{eff} = m(1 - \bar{v}\rho_0)$ or $\sum m_i(1 - \bar{v}_i\rho_0)$, of all protein components and complexes of the band 3/ankyrin/aldolase system conceivable. m_{eff} of band 3 includes bound detergent [14]. Hatched columns: m_{eff} of ankyrin and its complexes

Fig. 2 Search for a ternary band 3/ankyrin/aldolase complex: Determination of the effective molecular mass m_{eff}^* of that complex which, together with the free ankyrin and a binary band 3/ankyrin complex of stoichiometry 4:1, allows the best fit to the experimental $A(r)$-data. The plot shows the dependency of the sum of the squared residuals of the fit, σ, on m_{eff}^*. Initial protein concentrations: 110 μg/ml (ankyrin), 480 μg/ml (aldolase), 160 μg/ml (band 3). Rotor speed: 6000 rpm

Figure 2 shows that, in fact, the strategy described allows for an unambiguous detection of the ternary complexes and for a determination of their average stoichiometry. In the corresponding analysis, the parameters to be determined were the concentrations (at a fixed radial position r_0) of three different types of protein particles: unbound anykrin, the binary band 3/ankyrin complex, and a ternary complex of effective molecular mass m_{eff}^*. In the calculations, m_{eff}^* was systematically varied between that of the smallest and of the largest complex conceivable. The result of interest is the dependency of the sum of the squared residuals of the fit, σ, on m_{eff}^*. The rather well-defined minimum of σ corresponds to a ternary complex containing approx. 4 aldolase tetramers and is incompatible with the binary band 3/ankyrin complex. The corresponding $A(r)$-fit, together with the calculated local contributions of the unbound ankyrin and the two ankyrin-containing complexes, is shown in Fig. 3. As indicated by the absolute values of the deviations between fitted and experimental data and by their statistical distribution along the r-axis,

the fit is of excellent quality and confirms the conclusion suggested by the data in Fig. 2. It should be noted that omission of the ternary complex from the fit would increase σ by a factor of approx. 4. – The exact position of the minimum of the $\sigma(m_{eff}^*)$-curve depends on the concentration and molar ratio of the aldolase in the mixture. The curves representing the observed minimum and maximum values of m_{eff}^* corresponded, respectively, to 1.7 and 4.0 aldolase tetramers per complex. It is thus clear that the band 3/ankyrin complex can bind between one and four aldolase tetramers. This is in agreement with the results obtained with the binary band 3/aldolase system [15]. Except at very low and very high aldolase concentration, almost certainly a mixture of complexes of different aldolase content will be present [15].

Our interpretation of the data has disregarded the possibility that, after having bound 1–2 aldolase tetramers, the band 3/ankyrin complex could bind a second ankyrin molecule. This possibility cannot be ruled out by our data; there is, however, no support for it in the literature.

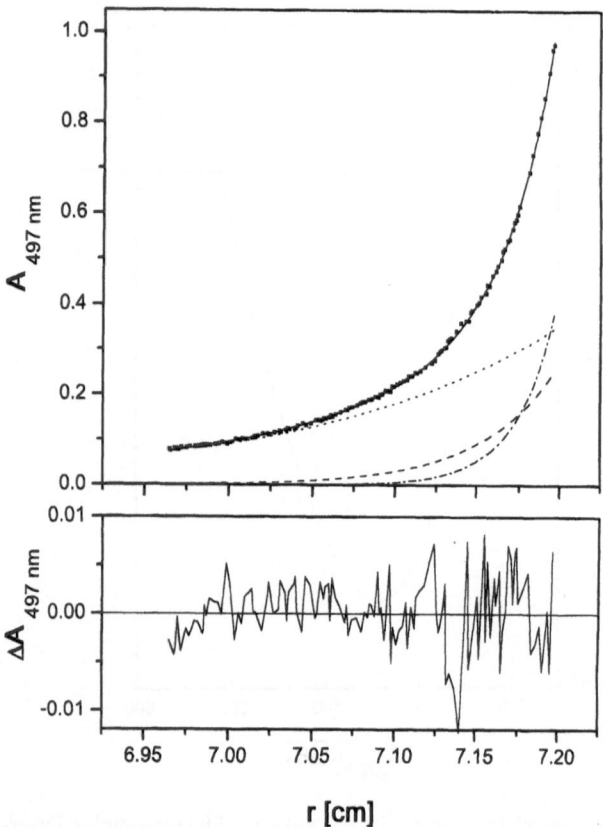

Fig. 3 Another analysis of the data of Fig. 2: (A) Experimental $A(r)$-data (■); curve fitted to the data under the assumption that the only dye-labelled protein particles are free ankyrin, the binary band 3/ankyrin complex, and a ternary complex of $m_{eff} = 320$ kDa (——); calculated contributions to $A(r)$ of the free ankyrin (....), the binary (- - - -) and the ternary (-.-.-) complex, respectively. The ternary complex contains approx. 45% of the total ankyrin in the cell. (B) Residuals of the fit to the experimental data

Stability of the complexes

Nonequilibrium centrifugation techniques were applied to establish whether the complexes described above are stable or in dynamic equilibrium with their constituents. The results of a sedimentation velocity run on a band 3/ankyrin/aldolase mixture are shown in Fig. 4. In this experiment, the ankyrin/band 3 ratio was chosen so low that, according to the results of sedimentation equilibrium experiments, approx. 70% of the ankyrin present should be complexed with band 3. The characteristic feature of the $A(r)$-diagrams of the figure is the lack of well-defined transitions between plateau regions. Instead, a single broad boundary of $s_{20,w} = 16S$ (as determined by the second moment method) is found. This shows that neither the band 3/ankyrin nor the band 3/ankyrin/aldolase complex are stable but take part in an association equilibrium. For the former complex, this finding is in agreement with earlier observations [10]. On the other hand, the average s-value found is higher than expected for a stable band 3 tetramer (12S [18]), contrary to the observations on the binary systems band 3/aldolase [15] and band 3/ankyrin [10] (and also band 3/hemoglobin [19] and band 3/band 4.1 [14]). Thus the ternary complex band 3/ankyrin/aldolase seems to have a larger lifetime than the binary band 3/ligand complex.

Discussion

In the present paper we have established, by sedimentation equilibrium analysis, the existence and stoichiometry of

Fig. 4 A sedimentation velocity run on a band 3/ankyrin/aldolase mixture. Protein concentrations: 710 µg/ml (band 3), 360 µg/ml (ankyrin), and 510 µg/ml (aldolase). The scans were recorded at intervals of 18 min after reaching a rotor speed of 35 000 rpm

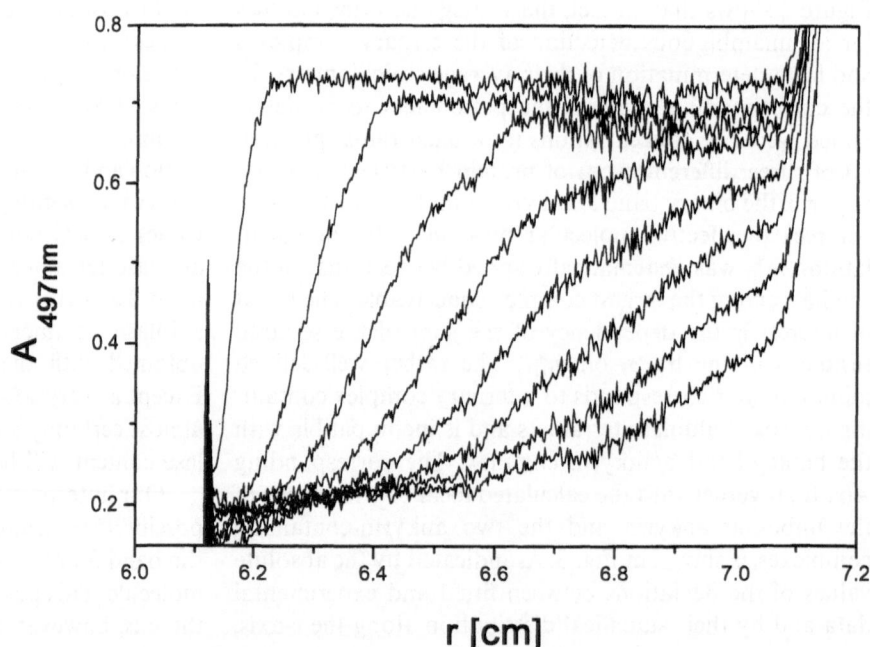

Progr Colloid Polym Sci (1997) 107: 77–81
© Steinkopff Verlag 1997

a complex between the erythrocyte proteins band 3, ankyrin, and aldolase in solutions of a nonionic, nondenaturing detergent. This achievement was possible despite the presence, in the solution, of 7–10 other proteins or protein aggregates (the molar mass of some is quite close to that of the complex) and of the transient character of the ternary complex, as demonstrated by sedimentation velocity runs. The success of the analysis depended on three prerequisites: (1) knowledge of the oligomeric structure of the binary aggregates which form the ternary complex, (2) dye-labelling of the most appropriate protein in the complex (i.e., the one which is present in the smallest number of aggregates), and (3) experimental proof that the label applied does not affect the association properties of the system. In the present case, the validity of condition (3) was confirmed in earlier studies [10].

The technique used in this paper is probably the only one available for analysis of the posed problem. In addition, it could be applied to clarify another aspect of ternary association, namely the competition for binding sites. With respect to the band 3/ankyrin/aldolase complex, e.g., competition between aldolase and glyceraldehyde-3-phosphate dehydrogenase for binding to band 3 [6] could be studied (labelling, however, one of the enzymes instead of ankyrin). This aspect is now being pursued in our laboratory.

Acknowledgements We are grateful to G. Mayer and B. von Rückmann for help with the preparations and to J.A. van den Broek for critically reading the manuscript.

References

1. Osborne JC, Powell GM, Brewer HB (1980) Biochim Biophys Acta 619: 559–571
2. Mulzer K, Kampmann L, Petrasch P, Schubert D (1990) Colloid Polym Sci 268:60–64
3. von Rückmann B, Huber E, Schuck P, Schubert D (1995) Prog Colloid Polym Sci 99:69–73
4. Gilligan DM, Bennett V (1993) Semin Hematol 30:74–83
5. Lux SE, Palek J (1995) In: Handin RJ, Lux SE, Stossel TP (eds) Blood: Principles and Practice of Hematology. Lippincott, Philadelphia, pp 1701–1818
6. Low PS (1986) Biochim Biophys Acta 864:145–167
7. Jenkins JD, Madden DP, Steck TL (1984) J Biol Chem 259:9374–9378
8. Pappert G, Schubert D (1983) Biochim Biophys Acta 730:32–40
9. Pinder JC, Smith KS, Pekrun A, Gratzer WB (1989) Biochem J 264:423–428
10. Mulzer K, Petrasch P, Kampmann L, Schubert D (1989) Stud Biophys 134:17–22
11. Schubert D, Schuck P (1991) Prog Colloid Polym Sci 86:12–22
12. Schuck P, Legrum B, Passow H, Schubert D (1995) Eur J Biochem 230: 806–812
13. Schuck P (1994) Prog Colloid Polym Sci 94:1–13
14. von Rückmann B, Jöns T, Dölle F, Drenckhahn D, Schubert D (1997) Biochim Biophys Acta 1325:226–234
15. Huber E, Bäumert HG, Spatz-Kümbel G, Schubert D (1996) Eur J Biochem 242:293–300
16. Schubert D, Boss K, Dorst H-J, Flossdorf J, Pappert G (1983) FEBS Lett 163:81–84
17. Schubert D, Huber E, Lindenthal S, Mulzer K, Schuck P (1992) Prog Cell Res 2:209–217
18. Dorst H-J, Schubert D (1979) Hoppe-Seyler's Z Physiol Chem 360:1605–1618
19. Schuck P, Schubert D (1991) FEBS Lett 293:81–84

Progr Colloid Polym Sci (1997) 107:82–87
© Steinkopff Verlag 1997

M.P. Jacobsen
D.J. Winzor

Studies of ligand-mediated conformational changes in enzymes by difference sedimentation velocity in the Optima XL-A ultracentrifuge

M.P. Jacobsen · Prof. Dr. D.J. Winzor (✉)
Centre for Protein Structure,
Function and Engineering
Department of Biochemistry
University of Queensland
Brisbane, Queensland 4072, Australia

Abstract Difference sedimentation velocity has provided an extremely convenient procedure for detecting and quantifying ligand-mediated conformational changes in enzymes by virtue of differences in hydrodynamic volume. However, the replacement of the Beckman model E instrument by the XL-A has necessitated reexamination of the existing method of analysis, which relied upon the comparison of simultaneously recorded distributions of solute in the two cells. After demonstration of the validity of the revised procedure by its application to simulated sedimentation velocity data, differential sedimentation velocity has been used to confirm the effect of phenylalanine on the sedimentation coefficient of rabbit muscle pyruvate kinase. Corresponding studies of the effect of glucose on the sedimentation coefficient of yeast hexokinase have demonstrated the substrate-mediated decrease in enzyme size that is evident from X-ray crystallographic studies, and identified this effect as the consequence of substrate perturbation of a preexisting enzyme isomerization rather than of substrate-induced isomerization of yeast hexokinase.

Key words Difference sedimentation velocity – enzyme isomerization – pyruvate kinase – hexokinase

Introduction

Difference sedimentation velocity [1, 2] was introduced nearly 30 years ago as a means of quantifying ligand-induced conformational changes in enzymes in terms of differences in hydrodynamic volume [3]. Such changes were estimated initially as the difference between values of the sedimentation coefficients obtained from simultaneous velocity runs on enzyme solutions with and without ligand, but more recent studies have employed a quantitative expression for the direct estimation of the difference in sedimentation coefficient [4–6]. The replacement of the Beckman model E instrument by the XL-A has necessitated reexamination of that analysis because of its reliance

upon the comparison of boundary positions in simultaneously recorded distributions of solute in the two cells. Those theoretical considerations have established that the same quantitative expressions retain validity for determination of the difference in sedimentation coefficients provided that the time interval between the recording of solute distributions for cells 1 and 2 is constant. However, despite the incorporation of this requirement into the programmed schedule of experiments, the delay period has been shown to fluctuate by as much as 8% (mean ±4%) during the course of difference sedimentation velocity experiments in the Optima XL-A. It has therefore been necessary to apply the analysis to simulated sedimentation velocity data in order to establish that assumed constancy of the time increment is an acceptable approximation under those circumstances. Application of the procedure is then illustrated by its use to confirm earlier reports [5, 6] of the effect of phenylalanine on the sedimentation coefficient of rabbit muscle pyruvate kinase, and to establish that the glucose-mediated decrease in the molecular dimensions of yeast hexokinase [7] reflects substrate perturbation of a preexisting enzyme isomerization equilibrium rather than of substrate-induced isomerization of yeast hexokinase – the earlier interpretation [8, 9] of the X-ray crystallographic findings [7].

Theory

Consider an experiment in which the sedimentation velocity behavior of an enzyme is compared with that of enzyme–ligand complex by simultaneous ultracentrifugation of two solutions of the enzyme. One ultracentrifuge cell is filled with a solution of enzyme in buffer, and the other with an identical concentration of enzyme in buffer containing a sufficiently high ligand concentration to ensure effective saturation of the enzyme site(s) for ligand. For an experiment in which the solute distributions in both cells are recorded simultaneously, the differential equation relating boundary position, r, to centrifugation time, t, is first written for each of the two solutions. Subtraction of one from the other then leads to the expression [4]

$$d(\ln r_1 - \ln r_2)/dt = \omega^2(s_1 - s_2) , \qquad (1)$$

where r_1 and r_2 are the respective radial positions of the boundary in cells 1 and 2 after centrifugation at angular velocity ω for time t. Although Howlett and Schachman [4] then developed an approximate solution of Eq. (1) with redefined radial variables, we shall proceed on the basis of the above rigorous expression for the difference between s_1 and s_2, the sedimentation coefficients of enzyme in cells 1 and 2, respectively.

Because the design of the Optima XL-A precludes the simultaneous recording of solute distributions, the two solute distributions being compared must be recorded consecutively. It is therefore appropriate to examine the integrated form of Eq. (1) in order to accommodate the different times of centrifugation (t_1, t_2) for the two cells. Specifically,

$$[\ln(r_1)_{t_1} - \ln(r_m)_1] - [\ln(r_2)_{t_2} - \ln(r_m)_2] = \omega^2(s_1 t_1 - s_2 t_2) , \qquad (2)$$

where $(r_m)_1$ and $(r_m)_2$ denote the respective positions of the air–liquid meniscus in cells 1 and 2. By expressing the larger centrifugation time (taken as t_2) as the sum of t_1 and

a time increment Δt, Eq. (2) may be written as

$$\ln(r_1)_{t_1} - \ln(r_2)_{t_2} = \omega^2(s_1 - s_2)t_1$$
$$+ [\ln(r_m)_1 - \ln(r_m)_2 - \omega^2 s_2 \Delta t] . \qquad (3)$$

Provided that the time increment separating the recording of the two solute distributions is constant, the term within square brackets is also constant, whereupon the differential form of Eq. (3) becomes synonymous with Eq. (1). On the basis that the maximum experimental fluctuation in Δt is likely to be in the vicinity of 3–4% about the mean, we explore the possibility of determining directly the difference in sedimentation coefficients from the slope of the dependence of $\{\ln(r_1)_{t_1} - \ln(r_2)_{t_2}\}$ upon t_1, the time of centrifugation for the first of each paired set of solute distributions.

Analysis of simulated difference sedimentation velocity data

To check the validity of the envisaged procedure, results were simulated for a system in which the presence of ligand effected a decrease in sedimentation coefficient of enzyme alone (s_1) from 10.0 to 9.7 S; this 3% change in s being pertinent to the effect of phenylalanine on the sedimentation coefficient of pyruvate kinase [5, 6]. The time dependence of boundary position in cell 1 was generated by assigning a magnitude of 5.990 cm to $(r_m)_1$ and calculating r_1 on the basis of $s_1 = 10.0$ S and values of $\omega^2 t_1$ (rad^2 s^{-1}) in the range $10^{10} \leqslant \omega^2 t_1 \leqslant 10^{11}$. The corresponding dependence of boundary position in cell 2 was first calculated on the basis of a meniscus position of 5.930 cm and values of $\omega^2 t_2$ based on those of $\omega^2 t_1$ and an identical time increment between each paired set ($\omega^2 \Delta t = 0.4 \times 10^{10}$ rad^2 s^{-1}). Solid lines in the major section of Fig. 1 summarize the individual dependences of the logarithm of radial distance upon $\omega^2 t$, and the solid line in the upper panel the plot of results according to Eq. (1) with the time of centrifugation taken as t_1. The slope of this difference plot, 0.3 S, exactly matches $(s_1 - s_2)$, there being no assumption inherent in the application of Eq. (1) to results for which there is a constant time increment Δt between each paired set of solute distributions.

As noted above, the maximum extent of fluctuation in Δt (or $\omega^2 \Delta t$) is likely to be 3–4% in an experimental situation. To assess the effect of such fluctuation on the validity of using Eq. (1) to determine $(s_1 - s_2)$, $\omega^2 \Delta t$ was assigned possible magnitudes of 0.38, 0.39, 0.40, 0.41 and 0.42×10^{10} s^{-1} (a maximum fluctuation of 5% from the mean of 0.40), the actual value added to obtain t_2 for a given t_1 being selected at random. From the resultant dependence (● in Fig. 1) of $\ln r_2$ upon $\omega^2 t_2$, taken as

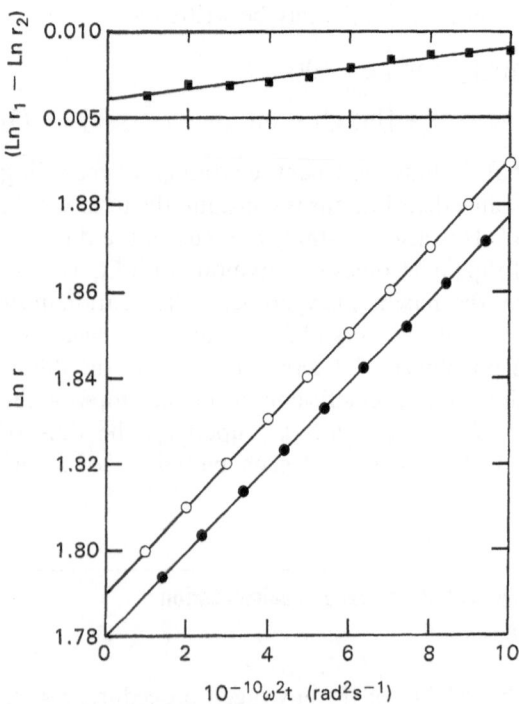

Fig. 1 Analysis of simulated sedimentation velocity patterns to test the validity of applying analyses developed for model E experiments to results obtained by difference sedimentation velocity in the XL-A ultracentrifuge. The main section depicts the $\omega^2 t$ dependence of the logarithms of the radial positions of boundaries in the two cells for a system with respective values of 10.0 and 9.70 S for s_1 (cell 1) and s_2 (cell 2) in an experiment with $\omega^2 \Delta t \approx 0.4 \times 10^{10}$ rad^2 s^{-1}, whereas the upper section is the difference plot of the simulated data in terms of Eq. (1). In this difference plot the simulated data obtained on the basis of a 5% random variation in the time increment (■) are compared with the theoretical plot (——) for the situation in which $\omega^2 \Delta t$ is exactly 0.4×10^{10} rad^2 s^{-1}

$(\omega^2 t_1 + 0.4 \times 10^{10})$, there is essentially no effect on the determined value of s_2, 9.68 (± 0.03) S compared with the input value of 9.70 S. Furthermore, although Eq. (1) now ceases to provide an exact description of the difference plot (upper panel in Fig. 1), the disparity between the calculated results (●) and the theoretical behavior (——) is extremely small. In other words, the random error associated with experimental scatter is likely to render insignificant this relatively minor departure from predicted behavior as the result of experimentally realistic fluctuations in the time increment Δt. We therefore proceed with application of the procedure to experimental systems.

Experimental

Prior to difference sedimentation velocity experiments, the preparations of rabbit muscle pyruvate kinase and yeast hexokinase (both Sigma products) were dissolved in Tris-chloride buffer (0.05 M Tris/HCl–0.072 M KCl–0.007

M MgCl$_2$), pH 7.5, and glycylglycine–chloride buffer (0.02 M glycylglycine/HCl–0.2 M KCl), pH 8.7, respectively, after which aliquots (50–100 μl) were subjected to zonal chromatography at 0.2 ml/min on a 600×7.8 mm Biosep-SEC3000 column (Phenomenex, Torrance, CA), preequilibrated with the same buffer. This exclusion chromatography step served not only to remove any contaminating material with markedly different size characteristics but also to provide an enzyme solution in dialysis equilibrium with the buffer to be used in the difference sedimentation velocity studies. For one series of experiments, the glycylglycine buffer was supplemented with 0.1 M poly(ethyleneglycol) 600 in order to examine the effect of this space-filling cosolute on sedimentation velocity characteristics of yeast hexokinase.

Solutions of pyruvate kinase (1 mg/ml) in Tris–chloride buffer (pH 7.5) and in the same buffer supplemented with 5 mM phenylalanine were centrifuged simultaneously at 20°C and 35 000 rpm in an Optima XL-A ultracentrifuge. Concentration distributions, monitored in terms of absorbance at 280 nm, were recorded at 10 min intervals for each cell, there being a delay (Δt) of 5 min between the recording of distributions for enzyme alone (cell 1) and enzyme–phenylalanine mixture (cell 2).

Difference sedimentation velocity experiments on hexokinase were conducted with 0.5 mg/ml enzyme in both cells, the solution in cell 2 being supplemented with 1 mM glucose. For these experiments at 60 000 rpm and 20°C the interval between records of the absorbance distribution (A_{280}) in a given cell was 9 min, whereas the time increment between distributions for cells 1 and 2 was 4.4 min. A final series of difference sedimentation velocity experiments examined the corresponding effect of glucose (1 mM) on hexokinase migration in buffer supplemented with 0.1 M poly(ethylene glycol) 600 (Merck): the time interval between absorbance distributions for the same cell was 10 min, the time increment between recorded distributions for cells 1 and 2 (Δt) again being 4.4 min in these experiments at 60 000 rpm and 20°C.

In view of the symmetry of the boundaries observed in all sedimentation velocity experiments, the boundary positions, r_t, were taken as the centroid (first moment) obtained by means of the VELGAMMA software, precautions having been taken to establish its essential identity with the correctly defined position – the square root of the second moment [10, 11]. In that regard the precision with which boundary positions may be located in solute distributions recorded by the absorption optical system of the XL-A is insufficient to justify concern about which moment of the boundary should be determined. Indeed, the position at which the absorbance attains half of its plateau value should probably suffice, given the limitations of the absorption optical system.

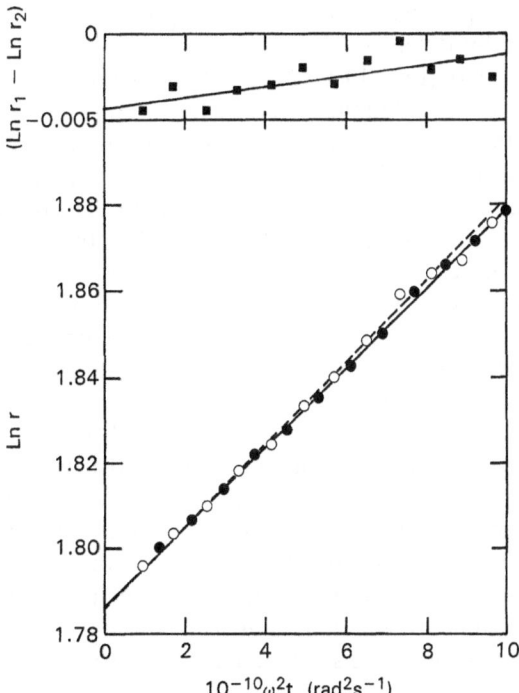

Fig. 2 Measurement of the effect of phenylalanine on the sedimentation coefficient of rabbit muscle pyruvate kinase (pH 7.5, I 0.13) by difference sedimentation velocity in the XL-A analytical ultracentrifuge. Major section: dependence of the logarithms of the radial positions upon $\omega^2 t$ for enzyme alone ($\bigcirc$, – – –) and enzyme solution supplemented with 5 mM phenylalanine ($\bullet$, ——) (cells 1 and 2, respectively) in experiments conducted at 35 000 rpm and 20 °C. Upper segment: difference plot of the results in accordance with Eq. (1), and an ordinate intercept based on Eq. (3)

Effect of phenylalanine on the sedimentation coefficient of pyruvate kinase

In order to illustrate the application of the present procedure, we choose initially an experimental system for which difference sedimentation velocity has already been used to establish the magnitude of $(s_1 - s_2)$. Inclusion of phenylalanine, an allosteric inhibitor of rabbit muscle pyruvate kinase, has been shown to effect a decrease of 0.3 S in the sedimentation coefficient because of its preferential interaction with the larger of two coexisting isomeric states of the enzyme [6].

Determination of sedimentation coefficients from the two separate time dependences of the logarithm of radial distance is summarized in the main part of Fig. 2, which clearly signifies a slightly larger migration rate for enzyme alone ($\bigcirc$) than for enzyme in the presence of 5 mM phenylalanine ($\bullet$). Least-squares calculations yield sedimentation coefficients ($s_{20,b}$) of 9.5 (± 0.3) and 9.2 (± 0.2) S for rabbit muscle pyruvate kinase under the respective conditions: numbers in parentheses denote the

uncertainty (± 2 S.D.) inherent in the estimates. Although these independent measurements have indicated a probable difference of 0.3 S between the sedimentation coefficients of enzyme and enzyme–inhibitor complex, the result, 0.3 (± 0.5) S, is clearly equivocal when the uncertainties in the estimates of s_1 and s_2 are taken into account. On the other hand, the difference plot of results according to Eq. (1) is far more definitive in that regard (Fig. 2, upper panel), least-squares calculations signifying a slope, $(s_1 - s_2)$, of 0.31 (± 0.08) S. Furthermore, the best-fit ordinate intercept of -0.0043 is the value predicted from the square-bracketed term in Eq. (3).

This conclusion that phenylalanine mediates a decrease of 0.3 S in the sedimentation coefficient of pyruvate kinase finds quantitative parallel in the previous report [6] of a 0.30 (± 0.03) S disparity. In this regard the smaller uncertainty associated with the earlier estimate reflects use of the schlieren optical system to record the sedimentation velocity patterns; and hence more definitive location of the boundary as the radial position corresponding to the peak in the optically recorded derivative of the concentration distribution. However, the results obtained with the new-generation analytical ultracentrifuge have sufficed to establish the existence of the phenylalanine-mediated change in the hydrodynamic characteristics of rabbit muscle pyruvate kinase.

Effect of glucose on the sedimentation coefficient of yeast hexokinase

The fact that complex formation between glucose and yeast hexokinase is accompanied by a decrease in the molecular dimensions of the enzyme was established many years ago by X-ray crystallography [7]. Indeed, advantage has subsequently been taken of this observation to account for the activating effect of a space-filling solute such as poly(ethylene glycol) on glucose binding and enzyme catalysis [8, 9]. It is therefore of interest to ascertain whether the glucose-mediated closure of the active-site cleft of the hexokinase [7] gives rise to a change in the hydrodynamic characteristics of the enzyme that is of sufficient magnitude for detection by difference sedimentation velocity. The conditions for this study (pH 8.7, I 0.22) have been chosen not only to duplicate those pertaining to the above-mentioned investigations of molecular crowding effects on enzyme affinity [8, 9] but also to avoid complications arising from hexokinase dimerization at neural pH [12, 13].

Analysis of the sedimentation velocity data in the absence of glucose yielded a sedimentation coefficient of 3.62 (± 0.12) S for yeast hexokinase, a value in keeping with the range of 3.5–3.9 S reported [12, 13] for the enzyme under

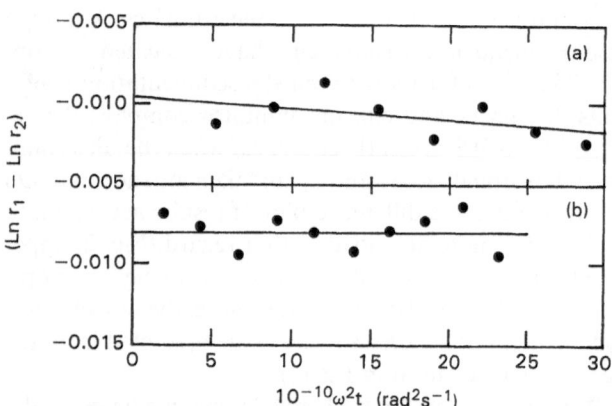

Fig. 3 Difference plots obtained in studies at 60 000 rpm and 20 °C of the effect of glucose (1 mM) on the sedimentation coefficient of yeast hexokinase (pH 8.7, I 0.22) in (a) the absence and (b) the presence of 0.1 M poly(ethylene glycol) 600. Results are plotted in accordance with Eq. (1) and ordinate intercepts based on Eq. (3)

comparable conditions. Inclusion of glucose (1 mM) in the enzyme solution led to a slight increase in the sedimentation coefficient to 3.68 ($\pm$ 0.12) S, the existence of that slight increase in s being substantiated by the plot of results in difference format (Fig. 3a), which signify a magnitude of -0.06 ($\pm$ 0.04) S for ($s_1 - s_2$), the difference between the sedimentation coefficients of hexokinase and its complex with glucose. This evidence that complex formation is accompanied by a 1.6% decrease in sedimentation coefficient is the hydrodynamic counterpart of the thermodynamic evidence that the mutually excluded volume between enzyme and poly(ethylene glycol 2000 is decreased by ES complex formation [9]. In fact, substitution of the respective Stokes radii of 3.22 and 3.18 nm for hexokinase and enzyme–glucose complex into Eq. (17) of Winzor and Wills [9] yields estimates of 258 and 251 l/mol for the protein–polymer covolumes, and hence to a predicted difference (7 l/mol) that is very similar to the experimental value of 6 l/mol inferred [9] from the measurements of hexokinase affinity as a function of poly(ethylene glycol) 2000 concentration [8].

Although the enhanced affinity of hexokinase for glucose in the presence of poly(ethylene glycol was interpreted in terms of the molecular crowding effect of the cosolute on a substrate-induced isomerization (9), the absence of an effect of cosolute on the maximal velocity of catalysis [8] is characteristic of substrate-mediated displacement of a preexisting isomerization equilibrium [14, 15]. In accordance with the protocol used to establish preexistence of the isomerization equilibrium for pyruvate kinase [6], the effect of glucose on the sedimentation coefficient of hexo-

kinase in the presence of poly(ethylene glycol) 600 was investigated by difference sedimentation velocity. From the difference plot (Fig. 3b) it is evident that the presence of cosolute has certainly diminished the effect of substrate on the magnitude of the sedimentation coefficient. Indeed, when cognizance is taken of the ordinate intercept predicted by Eq. (3), the results are adequately described by a horizontal line (Fig. 3b), which implies that the sedimentation coefficient of hexokinase is unchanged by complex formation with glucose under these conditions. Interpretation of this finding as signifying that the cosolute has therefore effected the conformational change in hexokinase brought about by glucose in its absence leads to the conclusion that the isomerization equilibrium was preexisting rather than substrate-induced. In other words, the change in sedimentation coefficient observed in the presence of glucose (Fig. 3a) reflects preferential interaction of the substrate with the smaller enzyme isomer; and the consequent change from a situation in which the larger, inactive isomer is the dominant species to one where all of the enzyme is in the smaller, active isomeric state.

Concluding remarks

The major outcome of this investigation is the demonstration that the expressions developed for the quantitative comparison of sedimentation coefficients by difference sedimentation velocity in the model E ultracentrifuge may also be used for interpreting corresponding results obtained with the Optima XL-A instrument, provided that the experiment is programmed for a constant time increment between the records of solute distributions for the two cells. The other outcome of note is the experimental use of difference sedimentation velocity to detect the glucose-mediated closure of the active-site cleft of yeast hexokinase as a change in the hydrodynamic characteristics of the enzyme that arises from preferential substrate interaction with the smaller of two coexisting isomeric states of the enzyme. This is the second enzyme system for which differential sedimentation velocity has been used in conjection with the molecular crowding effects of an inert cosolute to distinguish between the Monod [16] and Koshland [17, 18] mechanisms for ligand-mediated conformational changes.

Acknowledgements This investigation was supported by the Australian Research Council. Financial contributions by the National Health and Medical Research Council of Australia and the Ramaciotti Foundation toward purchase of the Beckman XL-A analytical ultracentrifuge are also gratefully acknowledged.

Progr Colloid Polym Sci (1997) 107:82–87
© Steinkopff Verlag 1997

References

1. Gerhart JC, Schachman HK (1968) Biochemistry 7:538
2. Schumaker V, Adams P (1968) Biochemistry 7:3422
3. Schumaker V (1968) Biochemistry 7:3427
4. Howlett GJ, Schachman HK (1977) Biochemistry 16:5077
5. Oberfelder RW, Barisas BG, Lee JC (1984) Biochemistry 23:458
6. Harris SJ, Winzor DJ (1988) Arch Biochem Biophys 265:458
7. Bennett WS, Steitz TA (1980) J Mol Biol 140:211
8. Rand RP, Fuller NL, Butko P, Francis G, Nichols P (1993) Biochemistry 32:5925
9. Winzor DJ, Wills (1995) Biophys Chem 57:103
10. Goldberg RJ (1953) J Phys Chem 57:194
11. Trautman R, Schumaker VN (1954) J Chem Phys 22:551
12. Shill JP, Peters BA, Neet KE (1974) Biochemistry 13:3864
13. Hoggett JG, Kellet GL (1976) Eur J Biochem 66:65
14. Bergman DA, Winzor DJ (1989) J Theor Biol 137:171
15. Bergman DA, Shearwin KE, Winzor DJ (1989) Arch Biochem Biophys 274:55
16. Monod J, Wyman J, Changeux J-P (1965) J Mol Biol 12:88
17. Koshland DE (1959) J Cell Comp Physiol Suppl 1:245
18. Koshland DE, Némethy G, Filmer D (1966) Biochemistry 5:365

Progr Colloid Polym Sci (1997) 107:88–93
© Steinkopff Verlag 1997

J. Vanhoudt
T. Aerts
S. Abgar
J. Clauwaert

Quaternary structure and interaction parameters of bovine α-crystallin: influence of isolation conditions

J. Vanhoudt · T. Aerts · S. Abgar
Dr. J. Clauwaert (⊠)
Biophysics Research Group
Department of Biochemistry
University of Antwerp
Universiteitsplein 1
2610 Antwerp, Belgium
E-mail: clauwaer@uia.ua.ac.be

Abstract The tertiary and quaternary structure of α-crystallin is still a matter of controversy. We have examined α-crystallins isolated at different temperature (4 °C, 20 °C and 33 °C), using equilibrium sedimentation and light scattering. Both techniques give the same structural and interaction parameters (molar mass, second virial coefficient) and complementary information (hydrodynamic radius, hydrodynamic volume).

The quaternary structure changes as a function of the temperature of isolation and processing. On cooling the cytoplasma below 30 °C, the quaternary structure of α-crystallin slowly changes to a larger particle which is unstable at 20 °C. On cooling further to lower temperatures (4 °C), the α-crystallin apparently recovers its stability, so it can be stored for longer times. The structural transition between 33 °C and 4 °C is reversible as we can conclude from our data of α-crystallin isolated and measured at 33 °C and α-crystallin isolated at 4 °C, stored at 33 °C for 24 h and measured at 33 °C.

The high hydrodynamic volume of α-crystallin suggests a very loose structure for this particle: a string of beads or a random coil. This loose structure suggests a rather limited interaction between the peptides and dramatically reduces the light scattering.

This structure can also explain the chaperone activity of the α-crystallin. The loose interaction between the crystallin peptides allows the interaction of the latter with the hydrophobic surfaces of the stressed proteins.

So both functions of α-crystallin, its chaperone activity and its low scattering capacity even at high concentration, are enhanced by its expanded quaternary structure.

Key words Eye lens – α-crystallin – quaternary structure – second virial coefficient – light scattering – equilibrium sedimentation

Introduction

The eye lens of mammalians is a biconvex, avascular, colorless and almost completely transparent structure. The major role of the cytoplasm of the vertebrate eye lens fibre cells is to form a high refractive transparent medium so that the lens can contribute to focus the images on the retina. This high refractive medium is obtained by a high concentration of soluble proteins. The lens crystallins are the main contributors to this high protein concentration. On a physical, biochemical and immunological basis, three main classes of crystallins can be distinguished in the eye lens of mammalians [1]. In order of decreasing molar mass, are the α-crystallins with a molar mass of about 6×10^5 g/mol, the β_H- and the β_L-crystallins with a molar mass of 2×10^5 and 5×10^4 g/mol and the γ-crystallins with a molar mass of about 2×10^4 g/mol. Their concentration in the lens is 45%, 20%, 20% and 15%, respectively. In spite of its high protein content, the eye lens is virtually completely transparent under normal healthy conditions. A theoretical explanation for this apparent contradiction was given by Benedek in the early 1970s [2]. He showed that a short-range order among the crystallins in the lens cytoplasm could account for the observed transparency. This was experimentally proven to be correct by Delaye and Tardieu more than a decade later [3]. α-crystallin is the largest protein in the cytoplasm and it is present in the highest concentration; it contributes for more than 90% to the light scattering. It is an oligomeric protein which mainly contains 4 peptides αA_1, αA_2, αB_1, αB_2 where the A peptides have an isoelectric point below pH 7 (acidic) and the B peptides have an isoelectric point above pH 7 (basic) [1]. αA_2, the major α-crystallin peptide, and αB_2 are the only primary gene products. αA_1 and αB_1 arise from these peptides by a specific postsynthetic phosphorylation [4, 5]. In addition to these "intact" peptides, α-crystallin also contains degraded peptides; these degraded peptides arise on maturing and/or aging by specific cleavages of the A or B peptides [6]. For example, $\alpha A_{2, 1-169}$ represents a peptide identical to αA_2 but only containing the first 169 amino acids.

The tertiary and quaternary structure is still a matter of controversy. The characterization of the native α-crystallin suggests a relation between peptide composition and quaternary structure. α-crystallin, isolated from the cytoplasm of newly synthesized fiber cells mainly contains the 4 undegraded A and B peptides: the young α-crystallin proteins also form a quite homogeneous population with a smaller molar mass. The α-crystallins, isolated from older cells, contain a broad pattern of peptides: the 4 undegraded peptides and a whole set of degraded and modified peptides, originating from the 4 peptides by quite specific degradation and chemical modification such as phosphorylation and deamidation. These α-crystallin solutions also contain a quite broad distribution of aggregates [7]. It is still an open question if this broad population of protein molecules is formed by a continuous collection of proteins each differing by one peptide, or if there are some discrete subclasses. Spectroscopic and light scattering studies have suggested a multi-layer tetrahedral model [8] which could give rise to a continuous set of sizes by filling up the different layers continuously or to a discrete set of molecules by giving the proteins with filled layers an extra stability. Some years ago, a micellar model has been suggested for the quaternary structure of α-crystallin [9]: this includes that hydrophobic forces are responsible for the

aggregation of the peptides and that polar interactions with solvent molecules keep the aggregates soluble and limit their size. Augusteyn and Koretz also claim arguments for a two-dimensional arrangement of the peptides instead of a more globular arrangement as suggested by hydrodynamic and light scattering methods [10, 11], but the experimental evidence for a two-dimensional arrangement is rather indirect [12]. A three layer structure model has been proposed for the native α-crystallin [13]: this model combines the multi-layer and the micelle model. The present uncertainty on the quaternary structure results in further new proposals for models of α-crystallin: a rhombododecahedric structure [14], a two-layer structure composed of annuli of α peptides [15] and an open micellar structure based on tertiary structure predictions of the α-peptides [16].

Any reliable model should include a broad distribution of sizes and of structures. Indeed experimental methods, which emphasize the presence of asymmetric structures such as transient-electric-birefringence and UV linear-dichroism, have proven the existence of a fraction of α-crystallin in a more extended asymmetric quaternary structure [17, 18]. The interest for α-crystallin recently reemerged when it became clear that α-crystallin has multiple important functions. The sequence homology with small heat shock proteins [19] was functionally confirmed by its chaperone activity [20]. The two subunits, the αA and mainly the αB peptides have been identified in several normal and abnormal cell types [21–23]. The αB peptide has been identified as the single immunodominant myelin antigen in multiple sclerosis-affected myelin [24].

We have examined α-crystallins isolated at different temperatures, using equilibrium sedimentation and light scattering in order to find out if differences in quaternary structure, as molar masses ranging from 3×10^5 [25], 5×10^5 [26] and 8×10^5 g/mol [8, 10] have been reported in literature, are related to the differences in isolation conditions of these proteins. As we have mimicked in *in vivo* conditions of temperature and ionic strength, we also hope to get an idea about the *in vivo* structure of the α-crystallin.

Materials and methods

Preparation of α-crystallin

The lenses of 6 month ($\pm$ 2 weeks) old calves were freshly obtained from a local slaughterhouse within 3 h after slaughtering and were subsequently stored at the temperature of the measurements 4 °C, 20 °C and 33 °C and all manipulations were performed at the appropriate temperature. The lens capsule was removed and the lenses were mixed with a sixth-fold quantity of buffer (containing 50 mM NaH_2PO_4, 50 mM Na_2HPO_4, 0.02% NaN_3, pH = 7.18 at 4 °C, pH 7.10 at 20 °C, pH 7.08 at 33 °C, ionic strength 0.200 M) and gently stirred for 20 min. In this way only the outer cortical fibre cells were dissolved. This suspension was centrifuged at 12 000 g for 30 min to remove the insoluble material.

About 2 ml of cortical protein solution, dissolved in the above-mentioned buffer (containing about $150A_{280\,nm}^{1\,cm}$ units), was loaded on a Bio-Gel A-5M column ($\varnothing$ 1.5 cm × 85 cm, Pharmacia). The eluent was monitored at 280 nm, using a LKB Uvicord II detection unit and collected in 2 ml fractions. The top fractions of the low molecular mass α-crystallin elution zone were collected and concentrated by using an XM-100 filter system (Amicon Corp.). For the measurement of the light scattering, we always started the measurements with the higher concentration and obtained the more diluted solutions by adding buffer.

Concentration determination

As the accuracy of most of the physicochemical methods does directly depend on the accuracy of the concentration measurements, we have paid special attention to the determination of the $A_{280\,nm,\,1\,cm}^{1\%}$ in order to use the absorbance at 280 nm as a method for the concentration measurements [27]. This resulted in a $A_{280\,nm,\,1\,cm}^{1\%}$ of 7.75.

Photon correlation spectroscopy

Photon correlation spectroscopy has been used for the determination of the diffusion coefficient D. Light scattered by the solutions was detected with an ITT FW 130 photomultiplier and the photocurrent output of the photomultiplier was analyzed using a Brookhaven BI-8000 AT correlator. The scattering setup was held at constant temperature (4 °C, 20 °C or 33 °C) by circulating dust-free water from a waterbath and the temperature was monitored directly in the scattering cell.

The quality of our setup was routinely checked by measurements at scattering angles from 40° to 140°. With a homodyne correlation setup the measured intensity correlation function of a diluted homogeneous solution, containing spherical particles which are small as compared to the wavelength of the light, becomes

$$g^2(t) = A + B\exp(-2Dk^2t)\,, \tag{1}$$

which is usually normalized to

$$g^2(t) = 1 + a\exp(-2Dk^2t)\,. \tag{2}$$

Light scattering measurements

The light scattered by α-crystallin solutions was measured using the same light scattering instrument as for photon correlation spectroscopy, at scattering angles from 40° to 140° in steps of 10°. The light scattered by a diluted solution of particles is commonly represented by the following equation:

$$\frac{Kc}{R_p(k)} = \frac{1}{P(k)}\left(\frac{1}{M_w} + 2Bc + \cdots\right)\,, \tag{3}$$

where $K = 4\pi^2 n^2 (dn/dc)^2/N_A\lambda_0^4$, with n being the refractive index of the solution, dn/dc the refractive index increment of the α-crystallin protein solution, 0.195 ml g^{-1} [7]; $\lambda_0 = 488$ nm the wavelength of the laser beam in vacuum, N_A the Avogadro's number. c is the concentration of the particles (mg/ml), $R_p(k) = (I_{sol}/I_{tol})R_{tol}(n/n_{tol})^2$ where I_{sol}/I_{tol} is the ratio of the scattered intensity by the protein solution to the reference solvent (toluene), R_{tol} the Rayleigh factor for toluene and n and n_{tol} are the index of refraction of the solution and the reference solvent, respectively. We have used the value $R_{tol} = 3.96 \times 10^{-3}$ m^{-1} [28] and $n_{tol} = 1.494$. $P(k)$ is the particle form factor, M_w the weight-average molar mass of the particles in solution and B the second virial coefficient.

At low concentration of particles, which are small relative to the wavelength of the incident beam so that $P(k) = 1$, Eq. (3) can be written in the following form:

$$\frac{Kc}{R_p(k)} = \frac{1}{M_w} + 2Bc\,. \tag{4}$$

Ultracentrifugation: equilibrium sedimentation

The Beckman Optima XL-A analytical ultracentrifuge was employed to perform the sedimentation equilibrium runs. The run conditions (angular velocity ω and duration of run) were calculated from the preset molecular parameters (sedimentation coefficient, molar mass range, 3 or 10 mm solution column), using the method proposed by Yphantis [29]. After reaching the equilibrium and having taken the equilibrium absorbance profiles, the angular velocity ω was increased to high speed (40 000 rpm) for another 24 h so that all the proteinous material were sedimented. The remaining absorbance profiles were considered as the best estimate of the residual blank absorbance and were subtracted from the sample absorbance profiles to obtain the c_r values as a function of r.

The standard equilibrium equation

$$c_r = c_0 \exp[(M_w(1 - v\varrho)\omega^2/2RT)(r^2 - r_0^2)]\,, \tag{5}$$

Progr Colloid Polym Sci (1997) 107:88–93
© Steinkopff Verlag 1997

where c_r and c_0 are the concentration at the distance r, respectively r_0, from the rotor center, ω the angular velocity, v the partial specific volume of the protein and ρ the density of the solution has been analyzed, using the Beckman software, which is based on nonlinear least-squares techniques [30] and the Equilibrium/Velocity Analysis Programs of Holladay [31, 32].

Results and discussion

The results of light scattering (Zimm plot and extrapolation to scattering angle zero) and equilibrium sedimentation should be comparable as both techniques give the same final relation

$$\frac{1}{M} = \frac{1}{M_0} + 2BC .$$ (6)

This is illustrated in Fig. 1 (measurements at 4 °C, 20 °C and 33 °C). It is clear that both methods give the same extrapolated $1/M_0$ value for all conditions. The second virial coefficient from equilibrium sedimentation at 4 °C and 20 °C is slightly more contaminated by noise than the light scattering value, whereas the equilibrium sedimentation measurements at 33 °C result in B values which are unreliable. This is probably the result of non-random fluctuations in the absorbance in the Beckman Optima XL-A analytical ultracentrifuge, when operating at higher temperatures.

Table 1 gives the molar mass M_0 and the hydrodynamic radius R, which is calculated from the extrapolated diffusion coefficient in the limit of zero concentration where the particle interaction can be considered negligible, at different temperatures. Two other parameters which can be calculated from these data, namely the hydrodynamic volume which can be deduced from the molar mass and the radius, and the second virial coefficient which can be calculated from the light scattering data, are also given.

Our data can explain the divergence in published molar masses. Indeed, the quaternary structure changes as a function of the temperature of isolation and processing. On cooling the cytoplasma below 30 °C, a slow structural transition takes place for the α-crystallin. This results in

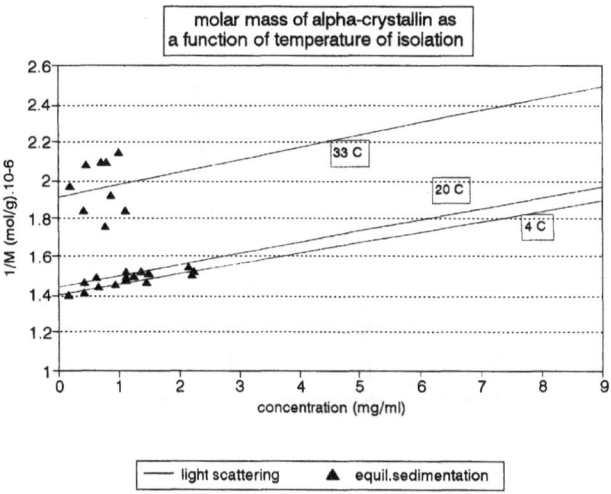

Fig. 1 Molar mass of α-crystallin solutions, isolated and measured at 4 °C, 20 °C and 33 °C as a function of concentration c. Full line: light scattering results; ▲: results from equilibrium sedimentation

a larger particle which is unstable at 20 °C. This instability explains our poor data at 20 °C: only measurements of solutions, obtained within 24 h after isolation by gel filtration, gave reliable results. On cooling further to lower temperatures (4 °C), the α-crystallin apparently recovers its stability and the solutions can be stored for longer times. The structural transition between 33 °C and 4 °C is reversible as can be seen from Fig. 2. Here the light scattering data are presented for α-crystallin isolated and measured at 33 °C (■) and α-crystallin isolated at 4 °C (▲), stored at 33 °C for 24 h and measured at 33 °C. It is clear that, within experimental error ($\pm$ 2%), both samples yield the same light scattering parameters (molar mass, second virial coefficient). As the molecules are smaller than $\lambda/20$, (see Table 1), the extrapolation to concentration zero in the Zimm plot does not yield a reliable value for the radius of gyration.

The high hydrodynamic volume of α-crystallin suggests a very loose structure for this particle: it almost can be considered as a string of beads or a random coil. This loose structure suggests a rather limited interaction between peptides. This limited interaction explains the labile quaternary structure of the α-crystallin particle. The high

Table 1 Physicochemical properties of bovine lens α-crystallin, isolated and measured at different temperatures

Temperature of processing and measurement	Molar mass g/mol ($\pm$ 10.000)	Hydrodynamic radius ($\pm$ 0.06) nm calculated from D using the Stokes–Einstein equation	Hydrodynamic volume ($\pm$ 0.12) ml/g	Second virial coefficient B ($\pm$ 0.1) $\times 10^{-5}$ ml mol g^{-2}
4 °C	725.000	9.61	3.09	2.6
20 °C	695.000	9.68	3.29	2.9
33 °C	515.000	8.92	3.47	3.2

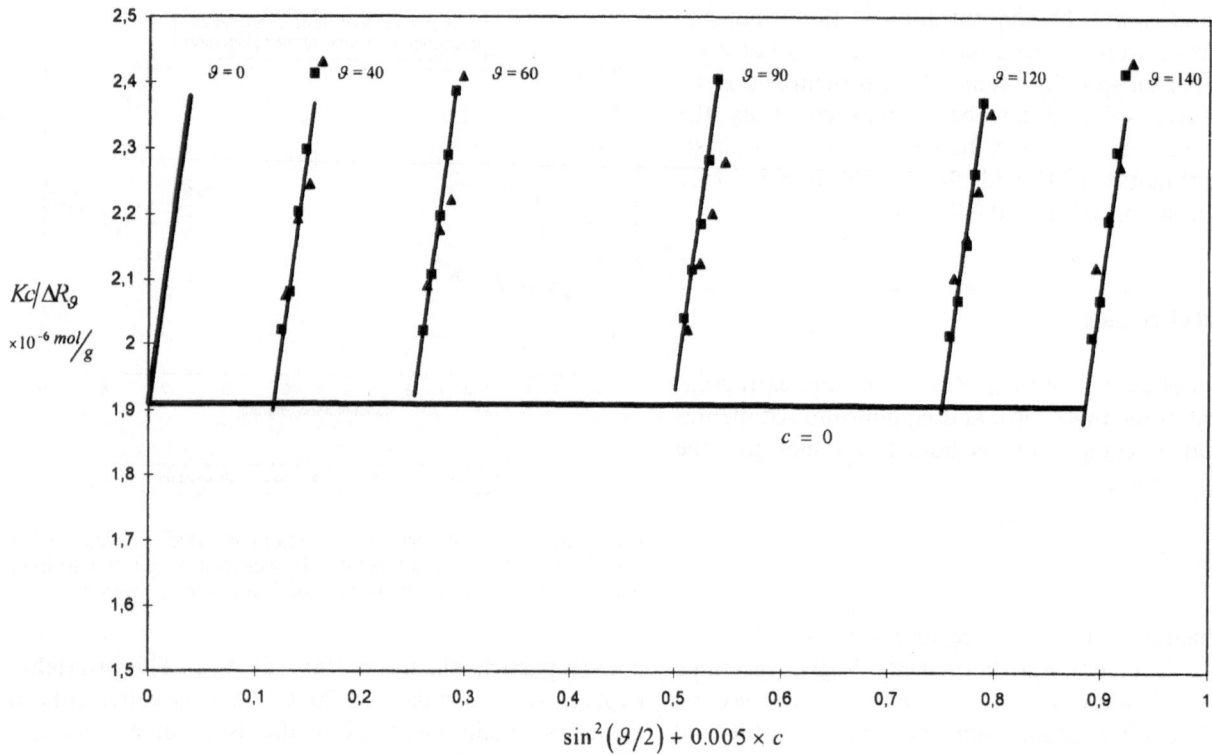

Fig. 2 Zimm plot of the light scattering data of α-crystallin solutions. ▲: α-crystallin isolated at 4 °C and then equilibrated and measured at 33 °C; ■: α-crystallin isolated at 33 °C and measured at 33 °C, Full line: linear regression of the data. Measurements were done in the range 40–140° in steps of 10° but only the 40°, 60°, 90°, 120° and 140° results are presented

hydrodynamic volume slightly increases at higher temperature, making the structure even more expanded. This increase in hydrodynamic volume can partially explain the increase in second virial coefficient [26], as can be concluded from the relation between B and the hard sphere volume v_2, i.e. $B = 4v_2/M_2$.

This loose structure dramatically reduces the light scattering at high concentrations. This is illustrated in Fig. 3. We have calculated the absorbance due to light scattering at 488 nm for solution of α-crystallin particles, having the hydrodynamic parameters of Table 1 (the 4 °C and 33 °C sample). We also included the calculation for α-crystallin particles having a more common hydrodynamic volume. The absorbance of a solution due to light scattering can be calculated via the turbidity τ [33]:

$$\text{Absorbance} = 0.434 \, \frac{32\pi^3}{3\lambda^3} \, n_0^2 \left(\frac{\delta n}{\delta c}\right)^2 \frac{M_w c S(c, 0)}{N_A} \,. \tag{7}$$

For the theoretical calculation of $S(c, 0)$, we have taken into account the relation between light scattering and osmotic pressure measurements,

$$\frac{Kc}{R(\theta)} = \frac{1}{RT} \frac{\delta\pi}{\delta C} \tag{8}$$

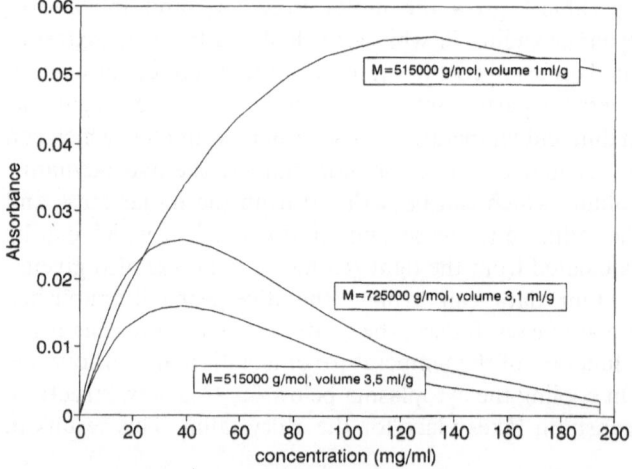

Fig. 3 Absorbance of α-crystallin solutions at 488 nm, as a result of light scattering in a concentration range 0–200 mg/ml, calculated from Eq. (7) using the appropriate light scattering constants of the α-crystallin particles, isolated at 4 °C and 33 °C. We also included the calculations for α-crystallins with a molar mass of 515 000 g/mol and an hydrodynamic volume of 1 ml/g, which is more common for globular proteins

Progr Colloid Polym Sci (1997) 107: 88–93
© Steinkopff Verlag 1997

which results in

$$S(c,0) = \frac{RT}{M_w(\delta\pi/\delta c)} \; . \tag{9}$$

Different expressions have been proposed for the osmotic pressure of hard spheres. Ree and Hoover have calculated an exact expression for the osmotic pressure of hard spheres as a function of the volume fraction up to the eight virial coefficient [34, expressions (4) and (21), Table II].

$$\pi = \frac{RT}{M_n} (c + 4vc^2 + 10v^2c^3 + 18.36v^3c^4 + 28.24v^4c^5$$
$$+ 39.53v^5c^6 + 56.52v^6c^7 + 87.65v^7c^8 + \cdots) \; .$$

Figure 3 illustrates that a collapse of the loose α-crystallin to a more dense structure can dramatically increase the light scattering. A concomitant increase of the polydispers-ity will further increase the light scattering [27] so that the absorbance reaches the critical level of 0.1 where the solution becomes opaque.

This loose structure can also explain the chaperone activity of the α-crystallin. The loose interaction between the crystallin peptides allows the interaction of the latter with the hydrophobic surfaces of the stressed protein [35]. So both functions of α-crystallin, its chaperone activity and its low scattering capacity even at high concentration, are enhanced by its expanded quaternary structure.

Acknowledgements This research was supported by grants from the Fund for Joint Basic Research (FKFO) and the Fund for Medical Scientific Research (FGWO). This research was performed within the framework of the EU Concerted Action "The Role of membranes in lens Aging and Cataract".

J. Vanhoudt holds a predoctoral fellowship from the Fund for Scientific Research (FWO).

References

1. Bloemendal H (1981) Molecular and Cellular Biology of the Eye Lens. Wiley, New York, p 469
2. Benedek GB (1971) Appl Optics 10:459–475
3. Delaye M, Tardieu A (1983) Nature 302:415–417
4. Spector A, Chiesa R, Sredy J, Garner W (1985) Proc Natl Acad Sci 82:4712–4716
5. Voorter CEM, de Haard-Hoekman WA, Roersma ES, Meyer HE, Bloemendal H, De Jong W (1989) FEBS Lett 259:50–52
6. Groenen PJTA, Merck KB, De Jong WW, Bloemendal H (1994) Eur J Biochem 225:1–19
7. Schurterberger P, Augusteyn RC (1991) Biopolymers 31:1229–1240
8. Tardieu A, Laporte D, Licinio P, Krop B, Delaye M (1986) J Mol Biol 192:711–724
9. Augusteyn RC, Koretz JF (1987) FEBS Lett 222:1–5
10. Siezen RJ, Berger H (1978) Eur J Biochem 91:397–405
11. Andries C, Backhovens H, Clauwaert J, De Block J, De Voeght F, Dhont C (1982) Exp Eye Res 34:239–255
12. Radlick JW, Koretz JF (1992) Biochim Biophys Acta 1120:193–200
13. Walsh MT, Sen AC, Chakrabarti B (1991) J Biol Chem 266:20079–20084
14. Wistow G (1993) Exp Eye Res 56:729–732
15. Carver JA, Aquilina JA, Truscott RJW (1994) Exp Eye Res 59:231–234
16. Groth-Vaselli B, Kumosinski TF, Farnsworth PN (1995) Exp Eye Res 61:249–253
17. van Haeringen B, Eden D, van den Bogaerde MR, van Grondelle R, Bloemendal M (1992) Eur J Biochem 210:211–216
18. van Haeringen B, van den Bogaerde MR, Eden D, van Grondelle R, Bloemendal M (1993) Eur J Biochem 217:143–150
19. Ingolia TD, Craig EA (1982) Proc Natl Acad Sci USA 79:2360–2364
20. Horwitz J (1993) Proc Natl Acad Sci USA 89:10449–10453
21. Bhat SJ, Nagineni N (1989) Biochem Biophys Res Commun 158:319–325
22. Renkawek K, Voorter CE, Bosman GJ, van Workum FP, de Jong WW (1994) Acta Neoropathol Berl 87:155–160
23. Head MW, Hurwitz L, Goldman JE (1996) J Cell Sci 109:1029–1039
24. van Noort JM (1996) J Mol Med 74:285–296
25. Thomson JA, Augusteyn RC (1983) Exp Eye Res 37:367–377
26. Wang X, Bettelheim FA (1989) Proteins 5:166–169
27. Xia J-Z, Wang Q, Tatarkova S, Aerts T, Clauwaert J (1996) Biophys J 71:2815–2822
28. Bender TMR, Lewis RJ, Pecora R (1986) Macromolecules 19:244–245
29. Yphantis DA (1964) Biochemistry 3:297–317
30. Johnson ML, Correia JJ, Yphantis DA, Halvorson HR (1981) Biophys J 36:575–588
31. Kelly L, Holladay LA (1990) Biochemistry 29:5062–5069
32. Shire SJ, Holladay LA, Rinderknecht E (1991) Biochemistry 30:7703–7711
33. van Holde KE (1985) In: Physical Biochemistry. Prentice-Hall, Englewood Cliffs, NJ, pp 209–304
34. Ree HF, Hoover GW (1967) J Chem Phys 46:4181–4195
35. Braig K, Otwinowski Z, Hedge R, Boisvert DC, Joachimack A, Horwich AL Sigler PB (1994) Nature 371:578–586

Progr Colloid Polym Sci (1997) 107:94–101
© Steinkopff Verlag 1997

BIOLOGICAL SYSTEMS

C. Fochler
H. Durchschlag

Investigation of irradiated eye-lens proteins by analytical ultracentrifugation and other techniques

C. Fochler
GSF
National Research Center for Environment and Health
85764 Neuherberg, Germany

Dr. H. Durchschlag (✉)
Institute of Biophysics and Physical Biochemistry
University of Regensburg
Universitätsstraße 31
93040 Regensburg, Germany

Abstract Crystallins, the major eye lens proteins, were investigated after preceding irradiation with X-rays or UV light by using analytical ultracentrifugation and other physicochemical and analytical techniques. Sedimentation velocity and sedimentation equilibrium runs of unirradiated and irradiated samples of the crude extract and individual fractions of calf-lens crystallins (α, β_H, β_L, γ) were performed under varying experimental conditions. While by the impact of radiation sedimentation coefficients remain essentially unchanged, high-speed sedimentation equilibrium experiments revealed significant changes of the molar masses of the respective crystallin species, depending on the crystallin class and the nature of the radiation used. The γ-crystallins turned out to be the most stable species against radiation-induced damages. UV absorption and fluorescence spectroscopy, size-exclusion chromatography, SDS electrophoresis, viscometry, and the analysis of SH and SS contents provided useful complementary information. In conclusion, several radiation damages could be registered: aggregation, crosslinking, dissociation, fragmentation, destruction of aromatic amino acids, cysteines and cystines. The ˙OH scavenger formate and the chaperone α-crystallin turned out to protect the crystallin crude extract to some extent against X-ray-induced aggregation phenomena.

Key words Eye-lens proteins – crystallins – radiation effects – radioprotection – analytical ultracentrifugation – comparative studies

Introduction

The lens of the eye provides the refractive index necessary to focus images on the retina. It is a highly transparent tissue made up mainly of soluble proteins called the crystallins, the remainder being essentially water. Three major classes (α, β, γ) and several subclasses of crystallins (e.g., β_H, β_L, γS) have been identified in mammalian lenses; bird and reptile lenses contain additionally δ-crystallins (cf. refs. [1, 2]). The molecular properties of the mammalian crystallins such as molar mass, number of subunits and thiol content vary considerably, α-crystallin and the β-subclasses β_H and β_L show a pronounced heterodispersity (Table 1). Crystallins have no significant turnover, they are not replaced during the lifespan of animals. They must therefore be exceptionally stable, withstanding all

---

Abbreviations AUC, analytical ultracentrifugation; HSSE, high-speed sedimentation equilibrium; SDS-PAGE, sodium dodecyl sulfate polyacrylamide gel electrophoresis; SEC, size-exclusion chromatography.

Table 1 Properties of
mammalian lens crystallins[a]

Property	α	β_H	β_L	γS	γ
Molar mass [kg/mol]	600–900	100–300	40–100	21	21
Molar subunit mass [kg/mol]	20	22–32	22–32	21	21
Number of subunits	30–45	4–14	2–4	1	1
Thiol content	Low	High	High	High	High
Aromatic content	High	High	High	High	High

[a] Values were excerpted from the literature (e.g., refs. [1, 2, 10] and references therein)

detrimental events which could lead to aggregation or denaturation, e.g., crosslinking reactions in proteins as a consequence of oxidative processes. The high content of aromatic and sulfur-containing amino acids (Trp, Tyr, Phe, Cys, Met), sometimes in close neighborhood, enables the interaction of aromatic groups and the polarizable sulfur, thereby contributing to protein stability and serving as electron-transfer system [3].

As a consequence of aging and irradiation, significant changes of the lens proteins and the lens can occur, finally leading to a loss of transparency and cataract formation (cf. refs. [4–6]). There is, indeed, some evidence that the occurrence of holes in the ozone layer of the atmosphere (enhanced UV radiation) as well as (too frequent) X-irradiation of persons may lead to an enhanced formation of human cataracts, even with young people. One possible reason may be due to a lack of free sulfhydryls, both in the crystallins and the medium. Therefore, the examination of lens proteins damaged by nonionizing and ionizing radiation and the search for radioprotective substances is of particular interest for many fields of research (radiation biology, biochemistry, medicine, pharmacology, etc.).

The investigation of radiation-induced damages of various other proteins has established the occurrence of several effects: change of aromatic and sulfur containing amino acid residues, fragmentation, dissociation, dimerization, aggregation, partial unfolding and exposure of hydrophobic residues, as well as inactivation [7]. In the case of X-irradiation, detrimental radiation effects are primarily caused by ·OH radicals, whereas photoionizations are responsible for damages caused by UV radiation [8, 9]. The damages of the protein structure caused by the impact of radiation can be monitored by a variety of analytical and physicochemical techniques. Alterations of the overall structure of proteins may be detected by scattering and hydrodynamic techniques. In this context, analytical ultracentrifugation (AUC) may be utilized as an efficient tool, especially by determining sedimentation coefficients and molar masses of the samples under analysis.

Our present investigations are concerned with the elucidation of possible radiation damages of the crystallin crude extract and individual fractions of crystallins in aqueous solution by means of AUC and other techniques of

structural investigation. The action of both X-rays and UV light resulted in several significant changes of their structure [10].

Materials and methods

A crystallin crude extract was prepared from fresh calf eye lenses by decapsulation, grinding and dissolving in buffer, followed by centrifugation [11]. Purified fractions of α-, β_H-, β_L- and γ-crystallins (the latter usually containing small amounts of γS) were obtained by gel filtration (Superdex 75 and 200 columns from Pharmacia Biotech).

For experiments, crystallins (0.1–100 mg/ml) were dissolved in 100 mM phosphate buffer pH 7.4 and irradiated with X-rays (0–20 kGy) or UV-C light (0–200 kJ m^{-2}) in the absence or presence of additives. For most experiments, protein concentrations of 0.1 mg/ml are required, in order to produce pronounced radiation effects on the one side, and measurable signals, on the other [7]. To allow a comparison of the action of X- and UV-irradiation on crystallins, doses were chosen which produced a comparable damage [10]. Conditions of typical experiments with crystallins were: protein concentration $c = 0.1$ mg/ml, dose $D = 0$ or 1 kGy or 60 kJ m^{-2}. All irradiations and investigations were performed at 25 °C.

Sedimentation velocity and equilibrium experiments were performed in a Beckman model E analytical ultracentrifuge equipped with a high-sensitivity ultraviolet scanner and multiplexer system and a 10 in recorder. Ultracentrifugal runs were performed in multi-hole rotors (AnF or AnG), using 12 mm double-sector cells (charcoal-filled epon) and sapphire windows. Apparent sedimentation coefficients of the proteins were obtained from ln r vs. t plots, and molar masses were determined by high-speed sedimentation equilibrium (HSSE) using the meniscus-depletion technique [12].

Ultracentrifugal analysis, especially HSSE, was performed under various experimental conditions (e.g., rotor speed and wavelengths of scanning [13]) in order to allow a discrimination between different particle species. Due to the different size of the proteins under consideration, optimum working conditions need to compromise between high and low centrifugation speeds and quite different times for reaching equilibrium. As a minimum, each sample was investigated at two rotor speeds and three scan wavelengths (230, 280 and 290 nm); achievement of equilibrium was checked by repeated scanning about 24 h after having reached equilibrium. Due to different absorption coefficients of the proteins at the mentioned wavelengths, scanning at 230 nm usually enhances the registered signals, while they are lowered at 290 nm, as compared to the absorbances at 280 nm. Finally, both measures (variation of rpm and scan wavelength) allow a modulation of the signals: in the case of heterogeneous systems the lowest molar mass estimate is obtained at 230 nm and high rotor speed, and the highest mass at 290 nm and low speed; homogeneous samples are characterized by the coincidence of all observed mass estimates. Of course, the use of low initial protein concentrations ($c_0 \leqslant 0.1$ mg/ml) leads only to marginal absorbances (especially at 290 nm) and causes a considerable noise in the signals. Especially in the case of low c_0, special care has to be taken to avoid superpositions of the signals of cell bottom with true sample absorbances, again most problematic when using 290 nm as scan wavelength.

Absorption and fluorescence spectra were recorded in Perkin–Elmer Lambda 5 and MPF-44A spectrometers, respectively. Nile Red fluorescence was determined according to [14]. For size-exclusion chromatography (SEC), high-performance liquid chromatography and a TSK G 2000 SW column (LKB) was used. SDS-PAGE was performed according to Laemmli [15]. Densities and viscosities of solvent and solutions were carried out in a Paar digital densimeter (DMA 02) and an Ostwald viscometer (6 ml, flow time for water about 70 s), respectively; intrinsic viscosities were calculated according to ref. [16]. Determination of the number of protein sulfyhydrils and disulfides was performed according to refs. [17] and [18], respectively.

Results and discussion

The primary goal of our experiments was the ultracentrifugal analysis of the effects of X and UV-irradiation on the crystallin crude extract (mainly consisting of α-, β_H-, β_L- and γ-crystallins), since the extract mimicks *in vivo* conditions most nearly. In this context, two compounds (the ·OH scavenger formate and the chaperone α-crystallin) were added to the crude extract prior to irradiation, in order to test their suspected radioprotective capability. In the following, also the individual crystallin species were investigated by AUC, in order to find out possible

sensitivities of certain crystallins against irradiation. The results obtained by AUC were complemented by several other techniques, in order to allow more detailed explanations of the observed effects on a molecular level.

Analytical ultracentrifugation

The investigation of crystallins is hampered by several obstacles such as low initial protein concentration ($c_0 = 0.1$ mg/ml) and the heterodispersity of α- and β-crystallins, and more seriously, of the crystallin crude extract which is composed of species of quite different molar masses ranging from about 20 to 900 kg/mol (cf. Table 1). In addition, heterodispersity still increases after preceding irradiation, due to the formation of fragments, dissociation products, and aggregates. This exaggerates the mentioned limits of masses still more. Of course, heterodispersity of the samples under investigation affects the accuracy of the obtained results markedly. Frequently, as a consequence of an insufficient separation of all the species during centrifugation, only mean values of parameters are obtainable.

Sedimentation velocity

The determination of the sedimentation coefficient s of the unirradiated crystallin crude extract (Table 2) showed the expected dependence on the experimental conditions: the observed s value decreases with elevated rotor speed, due to a spinning down of the large components of the heterogeneous crystallin mixture. The individual crystallin species show essentially sedimentation coefficients as reported in the literature (cf. ref. [10]). Surprisingly, under the given experimental conditions, neither X- nor UV-irradiation revealed any significant changes of this molecular parameter. To find out whether these findings are reality or

Table 2 Sedimentation coefficients of unirradiated and irradiated crystallins ($c_0 = 0.1$ mg/ml, $D = 0$ or 1 kGy or 60 kJ m^{-2})

Sample unirradiated/irradiated	Speed [rpm]	λ [nm]	$s_{20,w}$ [S]
Crystallin crude extract	20 000	280	2.9 ± 0.1
	34 000	230	2.8 ± 0.1
	40 000	280	2.4 ± 0.1
α-crystallin	30 000	230	21.4 ± 0.5
β_H-crystallin	34 000	230	4.6 ± 0.1
β_L-crystallin	34 000	230	3.8 ± 0.1
γ-crystallin	34 000	230	2.1 ± 0.1

Progr Colloid Polym Sci (1997) 107:94–101
© Steinkopff Verlag 1997

may be ascribed to a superposition of different effects, detailed equilibrium runs and other studies were performed.

Sedimentation equilibrium

High-speed sedimentation equilibrium experiments (Table 3) on the unirradiated crystallin crude extract reveals a very heterogeneous population of different crystallin species: the molar mass weight-average estimates span a wide range, from about 20 to 530 kg/mol. The masses at the extremes of the range (located in the table in the left- and right-hand columns, respectively) are obviously caused by the γ- and α-crystallins, respectively, and the β-crystallins lying between. At the first glance, the impact of both X-rays and UV light seems to have no pronounced

influence on the obtained mass averages, if compared to the unirradiated crude extract. A more thorough analysis (cf. 20 000 rpm: 280 nm; 10 000 rpm: 280 and 290 nm; Fig. 1), however, hints at a decrease (dissociation and/or fragmentation) of some of the larger crystallin species (β or $\beta + \alpha$), but does not reveal any change of the small γ-crystallins. Under the given experimental conditions ($c_0 = 0.1$ mg/ml), the registration of small amounts of very large aggregates is outside the limit of detection. On the other side, X-irradiation of the crystallin extract in the presence of the additives, formate or α-crystallin, yields molar mass averages significantly lower than found for the X-irradiated sample devoid of these additives (cf. 20 000 rpm: 290 nm; 10 000 rpm: 230 nm). By contrast, the crystallin extract, which has been UV-irradiated in the presence of the additives, did not show any comparable effect.

Table 3 Molar mass determinations of unirradiated and irradiated crystallin samples (crude extract and individual crystallin species, $c_0 = 0.1$ mg/ml) by HSSE. The crystallin crude extract was also irradiated after addition of 10 mM sodium formate (F) or 10% α-crystallin (10α) prior to irradiation, to test the possible influence of these additives on radiation-induced aggregation phenomena

Dose	Molar mass [kg/mol][a]					
	230 nm	280 nm	290 nm	230 nm	280 nm	290 nm
Crystallin crude extract		20 000 rpm			10 000 rpm	
0[b]	22 ± 2	57 ± 8	32 ± 3 / 153 ± 3	144 ± 7	218 ± 49	86 ± 16 / 532 ± 45
1 kGy	28 ± 1	30 ± 3	33 ± 3 / 183 ± 5	213 ± 29	27 ± 3 / 198 ± 15	21 ± 4 / 587 ± 15
1 kGy (F)	24 ± 7	41 ± 4	57 ± 4	40 ± 1	52 ± 11	309 ± 20
1 kGy (10α)[c]	29 ± 6	40 ± 2	41 ± 4	127 ± 17	381 ± 56	621 ± 79
60 kJ m^{-2}	26 ± 1	28 ± 2	20 ± 2	53 ± 4	68 ± 11	87 ± 20
60 kJ m^{-2} (F)	31 ± 1	41 ± 3	56 ± 8	45 ± 5	137 ± 24	279 ± 23
60 kJ m^{-2} (10α)[c]	27 ± 1	35 ± 1	39 ± 2	74 ± 4	132 ± 10	140 ± 12
α-crystallin		6000 rpm			4000 rpm	
0	581 ± 46	696 ± 80	963 ± 53	600 ± 69	857 ± 93	920 ± 72
1 kGy	706 ± 51	907 ± 79	893 ± 70	597 ± 48	1185 ± 193	1451 ± 158
60 kJ m^{-2}	669 ± 53	711 ± 53	839 ± 20	524 ± 63	759 ± 42	871 ± 77
β_H-crystallin		14 000 rpm			10 000 rpm	
0	59 ± 5	116 ± 9	106 ± 5	86 ± 5	129 ± 17	144 ± 5
1 kGy	53 ± 1	53 ± 5	75 ± 5	63 ± 2	66 ± 13	112 ± 17
60 kJ m^{-2}	38 ± 3	45 ± 5	37 ± 3	57 ± 3	59 ± 7	123 ± 11
β_L-crystallin		16 000 rpm			12 000 rpm	
0	48 ± 2	83 ± 2	76 ± 3	64 ± 3	89 ± 5	91 ± 4
1 kGy	36 ± 3	56 ± 4	55 ± 10	37 ± 2	94 ± 3	81 ± 10
60 kJ m^{-2}	38 ± 1	68 ± 1	51 ± 7	59 ± 4	68 ± 5	121 ± 7
γ-crystallin		26 000 rpm			20 000 rpm	
0	20 ± 1	22 ± 1	20 ± 1	25 ± 2	24 ± 1	21 ± 2
1 kGy	19 ± 1	20 ± 1	21 ± 1	24 ± 1	21 ± 1	24 ± 1
60 kJ m^{-2}	18 ± 1	20 ± 1	19 ± 2	20 ± 1	23 ± 3	20 ± 2

[a] Molar mass estimates are based on $\bar{v} = 0.735$ cm^3/g for all crystallins. In the case of heterogeneous samples, the obtained molar masses represent weight averages. Note that in some cases (crude extract) two averages are mentioned, belonging to the major part of the cell and the bottom region, respectively

[b] The presence of the additives yielded molar mass averages similar to those given in their absence

[c] Addition of 10% α-crystallin to the crystallin crude extract (which itself contains α-crystallin), so that a final concentration of 0.1 mg/ml protein results

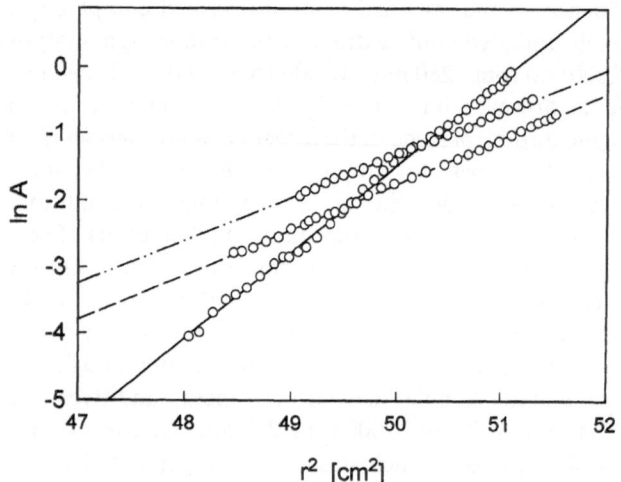

Fig. 1 High-speed sedimentation equilibrium of unirradiated and irradiated crystallin crude extract ($c_0 = 0.1$ mg/ml). The proteins were recorded at 20 000 rpm and 280 nm. ——: native protein; – – –: 1 kGy X-irradiation; – ·· –: 60 kJ m^{-2} UV-irradiation

The molar mass estimation of α-crystallin discloses some diversity (580–960 kg/mol), obviously due to some kind of heterodispersity in the isolated α-crystallin fractions. X-irradiation of α-crystallins causes a pronounced increase in the molar mass averages (cf. 4000 rpm: 280 and 290 nm), whereas UV-irradiation does not give evidence of aggregates of a detectable size.

The molar masses of the β-crystallin fractions exhibit marked heterodispersity, they range from about 50 to 150 kg/mol, the majority of masses of the β_H-fraction lying between about 100 and 150 kg/mol, and the β_L-fraction showing values between 50 and 90 kg/mol. Both subclasses of β-crystallins exhibit decreases of the observed masses upon X- and UV-irradiation (nearly at all rpm and scan wavelengths), thereby disclosing their dissociation and/or fragmentation.

The γ-crystallins exhibit molar masses of about 20–25 kg/mol at all experimental conditions, revealing that they are the only monodisperse crystallin species. They are very stable against both X-rays and UV light, apart from the observation of some fragmentation observed, for example, at 26 000 rpm and scanning at 230 nm.

Concerning the position of the baseline, data analysis revealed that all unirradiated samples exhibit a baseline near 0, whereas the irradiated crystallins caused a baseline > 0, due to the radiation-induced production of significant amounts of small fragments (amino acids, peptides). As a consequence of the enlarged absorption at 230 nm, the baseline effects were more pronounced when using this scan wavelength. Significant effects were monitored with the crude extract and the β-crystallins,

the most prominent one in the case of X-irradiated β_L-crystallin.

In this context, it should also be noted that several UV-irradiated samples (crude extract, γ-crystallin) exhibited prior to the centrifugation a turbidity of the solution. Obviously, as a consequence of UV-irradiation, very large aggregates are formed which sediment very rapidly after initiating the centrifugation process.

Other techniques

Spectroscopy

As evident from UV-absorption spectroscopy experiments (Fig. 2), crystallins ($c_0 = 0.1$ mg/ml) suffer from aggregation when irradiated with X-rays or UV light (cf. the absorbance at $\lambda > 330$ nm, i.e., outside the range of protein absorption). Such light-scattering phenomena are caused by aggregates of considerable size. Indeed, at high doses the turbidity of irradiated crystallin solutions is visible by the naked eye. The change/destruction of the aromatic amino acid residues of the crystallins can be concluded from the change of absorbance, and, above all, from the dose-dependent disappearance of the fluorescence signals (Fig. 3). At doses of about 15 kGy or 200 kJ m^{-2} the fluorescence of the aromatics of the crystallins is completely depleted. The dose–effect curves derived from the fluorescence emission spectra of individual crystallin species show a different susceptibility against radiation; γ-crystallin turned out to be relatively insensitive.

Application of Nile Red, an efficient fluorescent probe of hydrophobic domains of proteins, proves that new hydrophobic protein sites become exposed as a consequence of both X- and UV-irradiation (Fig. 4). The enhancement of the hydrophobic protein surface may be tentatively explained as an irradiation-triggered, slight unfolding of the protein.

Size-exclusion chromatography

SEC experiments allow the separation of the crude extract ($c_0 = 0.1$ mg/ml) into α, β and γ-crystallins, together with the registration of the damages of the separated species (Fig. 5). Obviously, X-rays and UV light induce different damages: under the given conditions, the α, β and γ-crystallin fractions are damaged severely and to a similar extent when using UV light, while X-irradiation destroys γ-crystallin extensively, but α-crystallin only moderately. Besides a different sensitivity of the individual crystallins against radiation, differences in their behavior may also be

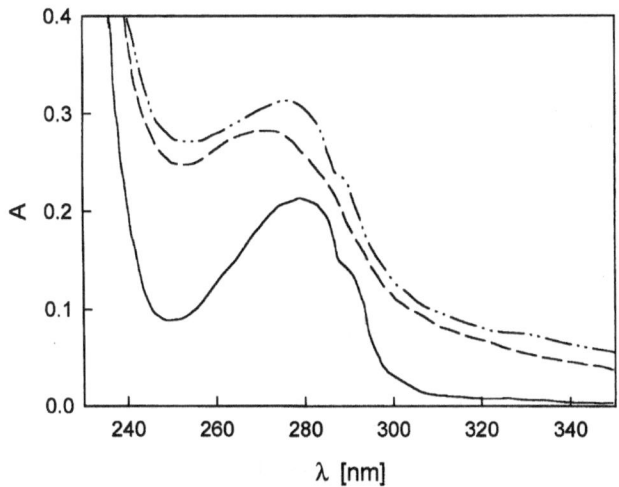

Fig. 2 UV-absorption spectra of unirradiated and irradiated crystallin crude extract ($c_0 = 0.1$ mg/ml). ———: native protein; – – –: 1 kGy X-irradiation; – ·· –: 60 kJ m^{-2} UV-irradiation

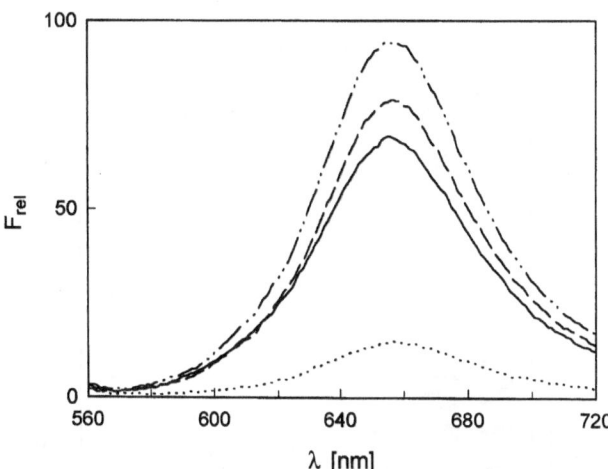

Fig. 4 Fluorescence of Nile Red in the presence of unirradiated and irradiated crystallin crude extract ($c_0 = 0.1$ mg/ml). ———: native protein; – – –: 1 kGy X-irradiation; – ·· –: 60 kJ m^{-2} UV-irradiation; ········: buffer

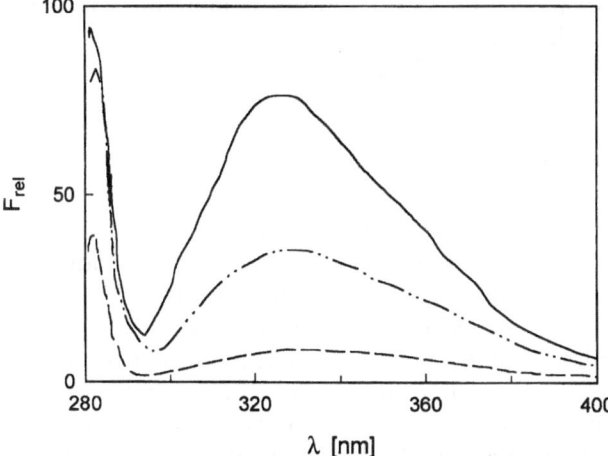

Fig. 3 Fluorescence emission spectra of unirradiated and irradiated crystallin crude extract ($c_0 = 0.1$ mg/ml). ———: native protein; – – –: 1 kGy X-irradiation; – ·· –: 60 kJ m^{-2} UV-irradiation

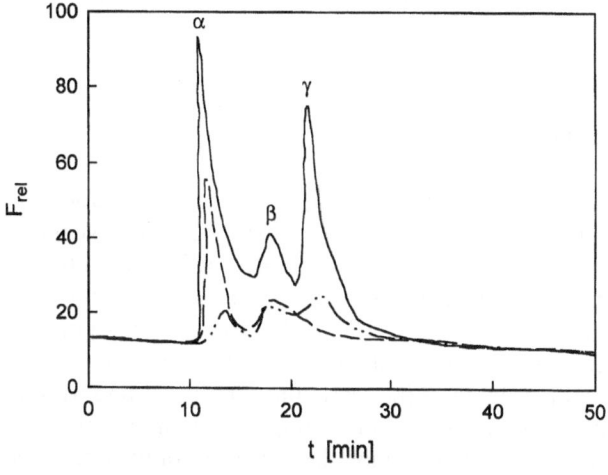

Fig. 5 Size-exclusion chromatography of unirradiated and irradiated crystallin crude extract ($c_0 = 0.1$ mg/ml), coupled to fluorescence detection. ———: native protein; – – –: 1 kGy X-irradiation; – ·· –: 60 kJ m^{-2} UV-irradiation

caused by both a mutual protective action of all crystallins in the crude extract and different concentrations of the individual species in the extract.

SDS electrophoresis

The SDS-PAGE of the crystallin crude extract ($c_0 = 1$ mg/ml) shows several subunit bands of molar masses between 20 and 30 kg/mol (Fig. 6), attributable to the various crystallin species and subspecies. Increasing X- or UV-irradiation induces a destruction of the protein subunits and, concomitantly, the formation of large, covalently crosslinked subunit aggregates (especially after UV-irradiation) which are unable to enter the pores of the gel. No significant amounts of large fragments are observed; however, small fragments cannot be detected by this method. With this type of experiment, the β-crystallins turned out to be less stable than the α- and γ-species, the γ-crystallins being the most resistant species.

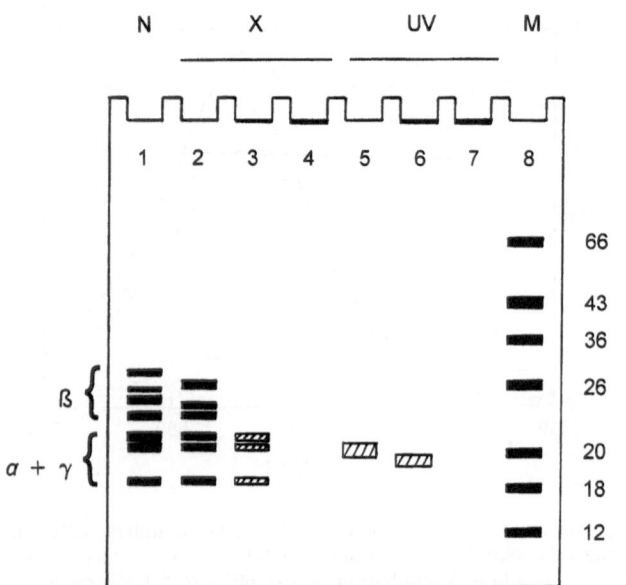

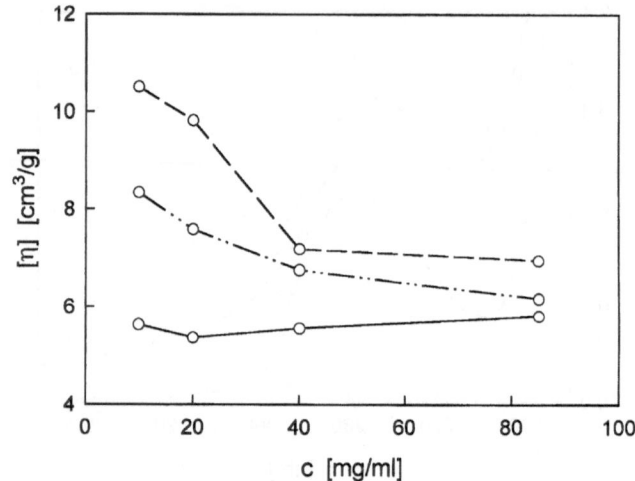

Fig. 7 Intrinsic viscosity of unirradiated and irradiated crystallin crude extract ($c_0 = 10-85$ mg/ml). ———: native protein; – – –: 20 kGy X-irradiation; – ·· –: 200 kJ m^{-2} UV-irradiation

Fig. 6 SDS-PAGE (reducing assay with 2-mercaptoethanol) of unirradiated and irradiated crystallin crude extract ($c_0 = 1$ mg/ml). Lane 1: native protein (N); lanes 2–4: 1, 5, 10 kGy X-irradiation; lanes 5–7: 60, 100, 150 kJ m^{-2} UV-irradiation; lane 8: unirradiated molecular weight standards (M given in kg/mol)

SH groups is caused by a partial unfolding of the protein, accompanied by an exposure of initially buried, integer sulfhydryls. Similar to the action of X-rays, UV light causes a decrease of SS bonds.

Viscometry

Viscometric studies on highly concentrated solutions of the crystallin crude extract ($c_0 = 10-85$ mg/ml) reveal that both X- and UV-irradiated samples lead to an enhancement of the intrinsic viscosity [η], the effects being more pronounced for X-rays (Fig. 7). The increase in [η] may be interpreted in terms of an aggregate formation and/or partial unfolding of the crystallins after preceding irradiation. While the unirradiated sample shows no pronounced concentration dependence, the viscosities of the irradiated samples are obviously dependent on the protein concentration. The largest effect can be found for the X-irradiated sample at the lowest concentration, in accord with the indirect action of X-rays in aqueous solution.

Determination of sulfhydryls and disulfides

An analysis of SH and SS groups discloses the presence of both groups in native crystallins ($c_0 = 0.1$ mg/mol), and their proceeding loss upon continuous X- and UV-irradiation. At high X-ray doses ($\geqslant 10$ kGy) a complete loss of SH groups, but only a decrease of SS bonds can be observed. While low UV doses induce a slight increase of available SH groups, high UV doses also deplete the number of SH groups considerably. The observed increase in

Conclusions

The application of analytical ultracentrifugation and complementary physicochemical and analytical techniques has proven that all crystallins suffer from a variety of radiation damages: aggregation, crosslinking, dissociation, fragmentation, partial unfolding and exposure of hydrophobic residues, destruction/changes of aromatics and sulfur-containing amino acid residues. However, the extent of the radiation effects is dependent on the type of radiation (X-ray or UV light, i.e. ionizing or nonionizing radiation) and the crystallin class (α, β, γ). X-irradiation of α-crystallin leads to a considerable aggregation, while the β-crystallins (β_H, β_L) essentially suffer from dissociation and fragmentation; the γ-crystallins turn out to be rather stable. The impact of UV light seems to yield rather large aggregates, and, again, give rise to dissociation products and fragments. As may be expected, the crystallin crude extract renders the behavior of all constituent species. However, the concerted action of all the crystallins in the extract may lead to some kind of stabilization against some of the detrimental effects of radiation. Though α-crystallins themselves suffer from radiation-induced aggregation, they may function as chaperones [19], thereby preventing the other crystallins from aggregation. This was shown by addition of extra α-crystallin to the crude

Progr Colloid Polym Sci (1997) 107:94–101
© Steinkopff Verlag 1997

extract prior to X-irradiation. Similarly, the presence of ˙OH scavengers, such as formate, may suppress aggregation and other radiation effects. On the other side, an effective UV protection may only be provided by substances ("chemical filters") exhibiting absorption in the UV range [7, 10].

Though the behavior of the crystallins upon X- and UV-irradiation is similar to those reported for other proteins, a comparison of the doses needed to produce comparable effects discloses that crystallins are more stable [7, 10]. Their stability against the impact of ionizing and nonionizing radiation is similar to their extraordinary behavior in the native state. Eye lens proteins exhibit a high stability against environmental changes, such as the presence of chaotropic agents [20, 21]. The resistance of crystallins against radiation may be further increased *in vivo* by favourable conditions: extremely high packing density of the crystallins in the lens (about 350 mg/ml), existence of α-crystallins in the assembly of lens crystallins, permanent presence of certain radioprotectives (thiols and other reductants) in the lens and the defense system in front of the lens (tear film, anterior chamber) as well as the UV-absorbing behavior provided by the cornea.

Acknowledgements The authors are much obliged to Profs. Jaenicke and Müller-Broich, Regensburg, for support and stimulating discussions, and to the University of Regensburg for a scholarship and financial support (granted to C.F.).

References

1. Hockwin O (ed) (1985) Biochemie des Auges. Ferdinand Enke Verlag, Stuttgart
2. Berman ER (1991) Biochemistry of the Eye, Plenum Press, New York
3. Summers L, Wistow G, Narebor M, Moss D, Lindley P, Slingsby C, Blundell T, Bartunik H, Bartels K (1984) Pept Protein Rev 3:147
4. Hockwin O (ed) (1976) Progress of Lens Biochemistry Research. Doc Ophthalmol Proc Ser, Vol 8. Dr W Junk bv Publishers, The Hague
5. Lerman S (1980) Radiant Energy and the Eye. Macmillan, New York
6. Harding J (1991) Cataract: Biochemistry, Epidemiology and Pharmacology. Chapman & Hall, London
7. Durchschlag H, Fochler C, Feser B, Hausmann S, Seroneit T, Swientek M, Swoboda E, Winklmair A, Wlček C, Zipper P (1996) Radiat Phys Chem 47: 501
8. von Sonntag C (1987) The Chemical Basis of Radiation Biology. Taylor & Francis, London
9. Kiefer J (ed) (1977) Ultraviolette Strahlen. Walter de Gruyter, Berlin
10. Fochler C (1996) Thesis, University of Regensburg
11. Bloemendal H, Zweers A (1976) In: Hockwin O (ed) Progress of Lens Biochemistry Research. Doc Ophthalmol Proc Ser, Vol 8. Dr W Junk bv Publishers, The Hague, pp 91–104
12. Yphantis DA (1964) Biochemistry 3:297
13. Durchschlag H, Durchschlag G (1977) Hoppe-Seyler's Z Physiol Chem 358: 228
14. Sacket DL, Wolff J (1987) Anal Biochem 167:228
15. Durchschlag H, Christl P, Jaenicke R (1991) Progr Colloid Polym Sci 86:41
16. Durchschlag H, Jaenicke R (1983) Int J Biol Macromol 5:143
17. Habeeb AFSA (1972) Meth Enzymol 25: 457
18. Thannhauser TW, Konishi Y, Scheraga HA (1987) Meth Enzymol 143:115
19. Borkman RF, McLaughlin J (1995) Photochem Photobiol 62:1046
20. Jaenicke R (1994) Naturwissenschaften 81:423
21. Jaenicke R (1996) FASEB J 10:84

Progr Colloid Polym Sci (1997) 107:102–114
© Steinkopff Verlag 1997

K.-J. Tiefenbach
H. Durchschlag
R. Jaenicke

Sedimentation analysis of SDS and albumin-SDS complexes

Klaus-Jürgen Tiefenbach (✉) · Dr. H.
Durchschlag · Rainer Jaenicke
Institute of Biophysics and Physical
Biochemistry
University of Regensburg
Universitätsstrasse 31
93040 Regensburg
Germany

Abstract Knowledge of the structure of protein–SDS complexes and the constituent free proteins and SDS micelles is important for our understanding of protein–detergent interactions and for the application and improvement of physicochemical techniques connected with the purification and characterization of proteins. Several reasons are responsible for present controversies regarding the size and structure of macrosolutes, e.g. presence of a variety of low- and high-molecular solutes, the fact that macrosolutes show heterogeneity and may exhibit equilibria that are strongly influenced by the environmental conditions, and the point that techniques available for their structural investigation refer to different global or local properties. Sedimentation velocity and equilibrium experiments of the detergent–protein complexes and their constituents were performed under a variety of experimental conditions, such as rotor speed, scanning wavelength, choice of the baseline, the solvent or additives, and the concentrations of the components. Monitoring the sedimentation profiles of micellar SDS is facilitated by labeling the detergent micelles using a fluorescent dye, and discriminating the species under analysis by scanning at specific wavelengths. Interpretation of the results is facilitated by monitoring absorption spectra at discrete radial distances in the centrifuge cells, in addition to spectra recording of mixtures and components outside the centrifuge. In order to estimate the amount of bound detergent, sedimentation data (sedimentation coefficients, particle weights and mass distributions) were complemented by size-exclusion chromatographic studies. In the case of nonreduced proteins about 0.4 g SDS/g protein are bound at low detergent concentrations, and about 0.8 to 1.2 g/g at elevated concentrations.

Key words Sodium dodecyl sulfate – protein–detergent complexes – micelles – analytical ultracentrifugation – absorption spectroscopy – size-exclusion chromatography

Abbreviations AC, ammonium chloride; AF, amplification factor; AUC, analytical ultracentrifugation; BSA, bovine serum albumin; ChA, chymotrypsinogen A; CMC, critical micelle concentration; CMT, critical micelle temperature; DB, dextran blue; HSSE, high-speed sedimentation equilibrium; βLg, β-lactoglobulin; Lys, lysozyme; MyG, myoglobin; NaP, sodium phosphate buffer; NPN, N-phenyl-1-naphthylamine; OvA, ovalbumin; PAGE, polyacrylamide gel electrophoresis; SDS, sodium dodecyl sulfate; SEC, size-exclusion chromatography.

Introduction

Detergents are soluble amphiphiles, possessing nonpolar as well as polar regions [1–11]. In the case of ionic detergents, the polar head groups bear positive or negative charges. Detergents tend to concentrate at surfaces and

Progr Colloid Polym Sci (1997) 107:102–114
© Steinkopff Verlag 1997

interfaces, thereby exhibiting organized assemblies including micelles. At low concentrations, detergents exist as monomers, but above a characteristic limit, the critical micelle concentration (CMC), micellization of detergent molecules occurs. The number of monomers necessary to form a micelle, usually spherical or ellipsoidal, is called the aggregation number n. Both the CMC and the micelle size depend on environmental conditions such as pH, temperature, ionic strength, presence of organic additives, and several impurities. Particularly, the concentration of ions affects ionic detergents and ionic detergent properties in a specific manner. Both CMC and size of the micelles may affect detergent performance. The conditions under which a particular detergent exists as a monomer, micelle, or crystal is commonly described by a temperature-composition phase diagram and a characteristic triple point ("Krafft point"). Micelle formation occurs only when the system is above a critical micelle temperature (CMT). At elevated temperature ("cloud point"), some nonionic detergents may form superaggregates which eventually separate from the solution as a detergent-rich phase.

The characteristic features of anionic detergents such as the alkyl sulfates, especially their pronounced destabilizing effect on proteins, are the reason for their widespread use in the biosciences (e.g. [12, 13]). Especially, sodium dodecyl sulfate (SDS) is applied both for the solubilization and analysis of simple and conjugated proteins: e.g. solubilization and fractionation of membrane, viral and ribosomal proteins; denaturation, dissociation, unfolding and reconstitution studies on many classes of proteins; examination of protein–detergent interactions; characterization of protein–detergent complexes with respect to their structure and binding behavior; mass determination of protein subunits.

The pronounced action of SDS is caused by the special features of this detergent [14–21]. It possesses a relatively high CMC (about 1–8 mM in water or aqueous solutions of low ionic strength [4, 19, 22–27]) and usually binds to proteins as a monomer. Finally, a layer of negatively charged SDS molecules is formed around the polypeptide chain (about 1.4 g SDS/g protein), causing protein dissociation, unfolding, inactivation and denaturation. The exact amount of SDS bound to proteins seems to be dependent on the conditions of protein, detergent and solvent (reduced or nonreduced polypeptide chains, SDS concentration, ionic strength).

Techniques available for the structural analysis of protein–SDS complexes and the constituent free proteins and SDS micelles comprise gel electrophoresis (preferably SDS-PAGE), size-exclusion chromatography (SEC), UV absorption, fluorescence emission, circular dichroism, viscometry, densimetry, analytical ultracentrifugation (AUC), light scattering, X-ray and neutron small-angle scattering,

etc. [27–41]. Though understanding protein–detergent interactions is of importance, and in spite of the fact that a variety of techniques has been applied in the past, the precise size and structure and the behavior of micelles and especially of protein–detergent complexes are still unknown and the discussions are controversial [16, 19, 21, 34, 37, 42–45].

In this study, analytical ultracentrifugation is the preferred tool for the characterization of the size of micellar SDS, free protein and protein-SDS complexes. Mainly serum albumin was chosen as a model protein [46]. Absorption spectroscopy, viscometry, densimetry, and size-exclusion chromatography were used as complementary methods. Data analysis yielded precise molar mass estimates which allowed the determination of the amount of bound detergent. This, however, required the development of new strategies in order to avoid artifacts encountered with the ultracentrifugal and chromatographic analysis of labile and heterogeneous systems.

Materials and methods

Bovine serum albumin (BSA), chicken egg-white ovalbumin (OvA), bovine β-lactoglobulin (βLg), bovine chymotrypsinogen A (ChA), sperm-whale myoglobin (MyG), and chicken egg-white lysozyme (Lys) as well as SDS were obtained from Serva (Heidelberg), and N-phenyl-1-naphthylamine (NPN) was obtained from Sigma (Munich); all other reagents were of analytical grade.

Proteins (initial concentration $c_0 = 0.5$–5 mg/ml) and detergent ($c = 0$–10%) were dissolved in bidistilled water or aqueous solutions containing 100 mM sodium phosphate pH 7.0 (NaP, $I \approx 0.2$ mol/l) or NaCl ($I = 0$–0.4 mol/l) for modulation of the ionic strength, and the fluorescent dye NPN for labeling (primarily micellar) SDS [25]. The concentration of SDS was determined by complexing the detergent with methylene blue [15].

Sedimentation velocity and equilibrium experiments were performed in a Beckman model E analytical ultracentrifuge equipped with a high-sensitivity photoelectric scanner and multiplexer system and a 10 in recorder; runs were performed in a six-hole rotor (AnG), using 12 mm double sector cells (charcoal-filled epon) and sapphire windows. Densities and viscosities of solvent were determined by means of a Paar digital density meter (DMA 02) and an Ostwald viscometer (5 ml, flow time for water about 320 s). Absorption spectra were recorded in a Perkin-Elmer Lambda 5 spectrophotometer. For size-exclusion chromatography (SEC), a Sephacryl S-300 High Resolution column (Pharmacia Biotech) was applied. Investigations were performed at 20–25 °C.

104

K.-J. Tiefenbach et al.
Sedimentation analysis of SDS and albumin-SDS complexes

AUC experiments made use of sedimentation velocity and equilibrium runs, allowing the characterization of all macrosolutes (proteins, detergent micelles, protein–detergent complexes) in terms of sedimentation coefficients and molar masses of the macromolecular species, as well as quantities derived therefrom (e.g. parameters characterizing detergent binding or overall shape) [41, 47–55]. Apparent sedimentation coefficients (s_{app}) were obtained from $\ln r$ vs. t plots; they were converted to standard conditions ($s_{20,w}$) making use of corrections for solvent viscosity and density as well as the partial specific volume of the macrosolute. Molar masses (M) were determined by high-speed sedimentation equilibrium (HSSE), using the meniscus depletion technique [50].

Experimental values of partial specific volumes of detergent and proteins were taken from the literature [56]. The volumes of protein–detergent complexes were calculated on the basis of the volume increments of the constituents (protein, detergent) and the conjectural weight fractions, thereby, however, neglecting preferential interactions.

Special problems with micellar detergents and protein–detergent complexes

The elucidation of the size and structure of detergent micelles and especially of protein–detergent complexes by means of AUC and SEC may be affected by various problems and pitfalls [41]:

(i) In general, investigations are performed in multicomponent systems consisting of several low-molecular and macromolecular solutes: low-molecular components such as buffer substances, salts, organic additives and dye labels, monomeric and micellar detergents, macromolecular proteins and protein–detergent complexes (including proteolytic fragments, crosslinks, aggregates and their complexes), in addition to unavoidable impurities, especially in the detergents.

(ii) Both detergent micelles and protein–detergent complexes may exhibit structural diversity and sensitive equilibria; both are influenced by environmental conditions, e.g. the concentration of all components, ionic strength, temperature, pressure. Evidently, the occurrence of heterogeneity prevents the analysis of structural details, so that the interpretation of the results is often ambiguous.

(iii) In the case of small proteins, the size of the protein–detergent complexes and the detergent micelles is of the same order of magnitude. This leads to mean values of the observed parameters and prevents a de-

tailed separation of the observed effects. This serious obstacle holds for both AUC and SEC experiments.

(iv) In heterogeneous systems with macromolecules of different sizes (micellar detergents on the one side and protein–detergent complexes on the other), the situation is less problematic for both types of experiments. However, AUC data analysis in terms of detailed mass estimates requires time-consuming experiments at different rotor speeds to separate all macrosolutes in solution; choice of a single, medium rotor speed turns out to be a rather unsuccessful compromise. Particle separation by SEC is optimal if the particles differ in size. However, a detailed interpretation needs the complementary determination of the detergent content.

(v) While proteins show pronounced absorbance in the UV range, the absorption properties of detergents such as SDS are rather moderate. This makes the detection of detergent micelles in the absence of proteins much more difficult that that of their complexes.

Despite obvious experimental deficiencies, some suggestions how to minimize errors and improve data analysis can be given:

(i) In order to discriminate between particles of different sizes and absorption behavior, and to study protein–detergent interactions in more detail, AUC runs need to be performed under different experimental conditions, e.g. different rotor speeds, scanning wavelengths, and concentrations of all solution components. In this context modern computer-aided evaluation methods and presentation techniques may help to process large amounts of data.

(ii) The correct choice of the baseline in AUC experiments is a necessary prerequisite for distinguishing between monomeric and micellar detergent molecules. To avoid erroneous baselines due to sedimentation of macrosolutes, the reference cell must be devoid of SDS micelles and proteins. Even if low detergent concentrations are present in the reference cell, small amounts of micelles may be formed as a consequence of the accumulation of detergent molecules near the cell bottom.

(iii) Labeling the detergent micelles may help to monitor micelles and to discriminate between the species to be analyzed by scanning at specific wavelengths. In the case of SDS, NPN labels SDS micelles more or less specifically. However, when using this fluorescent dye also for absorption measurements, higher concentrations of the label are required than for fluorescence studies. Since the dye seems to be incorporated into the micelles, labeling may interfere with micelle formation.

Progr Colloid Polym Sci (1997) 107:102–114
© Steinkopff Verlag 1997

(iv) In addition to the spectral characterization of the solutions and constituents before and after centrifugation, absorption spectra may also be monitored at selected positions of the AUC cells during a HSSE experiment, thereby relating mass estimates and definite particle species.

(v) If the free detergent concentration used for complexing proteins is slightly below the CMC, this may help to avoid heterogeneity caused by detergent micelles. In this context, however, it has to be noted that detergents bind to both the protein and the label; thus, the initial detergent concentration may be quite different from the free detergent concentration. Varying the protein and label concentrations at a constant detergent concentration causes different amounts of free detergent. Only the latter is responsible for the formation of protein-free micelles. Nevertheless, the saturation of the protein with detergent molecules is generally adequate, because detergents like SDS exhibit a high affinity of the detergent monomers to the protein moiety.

(vi) To avoid interference with crystal formation of detergents, the applied temperature has to be above the CMT. In the case of SDS, the CMT may be near 20 °C [57], its exact value, however, depending on the solvent conditions and the presence of impurities.

(vii) In order to prevent disturbancies due to insufficient detergent binding, the mixtures of proteins and detergents have to be incubated for some time (> 24 h at 20–25 °C). In the case of SEC experiments, the chromatographic columns have also to be equilibrated with buffer containing sufficient amounts of detergent (slightly below the CMC).

Results and discussion

This study is concerned with experiments on SDS micelles, native BSA and various BSA–SDS complexes, in addition to some experiments on other protein–SDS complexes, under various experimental conditions including a variety of pilot tests on how to perform such investigations efficaciously.

Absorption spectroscopy

The fluorescence intensity of the hydrophobic amine NPN has been exploited in the past in order to determine the CMC of SDS [25, 40]. Upon micelle formation, the fluorescence intensity increases precisely at the CMC, indicating that the fluorescent dye binds to the micellar detergent, presumably by inclusion into the micelle. In connection

with the use of the AUC absorption optics, the question was whether this dye can also be used as a label to increase the marginal absorbance of SDS micelles in the UV range [41].

Two conclusions may be drawn from the UV absorption spectra presented in Fig. 1: (i) There is a pronounced enhancement of the UV absorption of SDS solutions when using a concentration of 0.1 mM NPN. This concentration, however, is higher by a factor of 100 than commonly applied in fluorescence studies [25, 40]. (ii) There is a clear difference in the spectra of monomeric and micellar SDS: the spectra at SDS concentrations above the CMC exhibit distinct maxima at 250 and 346 nm, apart from the maxima in the far-UV region. In the present context the max-

Fig. 1 Absorption spectra of SDS solutions of different concentration (in NaP) in the absence (A) or presence (B) of 0.1 mM NPN. The inset shows the absorption of SDS–NPN solutions at 346 nm and varying SDS concentrations, measured against a SDS-free NPN solution

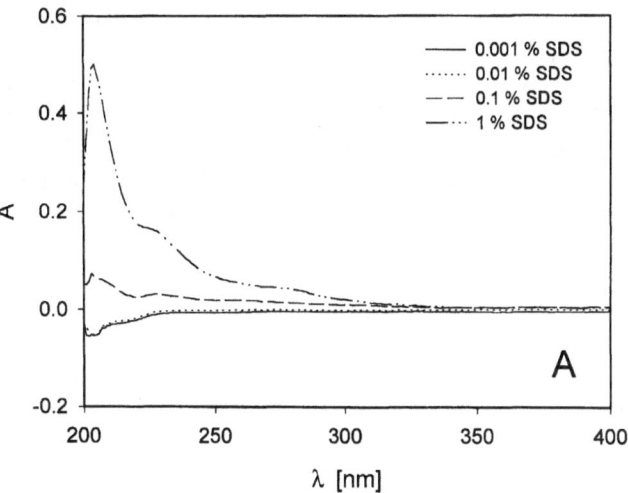

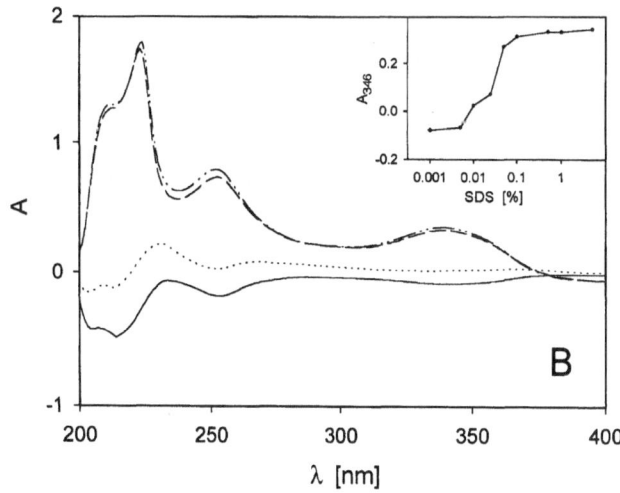

106
K.-J. Tiefenbach et al.
Sedimentation analysis of SDS and albumin-SDS complexes

imum at 346 nm is of special interest, because it is located outside the range of protein absorption.

The above findings indicate that absorption spectroscopy may be used as an additional means to roughly determine the CMC (cf. insert in Fig. 1B). Quantitative determination, however, is complicated by the limited solubility of NPN in detergent-free solution, and because difference spectra are affected by different amounts of free NPN in the cuvettes. Most importantly, the spectral behavior of labeled SDS micelles may be exploited in AUC experiments for monitoring SDS micelles, which otherwise exhibit only weak absorption.

In principle, the spectral characteristics of NPN-labeled micellar SDS may also be used for monitoring protein–SDS micelles, as may be concluded from the spectra in Fig. 2. In addition to the protein maximum at 280 nm, the spectra of the protein–SDS–NPN complexes show again pronounced maxima at 250 and 346 nm. Apart from the possibility to discriminate between different species, however, there is no real need for such a procedure, because proteins themselves absorb UV light appropriately.

Taken together, optimum scanning conditions are: (i) 250 or 346 nm for SDS micelles + NPN, (ii) 280 nm for proteins in the absence or presence of low or high SDS concentrations, and (iii) 250 or 280 or 346 nm for proteins in the presence of high SDS concentrations + NPN.

Analytical ultracentrifugation

Analytical ultracentrifugation yielded apparent sedimentation coefficients as well as masses and mass distributions of the macrosolutes. The results are strongly influenced by the environmental conditions such as detergent concentration, ionic strength, and presence of additives/reductants favoring protein unfolding and dissociation. Monitoring absorption spectra in the centrifuge cell, e.g. during an equilibrium run, allows the clear distinction between different macrosolutes.

Sedimentation velocity

The sedimentation coefficients of SDS micelles show a pronounced concentration dependence of both s_{app} and $s_{20,w}$, most pronounced at low detergent concentrations (Fig. 3). The sedimentation coefficient $s_{20,w}$ in NaP ($I \approx 0.2$ mol/l) decreases from about 1.9 to 1.4 S with increasing SDS concentration. If the mass of the micelles remains constant, this might be explained by a change in the shape of the micelles and/or altered interparticular interactions. Since parallel HSSE experiments, however, demonstrate a concomitant decrease in the mass of the micelle, such conclusions may fail.

The sedimentation coefficients observed for the BSA–SDS complexes reveal a strong dependence on the amount of added detergent ($4.3 \rightarrow 2.8$ S) (Fig. 3). However, the concentration dependence is quite different from the one of the SDS micelles in the absence of protein. Even at very low SDS concentrations there is a strong decrease of $s_{20,w}$, obviously caused by binding of the detergent to the protein as indicated by HSSE experiments. This is followed by (partial) protein unfolding at elevated detergent concentrations.

Fig. 2 Absorption spectra of BSA–SDS–NPN complexes and of the free components BSA (1 mg/ml), SDS (below and above its CMC), and NPN, in NaP

Fig. 3 Concentration dependence of s_{app} and $s_{20,w}$ of SDS micelles and BSA–SDS complexes (1 mg/ml BSA) in NaP. The micelles were monitored in the presence of 1 μM NPN

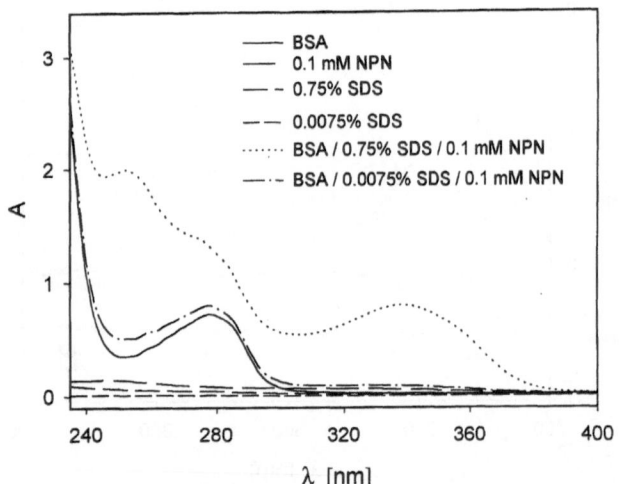

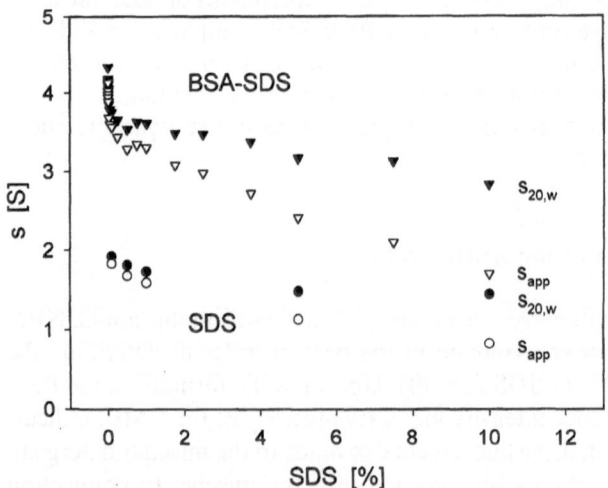

In general, the $s_{20,w}$ values of substantially liganded proteins are low compared to the s-values of the respective native proteins, a behavior obviously caused by increased particle anisotropy. If the solutions to be analyzed by AUC contain both protein–SDS complexes and SDS micelles (Fig. 4), the analysis is perturbed if both species are of similar size. This holds for the proteins of low molar mass. The extent of changes in s is also dependent on the pretreatment of the proteins; reduced proteins exhibit more pronounced detergent binding than their non-reduced counterparts.

Fig. 4 Sedimentation coefficients, s_{app} and $s_{20,w}$, of selected native proteins and their protein–SDS complexes (0.1% SDS) in NaP. Proteins were applied at $c_0 = 0.5$ mg/ml: BSA, bovine serum albumin; OvA, ovalbumin; βLg, β-lactoglobulin; ChA, chymotrypsinogen A; MyG, myoglobin; Lys, lysozyme

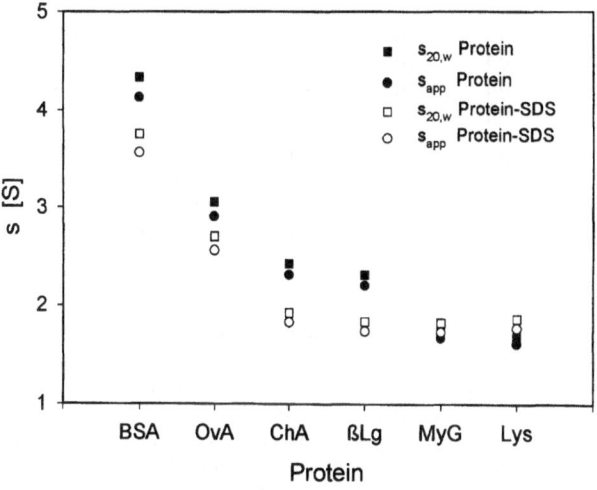

Fig. 5 Sedimentation coefficients, s_{app} and $s_{20,w}$, of SDS micelles (0.1% SDS + 0.1 mM NPN) and BSA–SDS complexes (1 mg/ml BSA + 0.1% SDS + 0.1 mM NPN) in NaCl solutions of varying ionic strength

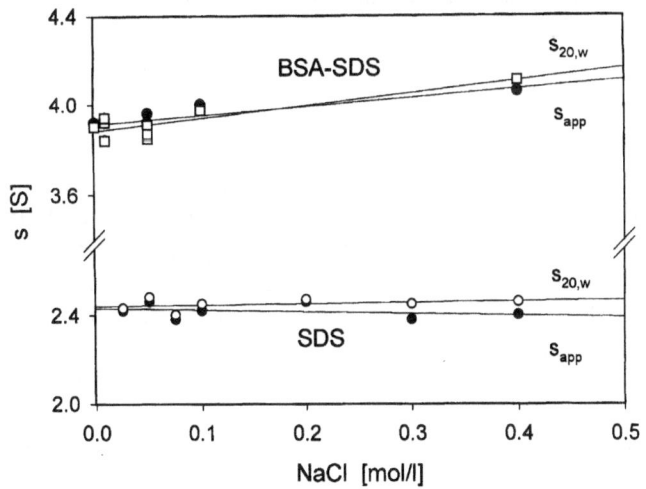

Determination of sedimentation coefficients of protein–SDS complexes in aqueous solutions of NaCl shows a significant increase of both s_{app} and $s_{20,w}$ upon increasing salt concentration (Fig. 5). This behavior is obviously dominated by the peculiarities of the protein, since a similar experiment with SDS micelles in the absence of the protein does not exhibit significant changes in s. The absolute values of micellar SDS in the presence of NaCl are higher than in the absence of this salt (for $c = 0.1\%$ SDS and $I = 0.2$ M: 2.4 vs. 1.9 S).

Sedimentation equilibrium

A great variety of HSSE experiments were performed with SDS micelles and protein–SDS complexes. Fig. 6 depicts typical data in A vs. r format for BSA and BSA–SDS complexes in the absence and in the presence of NPN, together with the usual linearizations. Some representative results are summarized in Tables 1 and 2. The comparison of the results obtained under different experimental conditions is facilitated by presenting the results in 3D-plots (Figs. 7 and 8).

An inspection of the mass estimates for SDS micelles (Table 1) reveals no significant changes of the molar mass obtained under differing conditions. The masses obtained at concentrations between 0.01 and 10% SDS vary between 34 and 20 kg/mol, with error limits of about ± 4 kg/mol; the masses correspond to aggregation numbers between 120 and 70. The value of about 24 kg/mol and the respective aggregation number of about 80, observed at intermediate SDS concentrations (0.1–1%), is in accord with literature data, mainly obtained by light scattering experiments [58, 59]. The values at low SDS concentrations point to slightly higher masses, similar to the behavior in the presence of NaCl. No significant differences were found for the mass estimates in the absence or presence of 1 μM or 0.1 mM NPN.

Equilibrium runs with native BSA demonstrate the high precision of such types of experiments. At 16 000 to 26 000 rpm, masses of 66 ± 1 kg/mol are obtained (Table 2, Fig. 7A), in full accord with amino acid sequence data of 66.3 kg/mol [46]. Because of the presence of several cross-links in BSA, at 10 000 rpm essentially higher molar mass weight-averages are found. The observed masses at this rotor speed are higher at a scanning wavelength of 295 nm as compared to 280 nm. The differences observed at these two wavelengths are caused by different extinction coefficients, which finally enable scanning in closer proximity to the cell bottom [60].

Molar mass determinations of BSA in the presence of low amounts of SDS ($c_0 = 0.001$–0.01%) yield masses corresponding to low detergent binding: < 0.1 g SDS per g

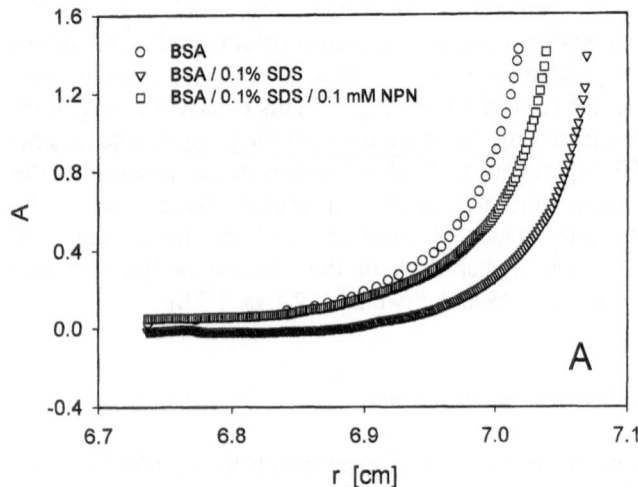

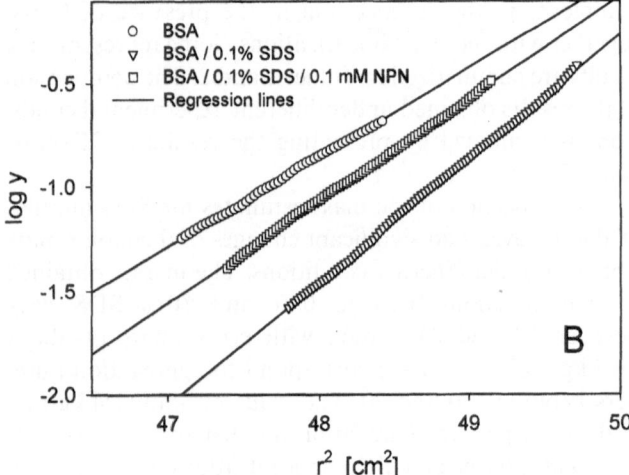

Fig. 6 HSSE of native BSA and BSA–SDS complexes in the absence or presence of NPN in A vs. r (A) and log y vs. r^2 (B) plots (1 mg/ml BSA, 0.1% SDS, 0.1 mM NPN, NaP). Recording of BSA and BSA–SDS complexes was performed at 16000 rpm and 280 nm. Molar masses (in kg/mol): BSA: 67.0 ± 0.8; BSA + 0.1% SDS: 124.7 ± 1.7; BSA + 0.1% SDS + 0.1 mM NPN: 103.1 ± 1.3. Because of their moderate absorbance, SDS micelles in the absence of NPN do not contribute significantly to the obtained mass averages; because of their small size, SDS–NPN micelles do not cause important perturbation at the given rotor speed

Table 1 Molar masses M and aggregation numbers n of micellar SDS in the absence and in the presence of the cosolutes NPN, NaP and NaCl, obtained by HSSE under various conditions of centrifugation and registration. *AUC-conditions*: 16000–40000 rpm, scanning at $\lambda = 280$ and 295 nm, and additionally at 250, 309 and 346 nm in the presence of NPN, AF = 1 or 4

Detergent[a]	Cosolutes			Ultracentrifugation	
SDS [%]	NPN [mM]	NaP [mM]	NaCl [mM]	M[b] [kg/mol]	n[c]
0.01	—	100	—	34.2 ± 8.4	119
0.1	—	100	—	28.7 ± 5.1	100
1.0	—	100	—	20.5 ± 6.0	71
10.0	—	100	—	24.9 ± 5.3	86
0.1	0.001	100	—	24.7 ± 1.1	86
0.5	0.001	100	—	23.8 ± 1.8	83
1.0	0.001	100	—	24.0 ± 2.1	83
5.0	0.001	100	—	25.4 ± 3.6	88
10.0	0.001	100	—	n.e.[d]	
0.01	0.1	100	—	28.3 ± 5.4	98
0.1	0.1	100	—	24.8 ± 1.0	86
1.0	0.1	100	—	22.7 ± 1.7	79
10.0	0.1	100	—	n.e.[d]	
0.1	0.1	—	0–10	n.e.[e]	
0.1	0.1	—	25	34.1 ± 3.8	118
0.1	0.1	—	50	29.1 ± 0.5	101
0.1	0.1	—	75	27.1 ± 0.5	94
0.1	0.1	—	100	28.6 ± 1.3	99
0.1	0.1	—	200	32.1 ± 1.5	111
0.1	0.1	—	300	30.3 ± 0.7	105
0.1	0.1	—	400	35.5 ± 2.3	123

[a] The given amounts for detergent and cosolutes are initial concentrations c_0. Due to the separation process in the centrifuge, the actual concentrations span a wide range: even if c_0 of the detergent does not exceed the CMC, at the bottom of the cell micelles may be formed
[b] Molar mass estimates are based on $\bar{v} = 0.870$ cm³/g for micellar SDS [56]. Different AUC conditions resulted in similar results; the values given in the table represent the averages of all mass estimates; n.e. signifies that the evaluation was not possible
[c] The determination of the aggregation number n is based on the molar mass of monomeric SDS of 288.4 g/mol
[d] Extremely high baseline
[e] At this low ionic strength the CMC is not reached

of protein (Table 2). At elevated SDS concentrations ($\geq 0.1\%$) considerable detergent binding is observed (up to 1.0 g/g for nonreduced BSA). In the case of 0.1% SDS, the results clearly demonstrate that protein–SDS complexes containing similar amounts of SDS are obtained at different rotor speeds (Table 2, Fig. 7B). By contrast, the results found for 1.0% SDS point to protein–SDS complexes containing different amounts of SDS (Table 2, Fig. 7C). However, the latter results may be influenced by two pitfalls: (i) For free SDS concentrations above the

CMC, the mass of the SDS micelles may contribute to the obtained mass averages of the protein–SDS complexes, though they exhibit only marginal absorption properties in the absence of a label (cf. Fig. 1A). (ii) If the reference cell contains appreciable amounts of micellar SDS, this may affect the baseline and may lead to erroneous mass estimates. This example also demonstrates the caution one must observe in calculating molar masses, without performing runs at different rotor speeds.

Table 2 Molar masses M of native (nonreduced) BSA and BSA–SDS complexes and amount of detergent binding δ to the protein in the absence and in the presence of the cosolutes NPN, NaP and NaCl, obtained by HSSE under various conditions of centrifugation and registration. *AUC-conditions*: c_0 of protein = 1 mg/ml, 10 000–26 000 rpm, scanning at $\lambda = 280$ and 295 nm, and additionally at 250, 309 and 346 nm in the presence of NPN, AF = 1 or 4

Detergent[a]	Cosolutes			Ultracentrifugation			
SDS [%]	NPN [mM]	NaP [mM]	NaCl [mM]	Rotor speed [rpm × 10^{-3}]	λ [nm]	$M^{b)}$ [kg/mol]	$\delta^{c)}$ [g/g]
Native BSA: $\bar{v} = 0.735$ cm³/g							
—	—	100	—	26	280, 295	65.3 ± 1.7	
—	—	100	—	22	280, 295	65.9 ± 2.6	
—	—	100	—	16	280, 295	66.8 ± 0.4	
—	—	100	—	10	280, 295	158 ± 59[d]	
—	—	100	—	10	280	107 ± 7[d]	
—	—	100	—	10	295	209 ± 17[d]	
BSA–SDS complexes: $\delta_{calc} = 0.01$ g/g; $\bar{v}_{calc} = 0.736$ cm³/g							
0.001	0.1	100	—	40	250	67.5 ± 1.8	0.02
0.001	0.1	100	—	26	250	60.6 ± 0.8	0.00
0.001	0.1	100	—	16	250	66.7 ± 5.0	0.01
BSA–SDS complexes: $\delta_{calc} = 0.1$ g/g; $\bar{v}_{calc} = 0.747$ cm³/g							
0.01	—	100	—	26	280, 295	66.8 ± 9.4	0.01
0.01	—	100	—	22	280, 295	70.2 ± 3.6	0.06
0.01	—	100	—	16	280, 295	71.0 ± 3.1	0.07
0.01	0.1	100	—	40	250	64.1 ± 1.2	0
0.01	0.1	100	—	26	250	76.2 ± 12.4	0.15
0.01	0.1	100	—	16	250	74.1 ± 5.5	0.12
BSA–SDS complexes: $\delta_{calc} = 0.93$ g/g; $\bar{v}_{calc} = 0.800$ cm³/g							
0.1	—	100	—	26	280, 295	126.5 ± 2.9	0.91
0.1	—	100	—	22	280, 295	132.6 ± 0.5	1.00
0.1	—	100	—	16	280, 295	120.5 ± 7.2	0.82
0.1	0.1	100	—	40	250	36.1 ± 5.1	
0.1	0.1	100	—	26	250	104.8 ± 4.8	
0.1	0.1	100	—	16	250	102.6 ± 1.7	
0.1	0.1	—	1	40	346	97.3 ± 4.1	
0.1	0.1	—	1	26	346	141.4 ± 9.0	
0.1	0.1	—	100	40	280, 346	20.1 ± 1.3	
0.1	0.1	—	100	26	280, 346	40.2 ± 9.2	
0.1	0.1	—	400	40	280, 346	18.0 ± 3.1	
0.1	0.1	—	400	26	280, 346	42.7 ± 1.9	
0.1	0.1	—	400	16	280, 346	48.7 ± 4.6	
1.0	—	100	—	26	280, 295	82 ± 10[e]	0.24
1.0	—	100	—	22	280, 295	87 ± 15[e]	0.31
1.0	—	100	—	16	280, 295	107 ± 3[e]	0.61
1.0	—	100	—	10	280, 295	137 ± 8[e]	1.07
1.0	0.1	100	—	40	250	23.3 ± 2.5	
1.0	0.1	100	—	26	250, 280, 309, 346	24.1 ± 2.2	

[a] The given amounts for detergent and cosolutes are initial concentrations c_0
[b] Molar mass estimates are based on the $\bar{v}$ values calculated for the predicted amounts of detergent binding [14]. If the solutions represent a population of different species, different rotor speeds result in quite different molar mass weight averages; n.e. signifies that the evaluation was not possible
[c] The determination of the amount of detergent binding δ to the protein is based on the molar mass of native BSA of 66.3 kg/mol [46]
[d] Due to crosslinks, BSA always contains a certain amount of dimers and higher oligomers [46]
[e] The reference cell contained 1% SDS

While NPN turned out to be most effective for scanning of the SDS micelles, the presence of significant amounts of NPN in protein–SDS complexes may lead to unexpected disturbancies (cf. Table 2, Figs. 6 and 7D). This especially holds for elevated concentrations of the detergent or the presence of high NaCl concentrations, where the dye may capture SDS molecules, thereby preventing normal SDS binding to the protein. Consequently, in this

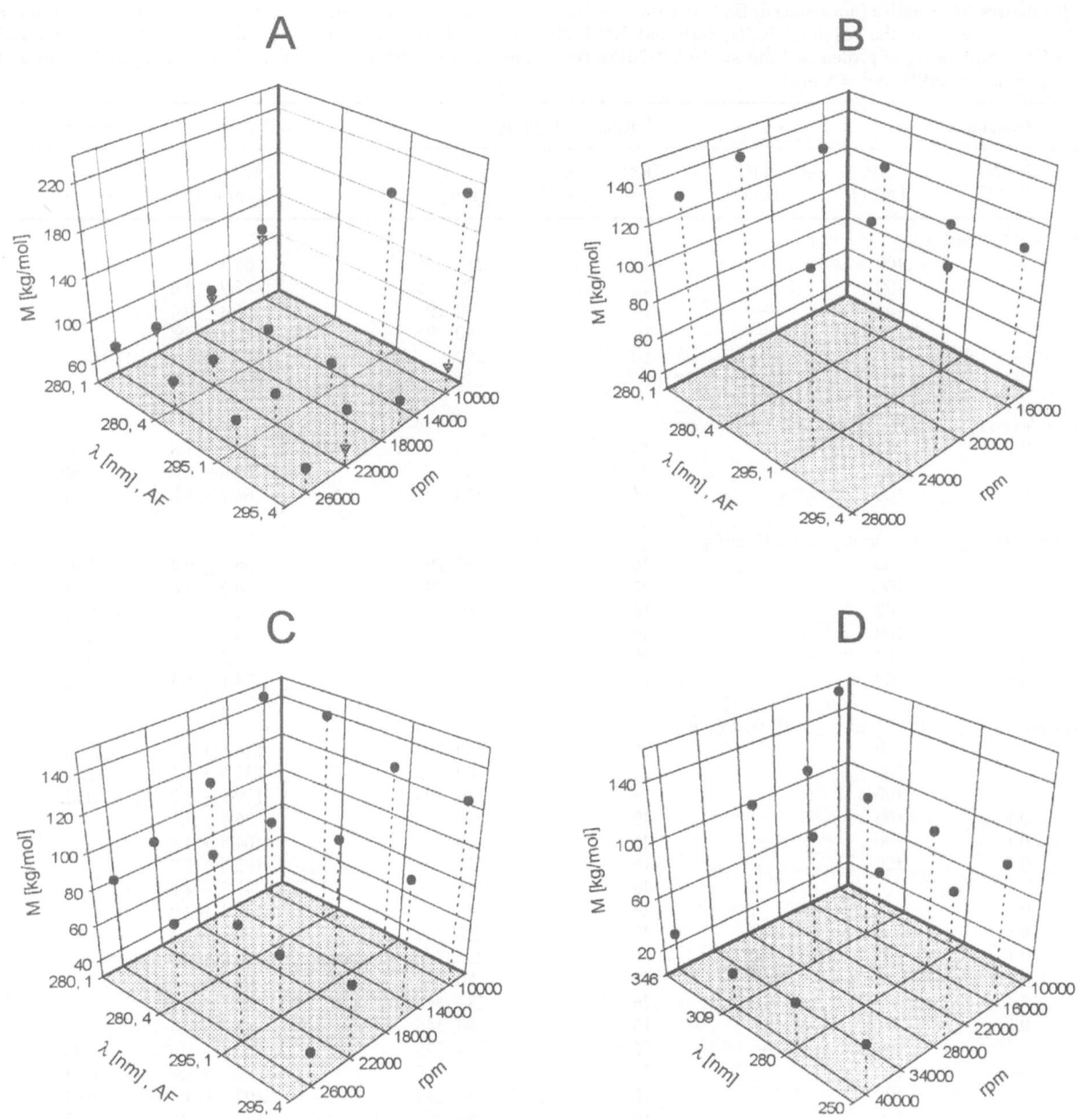

Fig. 7 Apparent molar masses of BSA and BSA–SDS complexes in NaP, as obtained from HSSE experiments at different rotor speeds (rpm), scan wavelengths (λ) and amplifications (AF), presented in 3D-plots. A: 1 mg/ml native BSA, the different symbols ($\bullet$, $\triangledown$) represent the results from scanning at different cell positions; B: 1 mg/ml BSA + 0.1% SDS; C: 1 mg/ml BSA + 1% SDS; D: 1 mg/ml BSA + 0.1% SDS + 0.1 mM NPN

case the SDS–NPN micelles are the preferred target of the scanning procedure, yielding essentially lower masses. In extreme cases, the masses of the micelles (about 20–25 kg/mol) are found.

A similar binding behavior as found for BSA, has also been found for the other proteins, the results are again influenced by the extent of reduction of the protein disulfides prior to SDS incubation. Mass estimates of protein–SDS complexes with low mass are commonly affected by

the simultaneous presence of SDS micelles, thus leading to erroneous binding data.

Recording absorption spectra in the ultracentrifuge

Monitoring absorption spectra inside the centrifuge cell (for example, during an equilibrium run) allows a clear distinction between different macrosolutes. Advantageously

Progr Colloid Polym Sci (1997) 107:102–114
© Steinkopff Verlag 1997

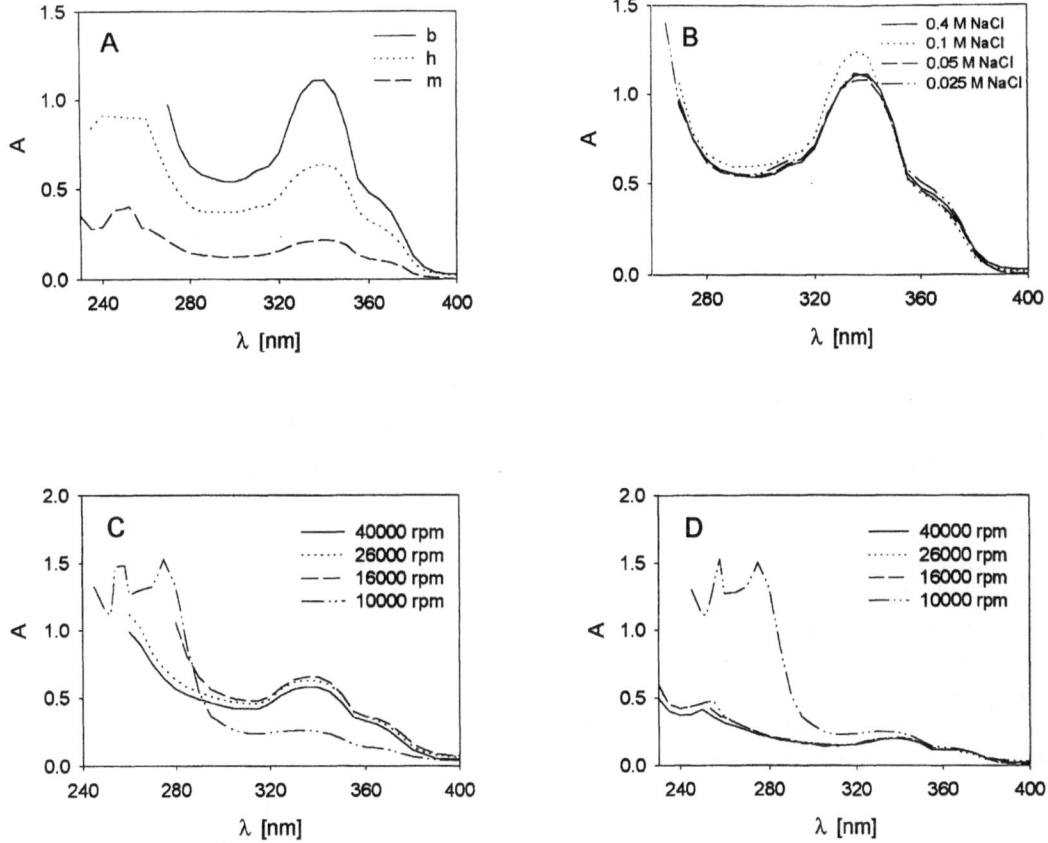

Fig. 8 Absorption spectra of NPN-labeled SDS micelles and BSA– SDS complexes, monitored in the AUC cell during an HSSE experiment. Recording was performed at different rotor speeds near the cell bottom (b), half-way (h) and near the meniscus (m). The substances under analysis can be identified at 346 nm as micellar SDS, and at 280 nm as protein plus SDS. A: 0.1% SDS + 0.1 mM NPN + 0.4 M NaCl, 40 000 rpm, positions b, h, m; B: 0.1% SDS + 0.1 mM NPN + variable amounts of NaCl, 40 000 rpm, position b; C: 1 mg/ml BSA + 1% SDS + 0.1 mM NPN, NaP, variable rpm, position h; D: 1 mg/ml BSA + 1% SDS + 0.1 mM NPN, NaP, variable rpm, position m

this may be performed at different rotor speeds and at selected radial positions. Recording at positions of the concentration gradient near the cell bottom may help to distinguish between different macrosolutes, thereby relating analyzed masses to difinite species. In the case of a flat concentration gradient or occurrence of an irregularly high baseline, recording near the meniscus and at intermediate positions may also be of interest. A comparison of the observed spectra with spectra recorded outside the centrifuge, before or after centrifugation, helps to identify the compounds to be analyzed.

In order to identify micellar SDS or to discriminate between protein, protein–SDS complex and micellar SDS, the use of NPN is required. Fig. 8 shows representative examples of spectra, recorded under varying experimental conditions.

Panel A shows spectra of SDS micelles in the absence of protein, monitored at three different positions. The spectra clearly reflect the concentration distribution of micellar SDS in the cell. The absorbance at 280 nm is low

compared to the maximum at 346 nm. A similar picture is obtained if variable rotor speeds and a fixed position near the cell bottom are used. The presence of variable ionic strength (NaCl) does not result in significant changes of the spectra of micellar SDS, if a constant rotor speed and a fixed position near the cell bottom are chosen (panel B).

In contrast to the spectra characteristics of pure SDS micelles, the spectra of the protein SDS complexes exhibit higher absorbancies at 280 nm compared to the signals at 346 nm. This is demonstrated in panel C for 10 000 and 16 000 rpm at an intermediate cell position; near the meniscus only at the lowest rotor speed considerable amounts of protein are present (panel D). On the other hand, at higher speeds (26 000 and 40 000 rpm) the majority of the large protein–SDS complexes has sedimented to the cell bottom. In these experiments micellar SDS can only be detected clearly at an intermediate cell position and speeds between 16 000 and 40 000 rpm (panel C), while at 10 000 rpm it is distributed over the whole cell.

A comparison of the spectra in the centrifuge (Fig. 8) with those outside (Figs. 1 and 2) shows slight differences. For example, in the spectra recorded in the centrifuge, a shoulder at about 370 nm occurs, which is not present in conventional spectra. This is presumably due to some deficiencies in the model E absorption optics which, however, do not influence the principle conclusions drawn from such experiments.

Size-exclusion chromatography

For SEC experiments of protein–SDS complexes, the column has to be equilibrated with SDS (1 mM). In order to test the amount of detergent that binds to the protein, two kinds of samples were applied to the chromatographic column: native protein and protein preincubated with SDS. Separation of excess of SDS from the protein–SDS complexes by means of SEC, and subsequent determination of the SDS content in the protein fractions, allow the amount of bound detergent to be estimated. In contrast to conductivity measurements, even in the presence of electrolytes, the methylene blue method may be applied successfully to determine the amounts of SDS with high accuracy.

Application of native proteins

Applying native proteins to a column equilibrated with SDS leads to the separation profile shown in Fig. 9A. The sequence of protein elution corresponds to their subunit masses, however, the polypeptide chains elute in a rather narrow range. The concentration of SDS determined in the respective protein fractions corresponds to about 0.4 g SDS per g of protein. This means that the amount of SDS in the buffer causes partial, but not complete binding of the detergent to the protein during the chromatographic run (about 6 h).

Application of protein–SDS complexes

Similar experiments with proteins preincubated with SDS for > 24 h (at room temperature), leads to elution profiles with the proteins distinctly separated (Fig. 9B). Parallel SDS determination reveals that the majority of proteins is saturated with SDS (about 0.8–1.2 g/g, depending on the number of disulfide bonds of the nonreduced proteins). The values for the smaller proteins (myoglobin, lysozyme), however, yield too high values (1.5–1.6 g/g). This latter behavior is obviously caused by the disturbancies of SDS

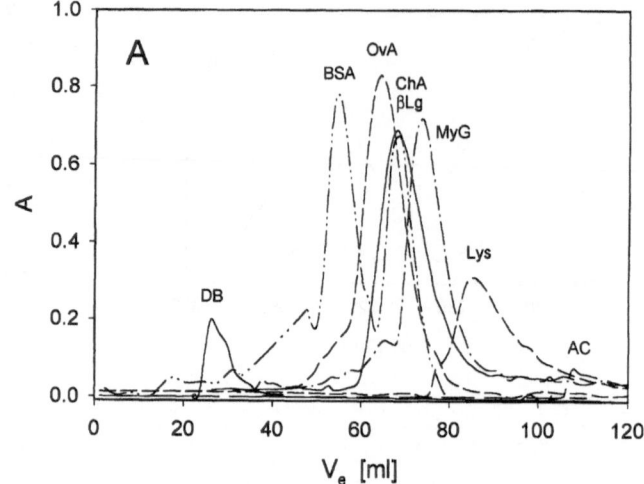

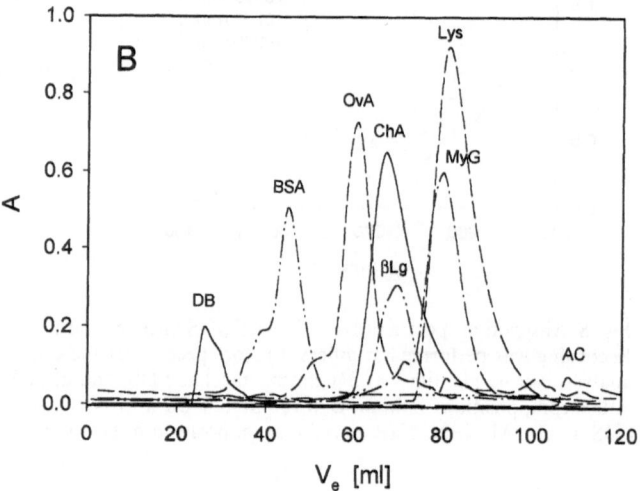

Fig. 9 SEC of protein–SDS complexes, using columns equilibrated with 0.1 M NaP pH 7.0, containing 0.1 M NaCl and 1 mM SDS. A: application of proteins in the native state; B: application of proteins preincubated for > 24 h with SDS (final concentration: 2%). Proteins and protein–SDS complexes were monitored in the eluted fractions at 280 nm. The abscissa gives the elution volume, V_e, in ml. The SDS content was determined using the methylene blue assay (in $CHCl_3$ at 650 nm). Proteins were applied at $c_0 = 3–5$ mg/ml: BSA, bovine serum albumin; OvA, ovalbumin; βLg, β-lactoglobulin; ChA, chymotrypsinogen A; MyG, myoglobin; Lys, lysozyme. References at the extremes of the separation range: DB, dextran blue; AC, ammonium chloride

micelles, which possess masses similar to the protein–SDS complexes of the small proteins.

Conclusions

In summarizing the above results, we may conclude that irrespective of a series of problems the investigation of

Progr Colloid Polym Sci (1997) 107:102–114
© Steinkopff Verlag 1997

SDS micelles and protein–SDS complexes by hydrodynamic methods is successful. The determination of both sedimentation coefficients and molar masses of the macrosolutes under consideration yields reproducible results, provided some precautions are taken into account. Combined with the analysis of the detergent-free proteins, changes of the overall structure upon detergent binding may be determined. Low detergent concentrations only cause moderate binding (about 0.4 g/g, presumably binding to ionic sites of the protein surface); they do not affect the native overall structure significantly. At elevated detergent concentrations, the electrostatic repulsion of charged detergent molecules initiates progressive protein unfolding and binding to initially buried hydrophobic regions. The hydrodynamic parameters now indicate appreciable protein–detergent complex formation (about 0.8–1.2 g/g in the case of nonreduced proteins, and about 1.4 g/g with reduced proteins) as well as alterations of the tertiary and quaternary structure. The occurrence of two binding steps is in accord with considerations of Tanford [5] and Takagi et al. [19].

References

1. Helenius A, Simons K (1975) Biochim Biophys Acta 415:29–79
2. Israelachvili JN, Mitchell DJ, Ninham BW (1976) J Chem Soc Faraday Trans II 72:1525–1568
3. Tanford C, Reynolds JA (1976) Biochim Biophys Acta 457:133–170
4. Helenius A, McCaslin DR, Fries E, Tanford C (1979) Meth Enzymol 56:734–749
5. Tanford C (1980) The Hydrophobic Effect: Formation of Micelles and Biological Membranes, 2nd ed. Wiley, New York
6. Birdi KS (1985) Progr Colloid Polym Sci 70:23–29
7. Hoffmann H, Ulbricht W (1986) In: Hinz H-J (ed) Thermodynamic Data for Biochemistry and Biotechnology. Springer, Berlin, pp 297–348
8. Luisi PL, Magid LJ (1986) CRC Crit Rev Biochem 20:409–474
9. Israelachvili JN (1992) Intermolecular and Surface Forces, 2nd ed. Academic Press, London
10. Myers D (1991) Surfaces, Interfaces, and Colloids: Principles and Applications. VCH, Weinheim
11. Moroi Y (1992) Micelles: Theoretical and Applied Aspects. Plenum Press, New York
12. Tanford C (1972) J Mol Biol 67:59–74
13. Creighton TE (1993) Proteins: Structures and Molecular Properties, 2nd ed. WH Freeman, New York
14. Pitt-Rivers R, Impiombato FSA (1968) Biochem J 109:825–830
15. Reynolds JA, Tanford C (1970) Proc Natl Acad Sci USA 66:1002–1007
16. Reynolds JA, Tanford C (1970) J Biol Chem 245:5161–5165
17. Nelson CA (1971) J Biol Chem 246:3895–3901
18. Tanford C, Nozaki Y, Reynolds JA, Makino S (1974) Biochemistry 13:2369–2376
19. Takagi T, Tsujii K, Shirahama K (1975) J Biochem 77:939–947
20. Maddy AH (1976) J Theor Biol 62:315–326
21. Makino S (1979) Adv Biophys 12:131–184
22. Mukerjee P, Mysels KJ (1971) Critical Micelle Concentrations of Aqueous Surfactant Systems. Nat Stand Ref Data Ser, Nat Bur Stand (USA), NSRDS-NBS 36, Washington DC
23. De Vendittis E, Palumbo G, Parlato G, Bocchini V (1981) Anal Biochem 115:278–286
24. Chattopadhyay A, London E (1984) Anal Biochem 139:408–412
25. Brito RMM, Vaz WLC (1986) Anal Biochem 152:250–255
26. Samsonoff C, Daily J, Almog R, Berns DS (1986) J Colloid Interface Sci 109:325–329
27. Sjöberg B, Pap S, Kjems J (1987) Eur J Biochem 162:259–264
28. Fish WW, Reynolds JA, Tanford C (1970) J Biol Chem 245:5166–5168
29. Nozaki Y, Schechter NM, Reynolds JA, Tanford C (1976) Biochemistry 15:3884–3890
30. Takagi T, Miyake J, Nashima T (1980) Biochim Biophys Acta 626:5–14
31. Kameyama K, Nakae T, Takagi T (1982) Biochim Biophys Acta 706:19–26
32. Jones MN, Midgley PJW (1984) Biochem J 219:875–881
33. Davis A (1984) In: Venter JC, Harrison LC (eds) Molecular and Chemical Characterization of Membrane Receptors. Alan R Liss, New York, pp 161–178
34. Ibel K, May RP, Kirschner K, Szadkowski H, Mascher E, Lundahl P (1990) Eur J Biochem 190:311–318
35. Durchschlag H, Christl P, Jaenicke R (1991) Progr Colloid Polym Sci 86:41–56
36. Hirai M, Kawai-Hirai R, Hirai T, Ueki T (1993) Eur J Biochem 215:55–61
37. Ibel K, May RP, Sandberg M, Mascher E, Greijer E, Lundahl P (1994) Biophys Chem 53:77–84
38. Durchschlag H, Binder S, Christl P, Jaenicke R (1994) Jorn Com Esp Deterg 25:407–422 and Anexo 26–27
39. Durchschlag H, Zipper P (1995) Jorn Com Esp Deterg 26:275–292
40. Durchschlag H, Weber R, Jaenicke R (1996) In: Proc 4th World Surfactants Congress, Vol. 1. AEPSAT, Barcelona, pp 519–534
41. Durchschlag H, Tiefenbach K-J, Jaenicke R (1997) Jorn Com Esp Deterg 27:185–196 and Anexo 35–36
42. Shirahama K, Tsujii K, Takagi T (1974) J Biochem 75:309–319
43. Mattice WL, Riser JM, Clark DS (1976) Biochemistry 15:4264–4272
44. Lundahl P, Greijer E, Sandberg M, Cardell S, Eriksson K-O (1986) Biochim Biophys Acta 873:20–26
45. Muga A, Arrondo JLR, Bellon T, Sancho J, Bernabeu C (1993) Arch Biochem Biophys 300:451–457
46. Peters T Jr (1985) Adv Prot Chem 37:161–245
47. Chervenka CH (1973) A Manual of Methods for the Analytical Ultracentrifuge. Spinco Division of Beckman Instruments, Palo Alto
48. Harding SE, Rowe AJ, Horton JC (eds) (1992) Analytical Ultracentrifugation in Biochemistry and Polymer Science. Royal Society of Chemistry, Cambridge (UK)
49. Schuster TM, Laue TM (eds) (1994) Modern Analytical Ultracentrifugation. Birkhäuser, Boston
50. Yphantis DA (1964) Biochemistry 3:297–317
51. Reynolds JA, Tanford C (1976) Proc Natl Acad Sci USA 73:4467–4470
52. Durchschlag H, Jaenicke R (1983) Int J Biol Macromol 5:143–148
53. Reynolds JA, McCaslin DR (1985) Meth Enzymol 117:41–53

114

K.-J. Tiefenbach et al.
Sedimentation analysis of SDS and albumin-SDS complexes

54. Schubert D, Schuck P (1991) Progr Colloid Polym Sci 86:12–22
55. Roxby RW (1992) In: Harding SE, Rowe AJ, Horton JC (eds) Analytical Ultracentrifugation in Biochemistry and Polymer Science. Royal Society of Chemistry. Cambridge (UK) pp 609–618
56. Durchschlag H (1986) In: Hinz H-J (ed) Thermodynamic Data for Biochemistry and Biotechnology. Springer, Berlin, pp 45–128
57. Becker R, Helenius A, Simons K (1975) Biochemistry 14:1835–1841
58. Mysels KJ, Princen LH (1959) J Phys Chem 63:1696–1700
59. Anacker EW, Rush RM, Johnson JS (1964) J Phys Chem 68:81–93
60. Fochler C, Durchschlag H (1997) Progr Colloid Polym Sci, this volume

Progr Colloid Polym Sci (1997) 107:115–121
© Steinkopff Verlag 1997

H. Wendt
R.M. Thomas

The self-association of basic helix–loop–helix peptides

H. Wendt
Department of Biological Chemistry
and Molecular Pharmacology
Harvard Medical School
240 Longwood Avenue
Boston, Massachusetts 02115, USA

Dr. R.M. Thomas (✉)
Institut für Polymere
ETH-Zentrum
8092 Zürich, Switzerland
E-mail: rthomas@ifp.mat.ethz.ch

Abstract As part of a study into the homo- and hetero-oligomerization properties of muscle-specific transcriptional factors, and their interation with DNA, sedimentation equilibrium studies, accompanied by circular dichroism measurements, have been made on peptides derived from the helix–loop–helix regions of MyoD and E47. In addition, a chimeric peptide, in which residues from the loop region of E47 were substituted into that of MyoD, a fluorescently labelled derivative of the MyoD–bHLH peptide and a disulphide crosslinked version of MyoD–bHLH have also been investigated. MyoD–bHLH has been found to form a monomer–tetramer equilibrium in the μM concentration range, while E47–bHLH exists as a highly associated dimer. The MyoD–bHLH derivatives appear to exhibit the same oligomerization behavior as their MyoD–bHLH parent. CD studies of the disulphide-crosslinked peptide show that a level of organization higher than that of the dimer is required for structural stability in the MyoD–bHLH system. The rôle of self-association in the context of the biological function of these proteins is discussed.

Key words Helix–loop–helix peptides – self-association – sedimentation equilibrium – circular dichroism

Introduction

The techniques of analytical ultracentrifugation are of central importance to the investigation and understanding of macromolecular interactions. Sedimentation equilibrium analysis provides the only method for the determination of interaction stoichiometry and affinity under the conditions of low sample concentration in aqueous solution that are often biologically relevant. Studies on the self-association of peptide systems using modern equipment and computational techniques have become increasingly successful (see, e.g., [1, 2] and references therein). Here, we describe peptides that not only self-associate but also form heterogenous associations and, in both states, interact with DNA. This study is concerned with the initial description of the nature of the homomeric association of the DNA-binding regions of a family of transcription factors. These proteins, which all rely on the basic-helix–loop–helix (bHLH) structural motif for their interaction with DNA, are exemplified here by two family members, E47 and MyoD. Complementary measurements, made with CD spectroscopy reinforce some of the conclusions on the nature of the oligomerization arrived at from the sedimentation equilibrium results.

bHLH transcription factors regulate key steps in early development by selective dimerization and binding to regulatory DNA sites. They regulate transcription in a variety of tissues including muscle [3, 4], nervous tissue [5, 6] and blood cell lineages. In general, a heterodimer of a tissue-specific and a widely expressed bHLH protein, known as an E-protein, performs the biological role of these factors. A well characterized example is MyoD, a master regulator of muscle development [7–9]. Binding of a heterodimer formed by MyoD and the E-protein, E47, to regulatory DNA sequences activates transcription of specific genes in muscle development [10–12]. The dimerization is mediated by the helix–loop–helix motif, a four-helix bundle that positions the N-terminal basic region in the major groove of DNA. A minimal fragment of MyoD spanning the 60 residues which comprise the bHLH domain, is sufficient to induce myogenic differentiation of cultured fibroblasts. Complexed with DNA, the separate bHLH domains of MyoD and E47 each form homodimers with very similar structures [13,14]. However, consistent with *in vivo* observations, mixing of MyoD, E47 and a specific DNA site at equimolar concentrations results in, predominantly, heterodimer formation. In addition, despite their similar structures in the DNA complexes, the stabilities of the MyoD and E47 protein dimers are very different [15, Wendt and Ellenburger, unpublished results]. This has led to a number of studies on the oligomerization behavior of different bHLH proteins, using the full-length proteins as well as the isolated bHLH domains. Here we report on an investigation into the MyoD–bHLH peptide and several of its derivatives, and on the E47–bHLH peptide.

Materials and methods

bHLH domains

Overexpression and purification

The MyoD and E47 bHLH domains were produced in *E. coli* from the pRSET T7 expression vector (Invitrogen, Inc.) using the expression host BL21(DE3) pLysS [17]. The MyoD–bHLH peptide used in our studies was identical to that used for the crystal structure determination [13], including the Met–Glu–Leu– sequence at its N-terminus and the substitution of Cys_{135} with serine. In addition, the MyoD peptide was appended with the C-terminal sequence Gln–Gly–Gly–Cys–COOH (see below). The E47–bHLH expression construct is the one used for the crystal structure. The bHLH peptides were purified from *E. coli* cultures as described previously for E47 [14]. Mass spectrometry of the purified bHLH domains confirmed that they comprised the residues shown in Fig. 1. Purified peptides were quantitated by amino acid analysis, and a molar absorptivity of $1434 \, M^{-1} \, cm^{-1}$ at 280 nm was calculated for MyoD. Because E47 lacks Tyr and Trp residues, protein concentrations were routinely determined by a Bradford protein assay that had been calibrated with concentrations of E47 determined by amino acid analysis.

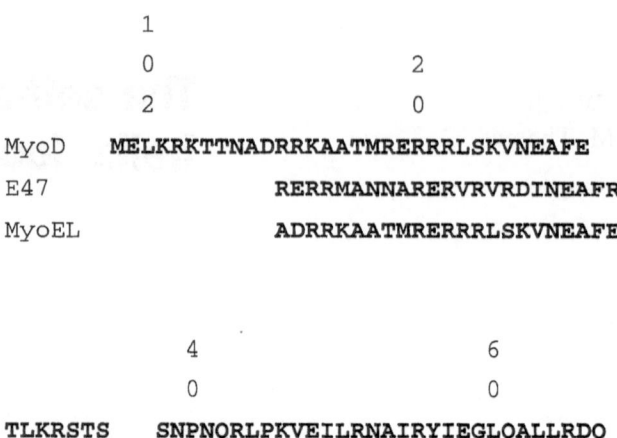

Fig. 1 The amino acid sequences of the peptides used in this study. The sequences are aligned with that of MyoD and are numbered accordingly. MyoDGGC has the same sequence as MyoD–bHLH with the C-terminal extension -GGC. (MyoDGGC)₂ is formed by the oxidation of this peptide, while in fMyoD the cysteine residue is labelled with fluorescein

Generation of bHLH mutants

MyoD–bHLH was appended with the sequence –Gly–Gly–Cys–COOH in order to facilitate labeling of the C-terminus with thiol-specific reagents. Nucleotides encoding these residues were introduced by polymerase chain reaction (PCR) using appropriately modified 3'-primers. PCR products were inserted between the NdeI and BamH1 restriction sites of the expression vector pRSET, and the resulting recombinant DNAs were sequenced. A chimeric MyoD–bHLH incorporating residues from the E47 loop and named MyoEL, was generated by PCR in two steps. First, the C-terminal mutations were introduced by PCR with a mutagenic 3'-primer. This reaction product then served as template for a second PCR reaction using a 5' mutagenic primer that extended to a unique StuI site near the 5'-end of the bHLH domain. A StuI-BamH1 fragment of the resulting PCR product was ligated into the corresponding position of the wild-type MyoD bHLH expression construct. The resulting MyoEL peptide lacks a C-terminal Gly–Gly–Cys extension and it was purified by methods used for the wild-type bHLH peptides.

Modification of C-terminal cysteine residues

Four milligrams of protein were dissolved in 6 ml of 6 M guanidinium hydrochloride in 50 mM phosphate buffer (pH 7.4)/10% glycerol/1 mM DTT, yielding a peptide concentration of 90 μM at pH 5.5. The mixture was stirred at room temperature for 45 min, the reaction was then adjusted to pH 8 and iodoacetamidofluorescein (IAF) or iodoacetamide was added at a calculated concentration of 2 mM. However, IAF is sparingly soluble and it continued to dissolve during the course of the reaction. The mixture was stirred at 37 °C in the dark for 2 h. The labeling reaction was stopped by the addition of 6 mM DTT and the solution was concentrated in a filtration device (Macrosep 3 K, Filtron Technology Corporation). Excess IAF or iodoacetamide was removed by gel filtration of the labeled protein on a Sephadex G25 column, equilibrated in reaction buffer at pH 5.5. Oxidation of the terminal cysteines was achieved by stirring the proteins in a phosphate buffer pH 8.0 for 2 h at 41°C. Final purification of all C-terminally modified bHLH peptides was carried out by reverse-phase high performance liquid chromatography using a semipreparative C18 column (Waters), eluted with binary gradients of acetonitrile/H_2O containing 0.1% trifluoroacetic acid.

Circular dichroism

Circular dichroism (CD) spectra were measured on an AVIV 62DS spectropolarimeter. Measurements were made at 25 °C and cells with pathlengths of between 0.1 and 1 cm were used, depending on peptide concentration.

Analytical ultracentrifugation

A Beckman XL-A analytical ultracentrifuge equipped with absorbance optics was employed. Conventional sedimentation equilibrium experiments using six-channel and standard double sector charcoal-filled Epon centerpieces were performed, relevant volumes of the fluorocarbon F47 being added to the samples in order to provide a false bottom. Each sample was run at varying speeds and data was collected at a radial spacing of 0.001 cm at wavelengths appropriate to the protein concentration under study, 20 scans being averaged in each case. All samples were dialysed against 20 mM ammonium acetate, pH 7.1, containing 100 mM potassium chloride, 2 mM dithiothreitol and 0.2 mM EDTA, and were run at 20°C (as reported by the ultracentrifuge). Data were analysed using software provided by the instrument manufacturer.

Partial specific volumes were calculated in the usual way [18] using an *ad hoc* computer program. Values of $\bar{v}_{20,c} = 0.73_3$, 0.73_5 and 0.74_0 cm^3 g^{-1} at 20°C were obtained for MyoD-bHLH, E47-bHLH and MyoEL, respectively. A value of 0.73_1 was calculated for fMyoD,

ignoring the contribution of the fluorescein group. Possible electrostrictive effects were not considered in these calculations [19]. Solvent density was measured on a Paar DSA48 density and sound analyser.

Results

Peptides

As well as MyoD-bHLH, which is identical to the construct used in the structure determination [13] except for the addition of an N-terminal glutamine residue, two other derivatives were investigated. MyoEL is a mutant in which loop residues and charged residues facing the interior of the four helix bundle have been grafted from E47 on to the MyoD sequence. These residues are unique in different bHLH proteins, and MyoEL was designed to probe the rôle of these regions in the association of bHLH domains. A construct with a glycine–glycine–cysteine spacer sequence which could be further modified by a cysteine-specific fluorescent label, facilitating the monitoring of oligomerization at low peptide concentration, has also been characterized.

Conventional sedimentation equilibrium

All samples were run at a minimum of three speeds in the range 20 000–35 000 rpm. Computed apparent whole cell mass averages, $M_{w,app}$, assuming the presence of a single ideal species, obtained from data measured at 30 000 and 35 000 rpm and at a variety of protein concentrations are shown in Table 1. There is a clear dependence of $M_{w,app}$ on both concentration and run speed for MyoD-bHLH and its derivatives and the lowering of $M_{w,app}$ as the concentration falls or as run speed increases is characteristic of the behavior of a self-associating system. Inspection of the fits to the data (an example is given in Fig. 2A) make this point clear, and suggest that an associating model for the MyoD-bHLH peptides is more appropriate, depending on the sample under consideration. The measured molar mass of E47 is, on the other hand, relatively insensitive to either speed or concentration. The two types of behavior will, therefore, be considered separately.

MyoD-bHLH, MyoEL and fMyoD

MyoD-bHLH has an apparent whole cell mass of approximately 23 000 g mol^{-1} at a protein concentration of 50 μM although the computed fit (Fig. 2A) clearly does not represent the data adequately. Bearing in mind that the

Table 1 $M_{w,app}$ for bHLH peptides at different concentrations and at 30 000 and 35 000 rpm, 20°C

Peptide	Conc [μM]	Speed [rpm]	$M_{w,app}$
MyoD	50	30 000	22930 ± 520
		35 000	21640 ± 550
	18	30 000	18150 ± 830
		35 000	15660 ± 550
	12	30 000	10650 ± 1250
		35 000	10670 ± 930
fMyoD	50	30 000	19440 ± 1240
		35 000	17840 ± 1540
E47	60	30 000	15540 ± 160
		35 000	15760 ± 160
	20	30 000	12780 ± 560
		35 000	12000 ± 300
	12	30 000	11200 ± 880
		35 000	15200 ± 580
MyoEL	80	30 000	21400 ± 760
		35 000	18500 ± 850
	20	30 000	16400 ± 530

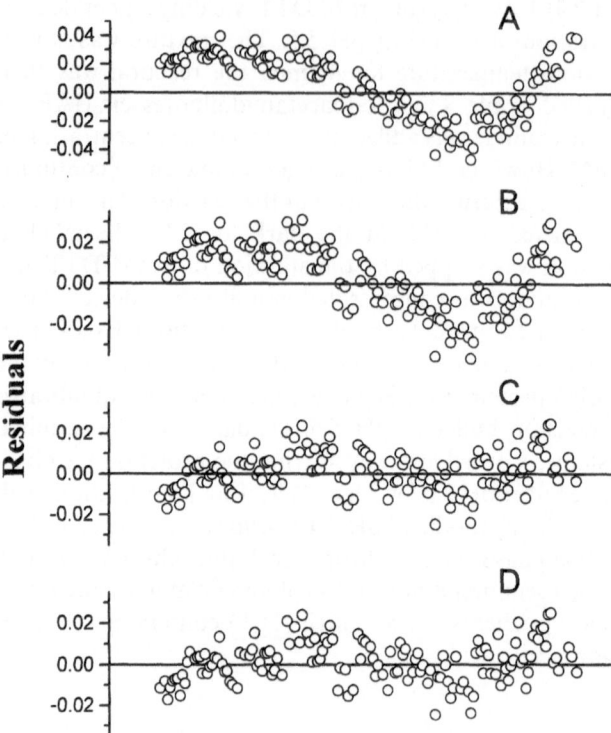

Fig. 2 MyoD–bHLH: Residuals to the computed fits for different self-associating models. The data shown here, which are representative of the complete set, were measured at a peptide concentration of 50 μM and at 30 000 rpm, 20°C. (A) Single ideal species, (B) dimer–tetramer equilibrium, (C) Monomer–tetramer and (D) Monomer-dimer-tetramer. The rms deviations of the residuals were 0.016, 0.010 and 0.010 for fits B, C and D, respectively

calculated monomer mass of MyoD–bHLH is 8169 $g\,mol^{-1}$, formation of a trimeric or tetrameric state immediately suggests itself, without excluding the possibility of higher levels of association. The best fit to the data is, in fact, achieved by assuming a monomer–tetramer equilibrium (Fig. 2C), there being no evidence of the presence of any species larger than the tetramer. The point average mass as a function of concentration increases smoothly from a value of 8500 and starts to level off at about 25 000 $g\,mol^{-1}$ although the data is, by its nature, rather noisy. Alternative models do not provide anything like as good descriptions of the data. In view of the DNA binding property of this peptide a model involving a dimeric state would seem reasonable. If the tetramer is regarded as a dimer-of-dimers (i.e., the base molar mass is taken to be 16 338 $g\,mol^{-1}$), treating the data as a dimerization process leads to a noticeable deterioration in the computed fit (Fig. 2B). Employing a monomer–dimer–tetramer model produces no improvement on the monomer–tetramer fit (Fig. 2D) and this model is therefore excluded on the principle of parsimony, and it is concluded that the dimeric state is not significantly populated at equilibrium.

fMyoD appears to exhibit the same properties as MyoD–bHLH, forming a tetramer in the micromolar concentration range. A detailed description of the equilibrium is, however, not possible due to an apparent slow deterioration of the sample with time during the runs. This manifests itself by progressive deviation of the fits to the data in the upper regions of the cell as the experiment progresses and is thought to be due to photolysis of the fluorescein moiety. The equivalently labelled form of E47–bHLH, fE47, exhibits this behaviour even more clearly

and, in this case, the presence of breakdown products has been confirmed by electrospray mass spectrometry (data not shown).

Relatively few measurements have been made on MyoEL as yet, and, while it clearly behaves like its parent, MyoD–bHLH, the data are not of the same quality. It is not possible to produce entirely convincing fits to the data, but the best produce a tetramer in the limit.

E47-bHLH

A typical sedimentation equilibrium experiment is shown in Fig. 3 and $M_{w,app}$ values listed in Table 1. The data can consistently be best fit as single species with a mass in the range of approximately 13 000–15 000 $g\,mol^{-1}$. Simultaneous fits to data collected at 25 000, 30 000 and 35 000 rpm give masses of 15 500 and 12 100 at 60 μM and 20 μM, respectively. Any attempt to fit the data to another model resulted in considerable statistical deterioration. As the calculated monomer mass of this peptide is

Progr Colloid Polym Sci (1997) 107:115–121
© Steinkopff Verlag 1997

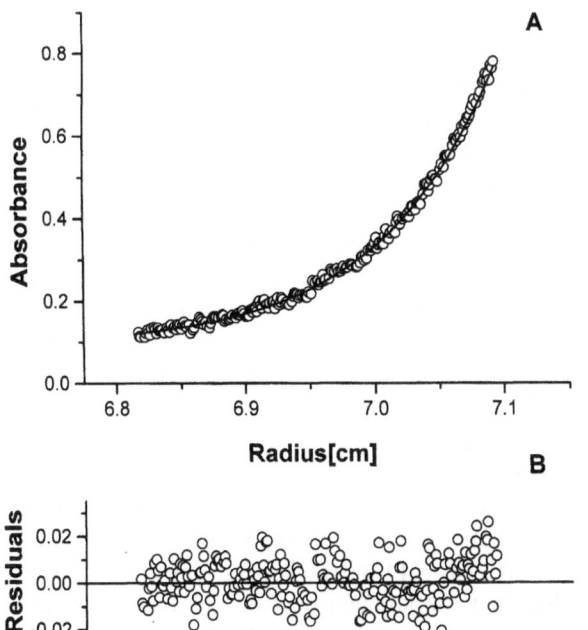

Fig. 3 E47–bHLH: (A) Concentration gradient, computed fit and (B) residuals to the fit at 60 μM, 30 000 rpm and 20°C. Similar results were found at the lower peptide concentrations investigated

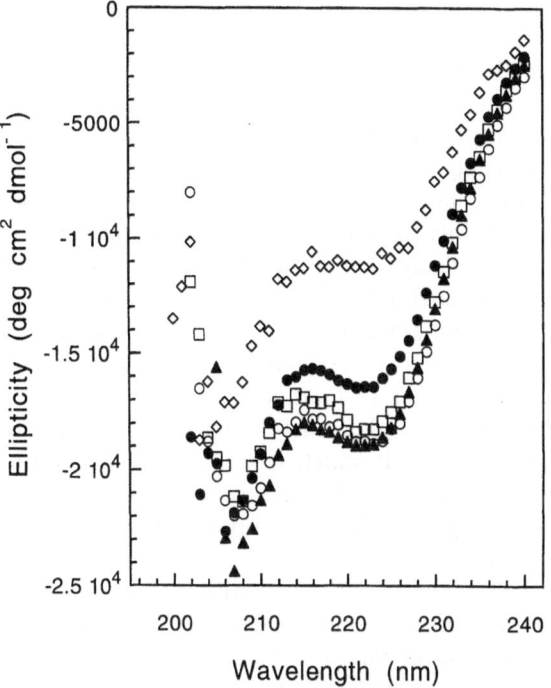

Fig. 4 The far UV circular dichroism spectrum of (MyoDGGC)$_2$ as a function of peptide concentration. ○ 36, □ 18, ▲ 10, ● 5 and ◇ 1 μM. 20 mM sodium phosphate, pH 7.4, 75 mM sodium chloride, 25°C

7022 $g\,mol^{-1}$, this implies that the material is present as a very highly associated dimer. The slight excess in mass at 60 μM might be accounted for by the presence of higher aggregates, although none were detected. Alternatively electrostrictive effects on $\bar{v}$ of the magnitude discussed by Starovasnik et al. [20] may be responsible: lowering $\bar{v}$ by about 2% would lead to a fall in $M_{w,\,app}$ of approximately 1000 $g\,mol^{-1}$. The apparent underestimate of mass at the lower concentrations is almost certainly due to the appearance of small, but measurable amounts of monomer, although at 12 μM the data are too noisy for detailed interpretation. Point averages for the data at 20 μM, when plotted against concentration, are again noisy but follow a shallow curve reaching a plateau at about 13 000 $g\,mol^{-1}$.

Circular dichroism

MyoDGGC can be crosslinked by oxidation of the C-terminal cysteine residues to form the covalent (MyoDGGC)$_2$. This obligate dimer binds to DNA with essentially the same affinity as the carboxymethylated form of MyoDGGC (MyoCM). Suprisingly, it was found that the shape and intensity of the CD spectrum of the crosslinked peptide were concentration dependent (Fig. 4). At 10 μM and above, the spectra can be viewed as the sum of contributions from an α-helical bHLH domain and from an

unstructured basic region, as has been reported for MyoD–bHLH [15]. At lower (<10 μM) concentrations the negative intensities of the minima at 208 and 222 nm decrease and the ratio $\theta_{208}:\theta_{222}$ increases, although the measured spectrum clearly still reflects an appreciable ordered structural content. This is characteristic of a system in which a disordered species is in equilibrium with an ordered one, each contributing to the spectrum in a manner that represents their individual concentrations. This result shows that it is association to a higher order than the dimer that is responsible for the stabilization of the MyoD–bHLH structure. The ultracentrifuge results on MyoD–bHLH indicate that the preferred configuration is presumably the tetramer although the relevant experiments on (MyoDGGC)$_2$ have not been carried out as yet. Dissociation of the tetramer at lower peptide concentrations leads to dissociation into crosslinked dimers and their unfolding.

Discussion

The results presented here both confirm and extend previous physicochochemical studies on basic helix–loop–helix peptides. However, in order to review the results on these systems to date it is important to remember that there are differences between the species studied. The peptide spanning residues 102–166 of the MyoD sequence with a methionine-glutamate-leucine N-terminal extension and cysteine$_{135}$ intact, MyoD$_{rec}$, has been modelled as a dimerizing system with an association constant, K_{obs}, of approximately 5 μM, on the basis of CD experiments. While MyoD$_{rec}$ bound DNA as a dimer, at concentrations above K_{obs} it was predominantly tetrameric in solution [15]. In a more detailed hydrodynamic study on the same peptide at concentrations between 21 and 600 μM it was shown to have an apparent mass of 25 800 g mol^{-1}. The system was modelled as a reversible dimer–tetramer association with a K_D of 17.3 μM, no free monomer being detected in solution. Only models that incorporated a non-ideality component adequately described the data, probably due to the molecule's high net charge. Oxidizing, and thus crosslinking, the peptide prevented tetramer formation and the molecule behaved as a single species that had a mass of 16 400 g mol^{-1} [20]. Subsequent work suggested that the peptide formed a monomer–dimer–tetramer equilibrium system and emphasized the possible importance of the tetrameric form in the regulation of DNA binding [16]. A study with full-length MyoD revealed even higher order oligomerization and reported micelle formation above a critical concentration. It was suggested that micellization might provide a regulatory mechanism that controlled the availability of free MyoD *in vivo* [19]. The MyoD–bHLH variant that we have reported on in this study is in fact the one for which the three-dimensional structure has been determined, albeit with an additional glutamine residue at the C-terminus. It differs further from MyoD$_{rec}$ in that cysteine$_{135}$ has been substituted by serine, which aided in crystallizing the peptide (Fig. 1). It has been shown that this substitution does not affect the biological function of the protein [13]. The results of sedimentation equilibrium analysis show clearly that this particular peptide forms a monomer–tetramer system with no dimer detectable at equilibrium. The peptide dissociates to form the monomer in the low μM concentration range, confirming data from circular dichroism experiments (Wendt and Ellenberger, unpublished results), although insufficient data are available for an accurate estimation of the dissociation constant. These results are reinforced by the CD studies on (MyoDGGC)$_2$ which show that this disulphide crosslinked peptide dissociates and simultaneously unfolds at lower concentration, implying that MyoD–bHLH is unstructured in the dimeric state and that a higher order, presumably tetrameric, is required to stabilize the conformation.

There have been previous reports on the mass properties of 71 [16] and 96 residue [19] versions of the E47–bHLH sequence. The present report concurs with the finding that this system forms a very stable dimer. Further,

as the peptide reported on here has only 58 residues, our results indicate that the sequence length can be reduced without affecting the dimerization property. Due to the tightness of association, the 58 residue E47–bHLH behaves as a single ideal species in the ultracentrifuge (Fig. 3) precluding the estimation of thermodynamic parameters. K_D for dimer formation in this peptide has been calculated as 0.02 μM from denaturation experiments using CD spectroscopy (Wendt and Ellenberger, unpublished results), while that for the 71 residue variant has been reported to be 4.5 μM [16]. However, in the latter variant the cysteine residue in the loop region was changed to serine, a mutation which, in contrast with MyoD, affects both dimerization and DNA binding (Wendt and Ellenberger, unpublished results).

The two MyoD–bHLH derivatives investigated appear to have the same, or very similar properties to the parent peptide. In the case of fMyoD this is important because it establishes that the labelled peptide has the same oligomerization behavior as the unmodified peptide and can, therefore, be relied on accurately to reflect the behavior of MyoD–bHLH in spectroscopic studies. The substitution of residues from E47 into the loop region of MyoD–bHLH does not seem to confer E47-like hydrodynamic properties on the MyoEL peptide although it increases the stability of the molecule to an extent intermediate between that of the E47 and MyoD–bHLH oligomers (Wendt and Ellenberger, unpublished results).

The specific formation of a bHLH heterodimer at the target DNA site is a seminal event in a number of developmental processes [3–9]. Detailed understanding of the origin of this specificity requires the characterization of the different contributions of DNA–protein and protein–protein interactions. In contrast to their DNA bound forms, bHLH proteins adopt a variety of oligomeric states in the absence of a binding site or of a partner for heterodimerization [15, 16, 18, 19]. Thus, the specific formation of stable heterodimers is believed to be the basis for the biological function of these domains [11, 25]. In addition, it has been suggested that the association of bHLH proteins might not only regulate their own homo- and hetero-oligomerization but might also be important in their interaction with members of the Id family [16, 21, 22]. This class of HLH proteins lack a basic domain and are capable of sequestering bHLH factors as inactive oligomers [21–23]. It has been shown that these proteins preferentially target E47 [21, 22, 24], suggesting that MyoD may escape this process by exhibiting a different oligomerization behavior.

It should be noted that all of the experiments presented and discussed here were conducted at concentrations much greater than that required for DNA binding. bHLH proteins typically bind DNA with affinities in the nano-

Progr Colloid Polym Sci (1997) 107:115–121
© Steinkopff Verlag 1997

molar range whereas the ultracentrifuge and CD experiments are generally limited, by their detection sensitivity, to low micromolar concentrations. Under these conditions MyoD is essentially monomeric, and presumbly fully unfolded, suggesting strongly that contact with DNA might be an additional mediator of regulatory bHLH association. Thus the formation of the final heterodimer–DNA complex might be initiated and directed by DNA contacts made by either MyoD or E47 monomers or homodimers of the latter. A similar mechanism has been suggested recently for the yeast transcription factor GCN4 [26], highlighting the need for further study into the different contributions of DNA binding and protein interaction.

Acknowledgement We would like to thank Professor Tom Ellenberger for his generous help and his careful reading of the manuscript.

References

1. Thomas RM, Wendt H, Zampieri A, Bosshard HR (1996) Colloid Polym Sci 99:24–30
2. Thomas RM, Zampieri A, Jumel K, Harding SE (1997) Eur Biophys J 25:405–410
3. Davis RL, Weintraub H (1992) Science 256:1027–1030
4. Weintraub H (1993) Cell 75:1241–1244
5. Caudy M, Vassin H, Brand M, Tuma R, Jan LY, Jan YN (1988) Cell 55:1061–1067
6. Cabrera CV, Alonso MC (1991) EMBO J 10:2965–2973
7. Weintraub H, Davis RL, Tapscott S, Thayer M, Krause M, Benazra R, Blackwell TK, Turner D, Rupp R, Hollenberg S, Zhuang Y. Lassar A (1991) Science 251:761–766
8. Weintraub H (1993) Cell 75:1241–1244
9. Olson EN, Klein WH (1994) Genes Dev 8:1–8
10. Murre C, McCaw PS, Baltimore D (1989) Cell 56:777–783
11. Lassar A, Davis RL, Wright WE, Kadesch T, Murre C, Voronova A, Baltimore D, Weintraub H (1991) Cell 66:305–315
12. Blackwell TK, Weintraub H (1990) Science 250:1104–1110
13. Ma PCM, Rould MA, Weintraub H, and Pabo CO (1994) Cell, 77:451–459
14. Ellenberger T, Fass D, Arnaud M, and Harrison SC (1994) Genes Dev 8:970–980
15. Anthony-Cahill SJ, Benfield PA, Fairman R, Wasserman ZR, Brenner SL, Stafford WF, Altenbach C, Hubbell WL, DeGrado WF (1992) Science 255:979–983
16. Fairman R, Beran-Steed RK, Anthony-Cahill SJ, Lear JD, Stafford WF, DeGrado WF, Benfield PA, Brenner SL (1993) Proc Natl Acad Sci 90:10 429–10 433
17. Studier FW, Moffat BA (1986) J Mol Biol 189:113–130
18. Laue TM, Shah B, Ridgeway TM and Pelletier SL (1992) In: Harding SE, Rowe AJ, Horton JC (eds), Analytical Ultracentrifugation in Biochemistry and Polymer Science. Royal Society of Chemistry, Cambridge, U.K.
19. Laue TM, Starovasnik MA, Weintraub H, Sun X-H, Snider L, Klevit RE (1995) Proc Natl Acad Sci 92:11824–11828
20. Starovasnik, MA, Blackwell TK, Laue TM, Weintraub H, Klevit RE (1992) Biochemistry 31:9891–9903
21. Benezra R, Davis RL, Lockshorn D, Turner DL, Weintraub H (1990) Cell 61:49–59
22. Sun XH, Copeland NG, Jenkins NA, Baltimore D (1991) Mol Cell Biol 11:5603–5611
23. Wilson RB, Megerditch K, Shen C, Benezra R, Zwollo P, Dymecki SM, Desiderio SV, Kadesch T (1991) Mol Cell Biol 11:6185–6191
24. Jen Y, Weintraub H, Benezra R (1992) Genes Dev 6:1466–1479
25. Sun XH, Baltimore D (1991) Cell 64:459–470
26. Matallo SJ, Schepartz A (1997) Nat Struct Biol 4:115–117

Progr Colloid Polym Sci (1997) 107:122–126
© Steinkopff Verlag 1997

BIOLOGICAL SYSTEMS

C. Bartmann-Lindholm
M. Geisert
U. Güngerich
W.E.G. Müller
D. Weinblum

Nuclear DNA fractions with grossly different base ratios in the genome of the marine sponge *Geodia cydonium*

C. Bartmann-Lindholm · M. Geisert ·
U. Güngerich · W.E.G. Müller ·
D. Weinblum (✉)
Institute für Physiologische Chemie und
Pathobiochemie
Johannes-Gutenberg-Universität
Duesbergweg 6
55099 Mainz, Germany

Abstract The DNA of the marine sponge *Geodia cydonium* (*G.c.*), a member of the phylogenetically old phylum Porifera, was characterized by density gradient centrifugation and by determining its genetic complexity by reassociation kinetics. At least five subcomponents were identified by curve-fit analyses of analytical density gradient centrifugation profiles of total *G.c.*-DNA. Four of these subcomponents were isolated from total *G.c.*-DNA by preparative density gradient centrifugation. The GC-contents of the subcomponents were determined to be 36.4%, 44.0%, 58.7%, and 66.1%, respectively. To our knowledge, such an extreme heterogeneity of DNA composition has never before been observed for any organism. The genetic complexities within the subcomponents were determined by reassociation kinetics to 2.1×10^8, 2.8×10^8, 9.2×10^8, and 1.4×10^9 bp, respectively. The orders of magnitude of the genetic complexities clearly indicate that the DNA subcomponents mainly contain eukaryotic single copy DNA, since DNA of symbiotic prokaryotes should show significantly lower complexities.

Key words Marine sponges – *Geodia cydonium* – DNA-isochores – genetic complexity – analytical CsCl-density gradient centrifugation – reassociation kinetics analysis

Introduction

The DNA molecules of vertebrates are composed of segments greater than 300 kb in length which are homogeneous with respect to their base ratios [1]. These segments contain single-copy DNA and moderately repetitive DNA, known as isochores. Vertebrate genomes contain a small number of isochores with characteristic base ratios. Thus, when isolating DNA, fragments of different base ratios and hence different buoyant densities will be obtained. As a consequence, density gradient centrifugation profiles of DNA from higher eukaryotes are not Gaussian, but asymmetrically distributed, always being skewed to the denser side. DNA fragments belonging to different isochores can be isolated by preparative density gradient centrifugation as first shown by Filipski et al. [2]. Since these DNA fractions comprise more than 90% of the genome they are also called major components and are distinguished from minor components which may comprise several classes of highly repetitive DNA, also characterized by specific base ratios. These DNAs frequently appear as satellite DNA bands upon density gradient centrifugation.

The complexity of the isochore distribution increases with the complexity of an organism; for instance, warm-blooded vertebrates contain more DNA fractions with greater density differences than cold-blooded vertebrates. For review see refs. [1, 3, 4].

The genomic organization of lower eukaryotes has been reported for only a few species, which mostly contain just one major component [5, 6]. When studying the gross organization of the genome from a lower eukaryote, the marine sponge G.c., we found the most heterogeneously composed genome of any species, yet so far described. In this paper we will report our results.

Materials and methods

Isolation of DNA

G.c. [Porifera: Demospongiae: Geodiidae] specimens were collected in the northern Adriatic sea near Rovinj, Croatia. Sponges were cut into pieces of approximately 1 g in weight and frozen in liquid nitrogen. To isolate DNA, 5 g frozen sponge tissue were ground in liquid nitrogen and the powder dissolved in 20 ml buffer (NaCl 100 mM, EDTA 50 mM, pH 8.0). Sodium dodecyl sulfate (SDS) solution (20%) was added to a final concentration of 2% (w/v) and the mixture heated to 60°C for 10 min. Nucleic acids were extracted with phenol/chloroform, the RNA subsequently digested with RNase. After reextraction with phenol/chloroform, the DNA was precipitated several times with isopropanol and finally with ethanol [7]. The isolated DNA was dissolved in CsCl-buffer (CsCl 50 mmol/l, TRIS/HCl 15 mmol/l, pH 8.0) [8].

Alcaligenes faecalis strain B 520 was obtained from Prof. Radler, Institute of Microbiology, University of Mainz, and DNA was extracted by standard procedures [9]. Saccharomyces cerevisiae-DNA was isolated according to Cryer et al. [10].

Analytical density gradient centrifugation

Total DNA from G.c. was dissolved in 15 mmol/l TRIS/HCl, pH 8 containing CsCl at a density of 1.7010 g/ml. Centrifugation was performed at 44 000 rpm at 25°C for 24–36 h in a Beckman model E analytical ultracentrifuge equipped with UV-optics, monochromator, photoelectric scanner, multiplexer and digital data output. KelF double-sector cells, wedge windows and a six-hole titanium rotor An-GTi were used. Densities were determined according to Szybalsky and Szybalsky [11] with DNAs from Clostridium perfringens and Micrococcus lysodeicticus as internal density markers.

Curve fitting

A density gradient centrifugation profile, composed of n subcomponents, can be described by [12]

$$f(\rho) = \sum_{m=1}^{n} c_m \frac{e^{-(\rho - \rho_m)^2/2\sigma^2}}{\sqrt{2\pi\sigma^2}}, \qquad (1)$$

where ρ_m is the mean buoyant density and c_m is the concentration of the nth subcomponent. Since the molecular weight of eukaryotic DNA is an artefact of the preparation method, the molecular weight and hence the standard deviation σ is identical for all subcomponents; in our experiments σ equals 0.0034 g/ml. The profiles were fitted to Eq. (1) using the program Minuit (CERN library). Starting values for ρ were taken from the maxima of the deconvolution curve (not shown). Deconvolution of the density gradient profile was performed according to Weinblum et al. [13].

Preparative density gradient centrifugation

Total G.c.-DNA was dissolved in CsCl/TRIS buffer (see above) at a concentration of 100 μg/ml and centrifuged for 20 h at 41 000 rpm at 25°C in a Beckman VTi 50 rotor. The contents of the tubes were divided into fractions of 1 ml while reading the absorbance at 254 nm.

Reassociation kinetics

DNA from the purified subcomponents was dissolved in phosphate buffer (0.12 M, pH 7.0), sonicated and dialyzed against the same buffer for 24 h. Samples containing approximately 50 μg/ml of DNA were equilibrated with helium, filled into the thermocuvette of a Gilford 250 spectrophotometer and heated until no further increase in absorbancy at 260 nm was observed (100°C). Then the cuvette was cooled within 1.5 min to a temperature 25°C below the melting temperature of the sample. The absorbances of the samples at 260 nm were recorded for 10 days. Data were fitted to Eq. (2) [14] by the curve-fit program Minuit (CERN library).

$$\frac{S}{c_0} = f(1 + k_1 f c_0 t)^{-0.445} + (1-f)(1 + k_2 (1-f)c_0 t)^{-0.445}$$

$$(2)$$

where S is the concentration of single-strand DNA, c_0 is the total DNA concentration, f is the fraction of the first component, and k_1 and k_2 are the reassociation constants of the two components, respectively.

Results

Density gradient centrifugation of G.c.-DNA

The analytical density gradient centrifugation profile of G.c.-DNA is shown in Fig. 1. The profile obviously

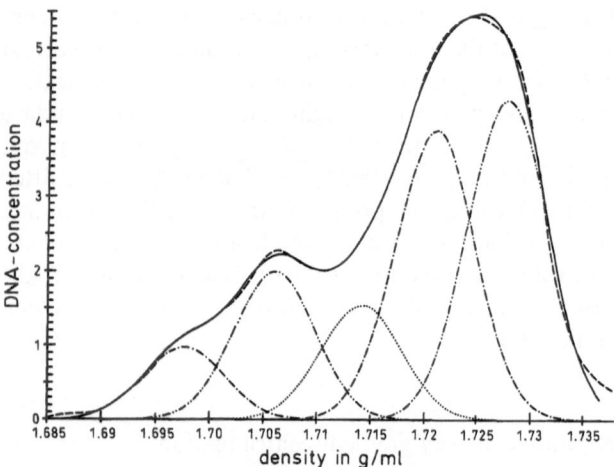

Fig. 1 Analytical density gradient centrifugation profile of total *G.c.* DNA. The curves represent: the measured profile (dashed line), the subcomponents, obtained from curve-fit calculations (dashed-dotted lines), the profile from the sum of subcomponents (solid line). The subcomponent, depicted with the dotted line, could not be isolated. DNA concentration in arbitrary units.

represents a superposition of several components of different buoyant densities. Digestion of the DNA sample with bovine pancreatic DNase prior to centrifugation led to the complete disappearance of the profile. Therefore, every component must be DNA. The deconvolution curve of the profile [13, 15] shows four well-resolved maxima, indicating four subcomponents. However, the profile can be described satisfactorily by superposition of at least five components (Fig. 1). This seems to be at variance with the results from the deconvolution procedure, but this method fails to resolve minor narrowly spaced components. Buoyant densities and the proportions of the fractions were obtained by curve fitting from independent DNA-preparations from four different *G.c.* specimens. The mean data are listed in Table 1.

Only such subcomponents are true isochores, which can be isolated by preparative centrifugation and will show Gaussian-distributed peaks upon recentrifugation. Therefore, from preparative gradients, fractions corresponding to the mean densities of the subcomponents were

cut and pooled. Upon repeated recentrifugations, Gaussian-distributed DNA of the appropriate densities were obtained for components 1, 2, 4 and 5. A component 3, albeit predicted by curve-fit analysis, could not be isolated by multiple preparative centrifugation. We assume that in this density region several DNA components with small density differences do exist. This assumption is supported by the fact that the deconvolution curve does not show a well-resolved maximum at this position. No attempts were made to analyze this region further.

Determination of genetic complexities by reassociation kinetics analysis

To demonstrate that DNA components of *G.c.* mainly contain nuclear single-copy DNA, the genetic complexities of the fractions were determined by reassociation kinetics. A genetic complexity of $> 10^8$ bp would unequivocally indicate eukaryotic nuclear single-copy DNA and would discriminate such DNA from repetitive satellite DNA or from the DNA of prokaryotic symbionts. For the latter types of DNA the genetic complexity would be smaller by at least two orders of magnitude.

Genetic complexities were determined by measuring the decreasing hyperchromicity of denatured DNAs at 260 nm as a function of time. With this simple method the order of magnitude of genetic complexities can be determined sufficiently accurate.

Eukaryotic DNA normally contains varying proportions of fast reassociating repetitive DNA and fold-back sequences besides the single-copy DNA; therefore a two-component curve (Eq. (2)) was selected which describes the experimental data quite well (Fig. 2). By a curve-fit procedure the reassociation constants and the proportion of the more slowly reassociating single-copy DNA were determined. These reassociation constants, which are inversely proportional to the genetic complexity, are listed in Table 2. Single-copy DNA from yeast, which was used as a reference, reassociates with a constant of 0.23. Using the size of the yeast genome of 1.4×10^7 bp [16] with a single-copy DNA content of 85%, we calculate the

Table 1 Subcomponents of *G.c.* DNA determined by curve fitting from analytical density gradient profiles. Values are means from 4 independent DNA preparations. The last column of the table contains the density values of the corresponding DNA fractions, isolated by preparative density gradient centrifugation

Number of fraction	Density [g/ml]	GC-content [%]	Proportion of total DNA [%]	GC-content of isolated fraction [%]
1	1.6972	37.2	8	36.4
2	1.7054	45.4	16	44.0
3	1.7128	52.8	12	Not found
4	1.7195	59.5	30	58.7
5	1.7262	66.2	34	66.1

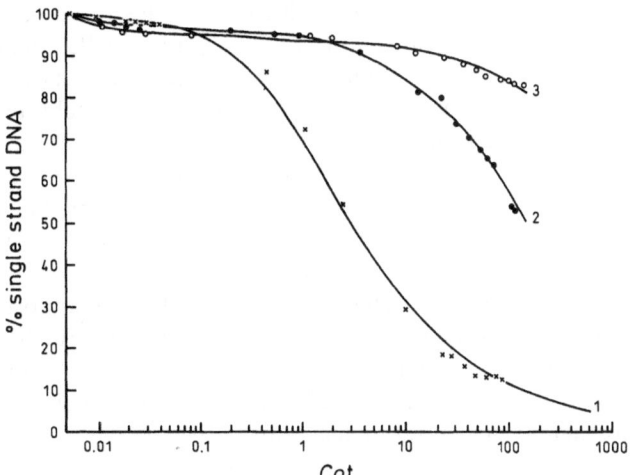

Fig. 2 Reassociation kinetics of subcomponents of *G.c.* DNA. Curve 1: *A. faecalis* DNA. Curve 2: second subcomponent (44.0% GC) of *G. cydonium* DNA. Curve 3: fifth subcomponent (66.1% GC) of *G.c.* DNA. c_0t-values in mol s

genetic complexity of the four *G.c.*-DNA fractions. The genetic complexities within the components increased from 2.1×10^8 bp (isochore with the lowest GC-content) to 1.4×10^9 bp (isochore with the highest GC-content). The amount of fast reassociating DNA was well below 10% for all fractions.

No corrections for different base ratios were applied. The influence of the GC-content upon reassociation velocity has been differently assessed: no change [17], a decrease [18], and an increase [19] have been reported for increasing GC-content. Therefore, the true value of the kinetic complexities for the components with higher GC content may be greater or smaller up to a factor of 2. To demonstrate that the genetic complexity of DNA with high GC-content can be determined sufficiently accurate in our system, we measured the reassociation of the DNA from the prokaryot *Alcaligenes faecalis*. This DNA has a GC-content of 65%, as determined from its melting temperature. As can be seen from Fig. 2 and from Table 2, it

reassociates approximately three orders of magnitude faster than the fifth subcomponent DNA of *G.c.* which has about the same GC-content.

Discussion

The patterns of distribution of major components or isochores in the DNA of vertebrates has been thoroughly investigated by Benardi et al. [1, 3, 4]. They occur regularly in the DNA of vertebrates. In the DNA of warm-blooded vertebrates, the number of GC-rich isochores is higher than in cold-blooded vertebrates, so these isochores were obviously generated during evolution. The GC-rich isochores contain proportionally more genes than the AT-rich. Bernardi [1] explained this by a selective evolutionary trend, since GC-rich DNA is more stable at elevated temperatures. Isochores have also been found in angiosperm genomes [20].

For invertebrate species only few data are found in the literature. Thiery et al. [5] investigated some invertebrates and found more than one major DNA component only for *S. cerevisiae*. The other species only contained minor components, which means satellite bands with very low genetic complexity, and showed symmetric profiles. Storck and Alexopoulos [21] determined the DNA base ratios from several hundred unicellular eukaryotes by density gradient centrifugation. They observed minor satellite bands in several species; however, no indications of skewed profiles were reported.

Therefore, the occurrence of isochores with grossly different base ratios in the genome of a lower eukaryote is surprising. Hence, we have to prove that none of the DNA-fractions are from symbiontic organisms, since symbionts from sponges have been described frequently [22]. Due to their high genetic complexity, none of the fractions can originate from DNA of prokaryontic symbionts. Furthermore, the occurrence of zooxanthella and zoochlorella in sponges has been reported; such sponges, however, display a yellow–green or a green color [23]. *G.c.* is

Table 2 Reassociation constants and genetic complexity of single copy DNA from subcomponents of *G.c.* in comparison with those of *Saccharomyces cerevisiae* and *Alkaligenes faecalis*. The different fractions representing the subcomponents of *G.c.* DNA are numbered according to Fig. 1 or Table 1

Source of DNA	GC-content [%]	Content of single copy DNA [%]	Reassociation constant $[\mathrm{mol^{-1}\,s^{-1}}]$	Genetic complexity [bp]
G.c., frac. 1	36.4	95.1	0.013	2.1×10^8
G.c., frac. 2	44.0	95.7	0.0097	2.8×10^8
G.c., frac. 4	58.7	93.0	0.003	9.2×10^8
G.c., frac. 5	66.1	94.4	0.002	1.4×10^9
S. cerevisiae	42.2	85.0	0.23	1.2×10^7
A. faecalis	65.2	99.0	1.28	2.1×10^6

colorless; moreover, pigment extraction and characterization according to described procedures [24] revealed no indication for the existence of e.g. chlorophyll. Therefore, it is highly unlikely that one of the DNA components is caused by an eukaryotic symbiont, particularly if one considers the high proportion of these subfractions. So we conclude that the subcomponents belong to the nuclear DNA of *G.c.*

The genetic complexities within the isochores increase approximately in the right proportion with the amount of DNA in the particular fraction, from 2.1×10^8 bp (smallest fraction comprising 8% of the total DNA) to 1.4×10^9 bp (largest fraction comprising 34% of the total DNA). The genetic complexity of the total DNA of *G.c.* amounts to 2.7×10^9 bp. If one adds another 10% for the fast reassociating DNA and for DNA, not contained in the four fractions, the diploid genome of *G.c.* contains approximately 6.6 pg DNA. This value seems to be high for a primitive organism, it is only slightly smaller than the value for the human genome. However, species from almost every major taxon of protists or invertebrates are described which contain even more DNA [25]. Imsiecke et al. [26] determined the DNA content of *G.c.* to 3.7 pg by DAPI staining. So by two independent methods the DNA content was determined in the same order of magnitude.

The described extreme heterogeneous DNA composition of the marine sponge *G.c.* genome is unique. Further studies will investigate if this feature is characteristic for other marine sponges as well and analyze in which isochore(s) the structural or regulatory genes occur.

Acknowledgement This work was supported by a grant from the Deutsche Forschungsgemeinschaft (SFB 169; A11 to WEGM).

References

1. Bernardi G (1993) Mol Biol Evol 10:186
2. Filipski J, Thiery JP, Bernardi G (1973) J Mol Biol 80:177
3. Bernardi G, Bernardi G (1990) J Mol Evol 31:265
4. Bernardi G, Bernardi G (1990) J Mol Evol 31:282
5. Thiery JP, Macaya G, Bernardi G (1976) J Mol Biol 108:219
6. Macaya G, Thiery JP, Bernardi G (1976) J Mol Biol 108:237
7. Ausubel FM, Brent R, Kingston RE, Moore DD, Smith JA, Seidman JG, Struhl K (1992) Current Protocols in Molecular Biology. Wiley, New York
8. Müller WEG, Zahn RK, Beyer R (1970) Nature 227:1211
9. Sambrook J, Fritsch EF, Maniatis T (1989) Molecular Cloning. Cold Spring Harbor Laboratory Press, Cold Spring Harbor
10. Cryer DR, Eccleshall R, Marmur J (1975) In: Prescott DM (ed) Methods in Cell Biology, Vol. 12. Academic Press, New York, p 39
11. Szybalski W, Szybalski EH (1971) In: Cantoni GL, Davis DR (eds) Procedures in Nucleic Acid Research, Vol. 2. Harper and Row, New York, p 311
12. Meselson M, Stahl FW, Vinograd J (1957) Proc Natl Acad Sci (Wash) 43:582
13. Weinblum D, Geisert M, Oswald E (1990) Colloid Polym Sci 268:55
14. Britten RJ, Davidson EH (1976) Proc Natl Acad Sci (Wash) 73:415
15. Medgyessy P (1961) Decomposition of Superpositions of Distribution Functions. Publishing House of the Hung Acad Sci, Budapest
16. Mortimer RK, Schild D (1985) Microbiol Rev 49:181
17. Britten RJ, Graham DE, Neufeld BR (1974) In: Grossman L, Moldave K (eds) Methods in Enzymology, Vol. 29. Wiley, New York, p 363
18. Gillis M, De Ley J, De Cleene M (1970) Eur J Biochem 12:133
19. Wetmur JG, Davidson N (1968) J Mol Biol 31:349
20. Matassi GL, Montero L, Salinas J, Bernardi G (1989) Nucleic Acids Res 17:5273
21. Storck R, Alexopoulos CJ (1970) Bacteriol Rev 34:126
22. Müller WEG, Zahn RK, Kurelec B, Lucu C, Müller I, Uhlenbruck G (1981) J Bacteriol 145:548
23. Simpson TL (1984) The Cell Biology of Sponges. Springer, New York
24. Müller WEG, Maidhof A, Zahn RK, Conrad J, Rose T, Stefanovich P, Müller I, Friese U, Uhlenbruck G (1984) Biol Cell 51:381
25. Cavalier-Smith T (1985) The Evolution of Genome Size. Wiley, New York
26. Imsiecke G, Custodio M, Borojevic B, Steffen R, Moustafa MA, Müller WEG (1995) Cell Biol Int 19:995

Progr Colloid Polym Sci (1997) 107:127–135
© Steinkopff Verlag 1997

Sedimentation equilibrium studies of synthetic polyelectrolytes by means of interference optical methods

E. Görnitz
M. Hahn
W. Jaeger
H. Dautzenberg

Dr. E. Görnitz (✉) · M. Hahn · W. Jaeger
Fraunhofer-Institute of Applied Polymer
Research
Kantstraße 55
14513 Teltow, Germany
E-mail: goernitz@iap.fhg.de

H. Dautzenberg
Max-Planck-Institute
of Colloid and Surface Science
Kantstraße 55
14513 Teltow, Germany

Abstract Synthetic quaternary ammonium polyelectrolytes have been investigated in aqueous 0.5 M NaCl solution by analytical ultracentrifugation using the "low speed" sedimentation equilibrium technique with interference detection. Lower molecular mass poly(diallyldimethylammonium chloride) (poly-DADMAC)-samples ($M_w < 10^5$ g/mol) and poly(methacryloyloxyethylbenzyldimethylammonium chloride) (poly-MADAMBQ) exhibit "ideal" sedimentation equilibrium behavior. Linear log $[\eta]$ vs. log M_w- and log s_0 vs. log M_w relations could be derived for poly-MADAMBQ. For higher molecular mass poly-DADMAC ($M_w > 10^5$ g/mol) and DADMAC-acrylamide copolymers a "non-ideal" solution behavior was observed resulting in deviations between molecular masses M_w obtained by sedimentation equilibrium and by elastic light scattering. The experimental conditions desired for the molecular characterization of synthetic polyelectrolytes by sedimentation equilibrium are discussed.

Key words Quaternary ammonium polyelectrolytes – molecular mass – analytical ultracentrifugation – sedimentation equilibrium – light scattering

Introduction

Synthetic polyelectrolytes find widespread application in many industrial processes, for the solution of environmental water and soil problems and in numerous products of our daily life. New homo- and co-polymerization techniques, including heterophase techniques like inverse emulsion and precipitation polymerization, have been developed for the synthesis of polyelectrolytes with improved application properties, especially with higher molecular masses. The molecular characterization of polyelectrolytes in general and especially of technical polyelectrolytes may be complicated for several reasons:

1. In aqueous solution the individual properties of the macromolecules are covered by long ranged intra- and intermolecular Coulomb interactions between the electrical charges of the chains. These so-called polyelectrolyte effects have to be suppressed by an appropriate amount of a low molecular weight salt. But even in salt containing solutions many polyelectrolytes behave non-ideal, "in the thermodynamic sense", exhibiting large excluded volumes due to high solvent affinity.

2. Synthetic polyelectrolytes are heterogeneous substances. The different contributions to sample heterogeneity may be classified as follows.
- polydispersity with respect to the molecular mass,
- heterogeneity of the molecular architecture (ranging from linear and branched chains to even globular micronetworks),
- chemical heterogeneity (copolymer composition)
- aggregation and self-association phenomena.

The common methods of molecular mass determination can be affected by these problems in a different manner [1]. Therefore, advantageously combinations of methods are applied in order to obtain reliable results.

The main advantages of analytical ultracentrifugation are that it is an absolute method and a fractionating one. The last property makes it, additionally, insensitive to the presence of aggregates within the solution.

In this paper we report on molecular mass determinations of a number of different laboratory-made cationic polyelectrolytes containing quaternary ammonium groups as charged units by means of the sedimentation equilibrium method. Because most of the synthetic polyelectrolytes are not absorbing in the UV/VIS region equilibrium concentration profiles were detected via the refractive index gradients by means of Rayleigh interference optics. The sensitivity of this method restricts the dilution of the polymer solutions to about 0.1 mg/ml.

Sample synthesis and preparation

The chemical structures of the investigated polymers are presented in Fig. 1. The synthesis of model samples of poly(diallyldimethylammonium chloride) (poly-DADMAC) was carried out by free radical cyclopolymerization of the monomeric DADMAC in aqueous solution. This reaction leads to soluble and preferably linear polymers with configurational isomers of pyrrolidinium rings (cis–trans-ratio 6:1) as structural units of the polymer chain [2]. The reaction proceeds via dimeric associates of the monomer with monomer concentrations above 1 mol/l resulting in an increase of the rate of the reaction but with no influence on the chemical structure of the polymer.

Fig. 1 Structural formulae of the polyelectrolytes investigated

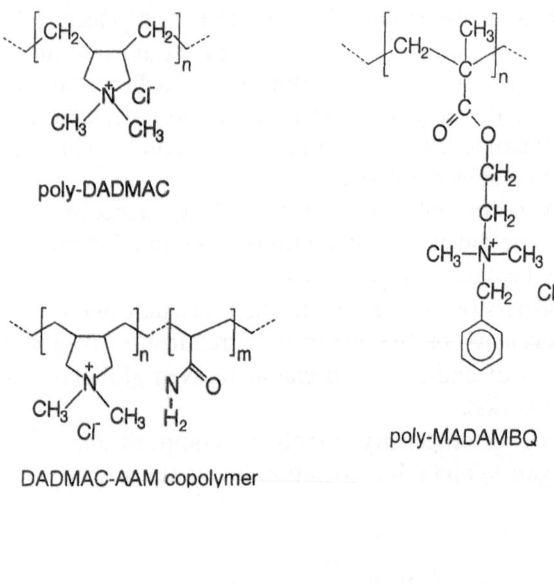

poly-DADMAC

DADMAC-AAM copolymer

poly-MADAMBQ

Using water soluble azo initiators like 2,2′-azobis(amidinopropane)dihydrochloride disturbing side reactions [2, 3] can be avoided. The kinetics of this process has been studied extensively [3, 4]. Therefore, reaction conditions for the synthesis of poly-DADMAC with broad variability of the molecular mass can be established by varying monomer and initiator concentrations as well as the reaction temperature. In any case polymerization was stopped at conversions < 10% by dilution and cooling. Due to the termination of the polymerization reaction by combination the described procedure resulted in a set of model polymers with different molecular masses in the range of 10^4–5×10^5 g/mol and with polydispersity ratios in the order of $M_n/M_w/M_z = 1/1.5/2$. In former ultracentrifugation studies [5] it was already shown that technical polymerizations of DADMAC often result in much broader distributions.

Polyelectrolytes with variable charge density were obtained by copolymerization of DADMAC with acrylamide (AAM) in water. A feeded polymerization technique with controlled dosage of the more reactive AAM was developed [6, 7] to synthesize copolymers with different charge densities but uniform charge distribution along the chain as well as nearly constant number averages of the molecular mass ($M_n \approx 5 \times 10^5$ g/mol) but broad molecular weight distributions.

A series of hydrophobically modified vinyl polyelectrolytes was synthesized by radical homopolymerization of benzyl-2-(methacryloyloxy)ethyldimethylammonium chloride (MADAMBQ) in aqueous solution [8]. Varying the reaction conditions the molecular masses of the products could be changed between 10^5 and 10^7 g/mol.

All samples were extensively purified via ultrafiltration immediately after polymerization and the polymers were isolated by vacuum freeze drying. A generalized scheme of the procedures necessary for purification and preparation of the polyelectrolytes for analysis is given in Fig. 2. Polymer solutions are prepared by dissolving the samples in pure water to achieve maximal decoiling of the charged macromolecules. An exact control of the polymer concentration is important especially for the determination of partial specific volumes. In the case of the homo-polyelectrolytes this was done by titration of the Cl$^-$ counterions in the aqueous solution. Otherwise the moisture content of the isolated polymers has to be measured immediately before dissolution. Then, the solutions were adjusted to 0.5 M NaCl. As a result of viscosimetric studies, this salt concentration was found to be sufficient to suppress the polyelectrolyte effects without remarkable desalting of the polymers. Dilution viscosimetry resulted in linear Huggins plots (η_{spec}/c vs. c) with Huggins coefficients in the range of $k_H = 0.25$–0.6. The salted polymer stock solutions were dialyzed extensively against 0.5 M NaCl and the dialysate

Progr Colloid Polym Sci (1997) 107:127–135
© Steinkopff Verlag 1997

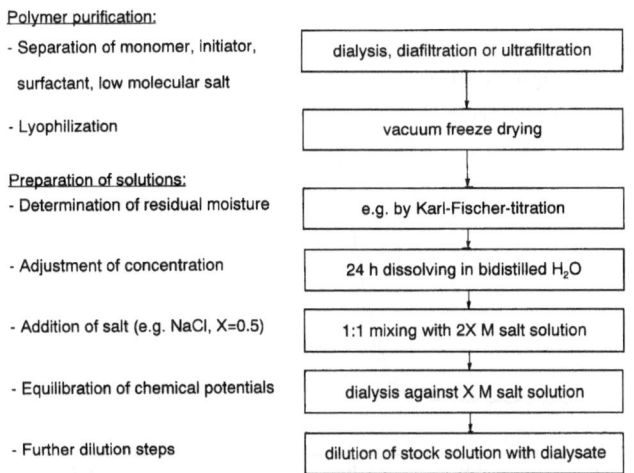

Polymer purification:
- Separation of monomer, initiator,
  surfactant, low molecular salt

| dialysis, diafiltration or ultrafiltration |

- Lyophilization

| vacuum freeze drying |

Preparation of solutions:
- Determination of residual moisture

| e.g. by Karl-Fischer-titration |

- Adjustment of concentration

| 24 h dissolving in bidistilled H_2O |

- Addition of salt (e.g. NaCl, X=0.5)

| 1:1 mixing with 2X M salt solution |

- Equilibration of chemical potentials

| dialysis against X M salt solution |

- Further dilution steps

| dilution of stock solution with dialysate |

Fig. 2 Preparation procedure of polyelectrolyte samples for molecular mass analysis (schematically)

was used for further dilution steps. Thus, problems arising from the Donnan ion exclusion effect can be avoided during the determination of the partial specific volume $\bar{v}$, the refractive index increment dn/dc and with respect to the reference solvent needed for the interference detection of the sedimentation equilibrium concentration profiles in double sector cells.

Experimental instrumentation and data analysis

The partial specific volumes $\bar{v}$ of the macromolecules have been derived from density–concentration plots obtained by the Paar density meter DMA 60/602 (Anton Paar K.G.,

Graz, Austria). Most of the sedimentation equilibrium experiments have been performed with the MOM 3180 analytical ultracentrifuge (Hungarian Optical Works, Budapest) with photographic registration of the interference patterns. A data analysis system for semi-automatic evaluation of the photographs was developed containing a computer coupled microscopic stage and a CCD camera, both controlled via OPTIMAS® 5.2 image analysis software. A fast Fourier transform algorithm was used to extract fringe displacement vs. radius data.

Recent results were obtained by the OPTIMA XL-I analytical ultracentrifuge (Beckman Instruments, Palo Alto, California, USA). The fast automatic on-line data acquisition of this instrument combined with the possibility of investigating several samples in one run by means of an 8-hole rotor and multi-channel cells enables to establish sedimentation equilibrium experiments as a routine method in polymer analysis.

We used 6 channel centrepieces with an optical path length of 12 mm and filling heights of 2.7 mm. The polymer concentrations were between 0.15 and 1.0 mg/ml. Because of the expected polydispersities of the investigated polymers the "low speed" or "non-meniscus depletion" equilibrium technique was used. Therefore, depending on the molecular mass under investigation, rotor speeds ranged between 2000 and 20 000 rpm with equilibrium times between 15 and 40 h. All measurements have been performed at 20 °C.

Data from the interference patterns of the equilibrium concentration profiles are used to study the relations of fringe displacement Δl vs. cell radius r (Fig. 3). In some cases, the radial position of the meniscus and especially of the cell bottom cannot be easily detected from the

Fig. 3 a) Interference pattern from equilibrium run in a 6-channel-cell from OPTIMA XL-I. b) Intensity scan at 280 nm from Optima XL-A absorption optics. c) Fringe displacement vs. radius derived from (a)

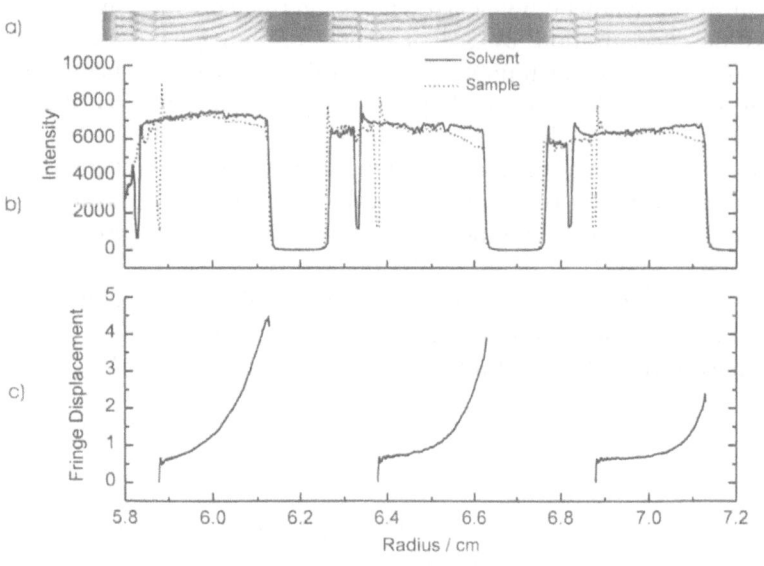

Fig. 4 Synthetic boundary
experiment for poly-DADMAC
(M_w(UC) $= 3.3 \times 10^5$ g/mol),
time between scans
$\Delta t = 20$ min

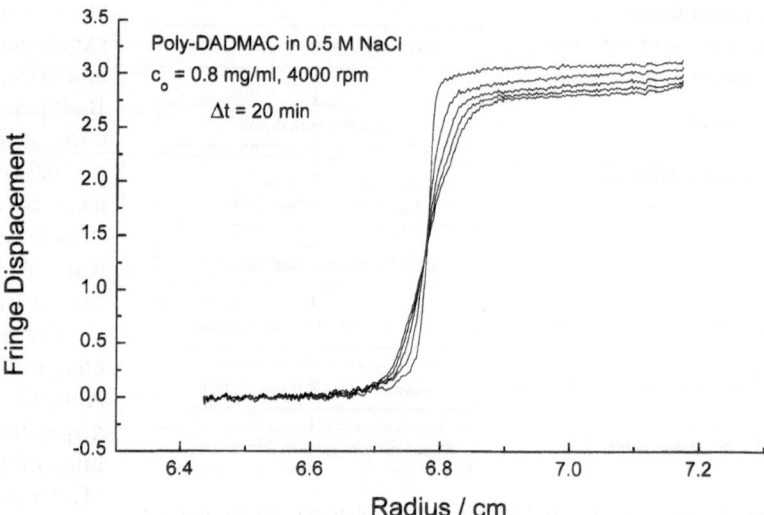

interference fringes. In this case the absorption scan of the OPTIMA XL-A offers an independent possibility for the determination of the meniscus and bottom positions which can then be used to determine a cut-off of the interference fringes (see Fig. 3).

For the calibration of the fringe displacement in units of the polymer concentration we prefer the direct determination by means of an independent synthetic boundary experiment producing a boundary between the solution of known polymer concentration and the solvent (dialysate) at low rotation speed (e.g., 4000 rpm). As has been mentioned already in former investigations [5] this type of experiment can, additionally, give some information about the presence of heavy aggregates within the solution. As it is demonstrated in Fig. 4 for some samples the upper plateau decreases with time without a remarkable movement of the synthetic boundary. This is interpreted as a loss of concentration by sedimentation of individual heavy aggregates which do not form a visible sedimentation front. Under the conditions of the equilibrium experiment these aggregates should be located at the cell bottom and will not contribute to the evaluated equilibrium concentration profile. For the calibration of fringe displacement vs. concentration this effect was taken into account.

The next step of data evaluation is the calculation of the meniscus concentration c_m by means of the condition of conservation of the cell concentration

$$c_m = c_0 - \frac{1}{r_b - r_m} \int_{r_m}^{r_b} c_{rel}(r)\, dr \tag{1}$$

with c_0 the cell loading concentration, $c_{rel}(r)$ the relative concentration displacement and r_b, r_m the radial positions

of bottom and meniscus (for sector shaped cell sections r-values have to be replaced by r^2).

The absolute equilibrium concentration profile is then given by

$$c(r) = c_m + c_{rel}(r) . \tag{2}$$

The radial position r can be transformed into the dimensionless radial displacement squared parameter

$$x = (r^2 - r_m^2)/(r_b^2 - r_m^2) . \tag{3}$$

The resulting $c(x)$-profiles are fitted by a sum of positive exponential functions

$$c(x) = \sum_{i=1}^{n} k_i \exp(a_i x) \quad \text{with } a_i > 0 , \tag{4}$$

where a number $n = 3$ was found to be sufficient. The apparent cell weight- and z-average of the molecular masses $M_{w,app}$ and $M_{z,app}$, which depend on fringe displacement only in terms of concentration c and slope $d\ln c/dx$ at meniscus ($x = 0$) and bottom ($x = 1$) position [9, 10], can easily be calculated from the parameters k_i and a_i of the fit:

$$M_{w,app} = \frac{1}{\lambda} \cdot \frac{\sum k_i(\exp a_i - 1)}{c_0} \tag{5}$$

and

$$M_{z,app} = \frac{1}{\lambda} \frac{\sum k_i a_i(\exp a_i - 1)}{\sum k_i(\exp a_i - 1)} \tag{6}$$

with

$$\lambda = (1 - \bar{v}\rho_0)\, \omega^2 (r_b^2 - r_m^2)/2RT \tag{7}$$

(ρ_0 – solvent density, ω – rotor speed in rad/s, R – gas constant, T – absolute temperature).

Extrapolation of $M_{w,app}$ and $M_{z,app}$ to zero concentration was performed according to

$$\frac{1}{M_{app}} = \frac{1}{M} + Bc_0 . \tag{8}$$

M_w- and M_z-values obtained by the described fit procedure are quite insensitive to errors in the determination of fringe displacement at bottom and meniscus position. Additionally, the quality of the exponential fit (Eq. (4)) can be discussed in terms of the thermodynamic ideality of the solution behavior.

Light scattering experiments were carried out with a Sofica 42 000 instrument (Wippler and Scheibling, Strasbourg, France), which was equipped with a 5 mW He–Ne laser as light source and a PC for data recording. The accuracy of measurements was better than 1%. The scattering data analysis was performed by a Zimm plot corresponding to

$$\frac{Kc}{R(q)} = \frac{1}{M_w} + 2A_2c + \frac{1}{3M_w}R_G^2 q^2 , \tag{9}$$

where $R(q)$ is the Rayleigh ratio of the scattering intensity, c the mass concentration of polymer, K contrast factor, containing the optical constants of the system, M_w the weight average of the molar mass, A_2 the second virial coefficient ($A_2 = B/2$, compare Eq. (8)), R_G^2 the z-average of the square of the radius of gyration, $q = (4\pi/\lambda)\sin\theta/2$ (with λ the wavelength in the medium and θ the scattering angle).

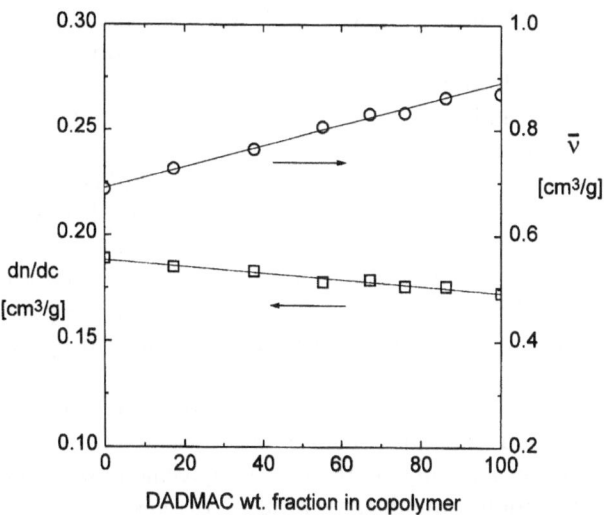

Fig. 5 Partial specific volume $\bar{v}$ and refractive index increment dn/dc for DADMAC-AAM copolymers is 0.5 M NaCl dependent on copolymer composition [7]

Results and discussion

Partial specific volumes in 0.5 M NaCl were $\bar{v} = (0.856 \pm 0.014)$ cm^3/g for poly-DADMAC and $\bar{v} = (0.783 \pm 0.014)$ cm^3/g for poly-MADAMBQ, independent of the molecular mass of the polymers. For the series of DADMAC-AAM-copolymers, $\bar{v}$ as well as the refractive index increment dn/dc vs. copolymer composition are given in Fig. 5 (from [6, 7]).

Fig. 6 Equilibrium concentration profiles from a 6-channel cell experiment. Example for "ideal" equilibrium behavior of a polydisperse system: poly-DADMAC in 0.5 M NaCl, rotor speed 18 000 rpm, ($M_w = 0.98 \times 10^5$ g/mol, $M_z = 1.6 \times 10^5$ g/mol, $B = 2 \times 10^{-3}$ ml mol/g^2)

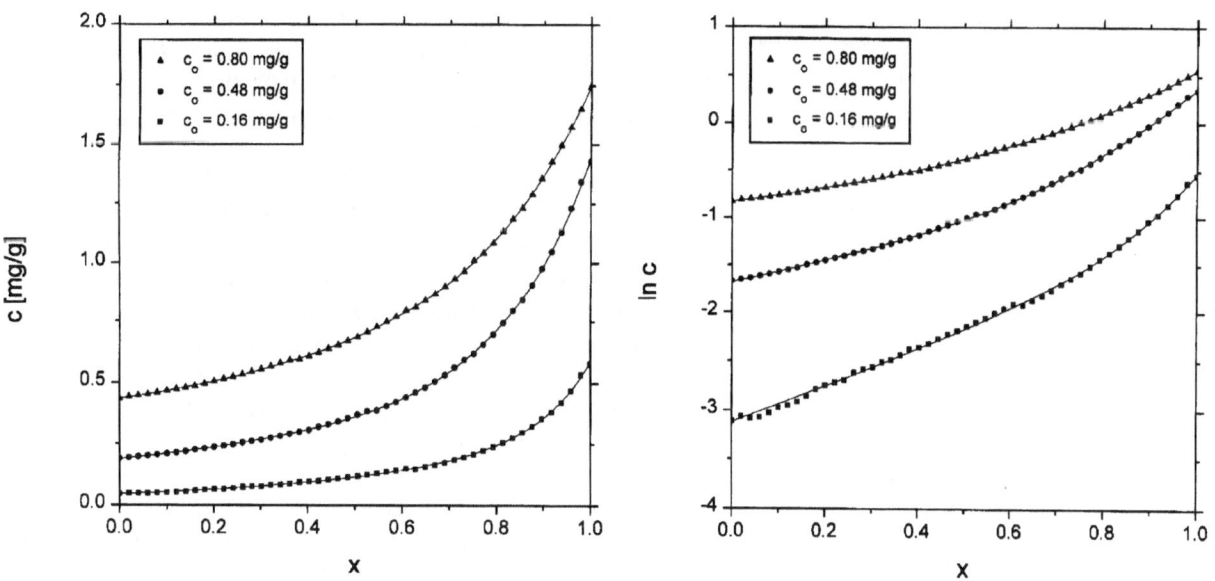

Figure 6 shows a typical set of "low-speed" sedimentation equilibrium profiles (from one run with three loading concentrations c_0 in a 6-channel cell plotted in the form $c(x)$ and $\ln c(x)$). This type of "ideal" equilibrium concentration distribution was observed in the case of hydrophobic modified polyelectrolytes poly-MADAMBQ as well as for lower molecular mass poly-DADMAC's ($M_w < 10^5$ g/mol). The experimental points can be fitted well by positive exponential functions (Eq. (4)). The upward bending of the $\ln c$ vs. x graph indicates the polydispersity of the sample, which may be expressed by the M_z/M_w-ratio.

For those samples the results of other characterization methods (viscosity, sedimentation velocity, light scattering) are generally in reasonable agreement with the sedimentation equilibrium results. In case of poly-MADAMBQ regardless of different polydispersities of the samples, M_z/M_w ranging from 1.3 to 2.3, linear plots of

Fig. 7 Mark-Houwink $[\eta]$–M_w and s_0–M_w relations for poly-MADAMBQ in 0.5 M NaCl

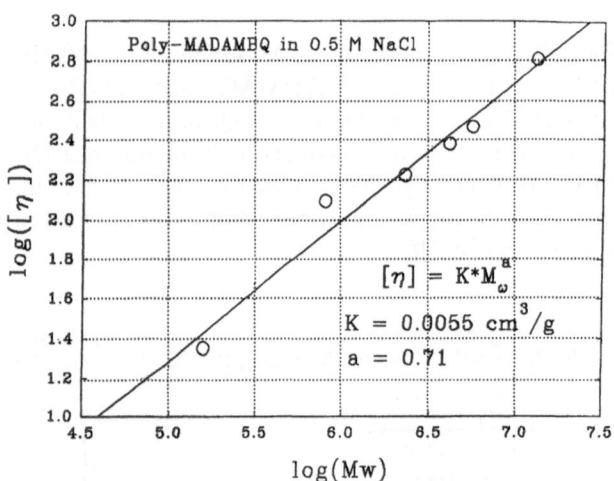

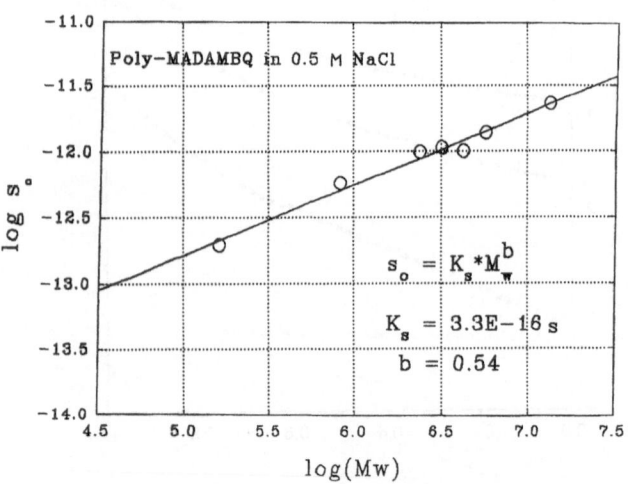

Table 1 Exponents a and b of the relations $[\eta] = K M_w^a$ and $s_0 = K_s M_w^b$ for poly-MADAMBQ compared with poly-MADAM from [11]

	poly-MADAMBQ in 0.5 M NaCl	poly-MADAM [11] in	
		0.1 M NaCl	1 M NaCl
a	0.71	0.76	0.71
b	0.54	0.35	0.41

$\log [\eta]$ vs. $\log M_w$ and $\log s_0$ vs. $\log M_w$ can be observed [8] (Fig. 7). The resulting parameters of the intrinsic viscosity respectively sedimentation coefficient vs. molecular weight relations are presented in Table 1 and compared with those from Stickler [11] obtained for the unmodified cationic polyelectrolyte poly-(methacryloyloxyethyltrimethylammonium chloride) (poly-MADAM). Comparing the results one can see that the sedimentation behaviour is more sensitive to the hydrophobic modification of the polymer than the intrinsic viscosity.

A qualitatively different behavior of equilibrium concentration distributions was observed in the case of higher molecular mass poly-DADMAC ($M_w > 10^5$ g/mol, see Fig. 8). For higher concentrations the exponential fit is very poor. Nevertheless, reliable results for M_w can be obtained via the extrapolation to zero concentration. The $\ln c$ vs. x plot is downward bent in the bottom region indicating a non-ideal behavior of the solution and covering the polydispersity effect. In such cases a reliable determination of M_z is mostly impossible. The non-ideal solution behavior may be explained by the intensive interaction of the strong polyelectrolyte with water as a highly polar solvent resulting in large excluded volumes even in the presence of the low molecular salt. This find an expression in high virial coefficients B obtained from the concentration dependence of M_w according to Eq. (8). For the poly-DADMAC samples we found $B = 2$–5×10^{-3} ml mol/g^2 compared with $B = 0.08$–0.13×10^{-3} ml mol/g^2 for the hydrophobically modified polyelectrolyte Poly-MADAMBQ.

Another expression of the non-ideality of the solutions of higher molecular weight poly-DADMAC in 0.5 M NaCl is the inconsistency of the results of different characterization methods. For 15 poly-DADMAC samples with molecular masses between 10^4 g/mol and about 5×10^5 g/mol the results of M_w from sedimentation equilibrium experiments have been compared with those from elastic light scattering (obtained after filtration through 0.45 μm) (Fig. 9). Only for the samples with the lowest molecular masses ($M_w < 10^5$ g/mol) the results agree satisfactorily. With increasing molecular mass sedimentation equilibrium results in much lower M_w-values than light scattering. In order to clear up this discrepancy for the case of the

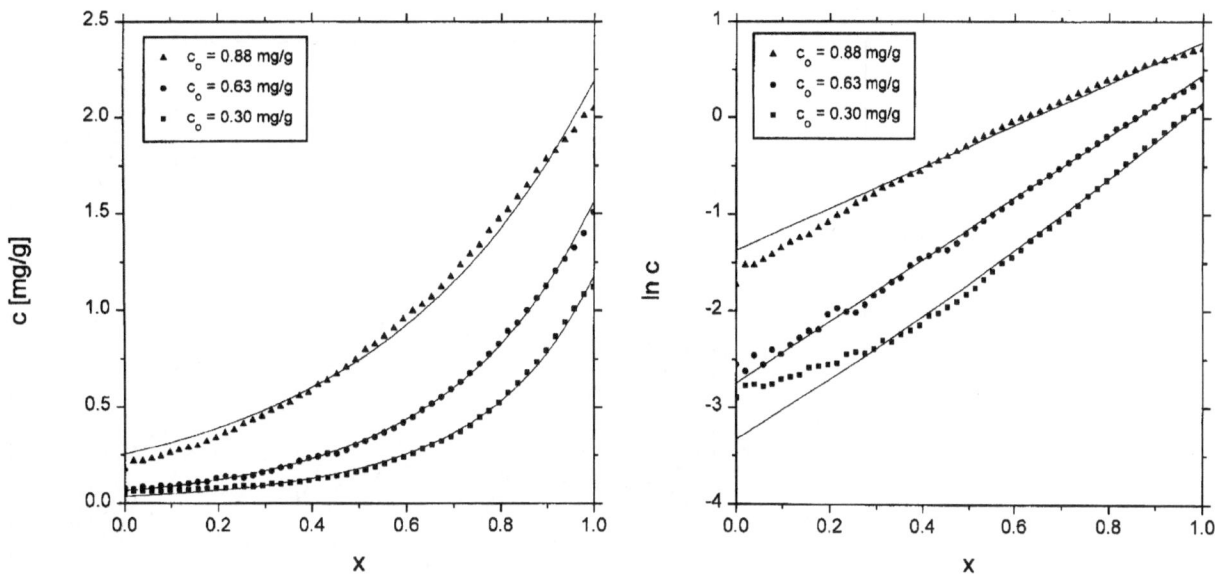

Fig. 8 Equilibrium concentration profiles from a 6-channel cell experiment. Example for "non-ideal" equilibrium behaviour at higher concentrations: poly-DADMAC in 0.5 M NaCl, 12 000 rpm ($M_w = 3.3 \times 10^5$ g/mol, $B = 3 \times 10^{-3}$ ml mol/g^2)

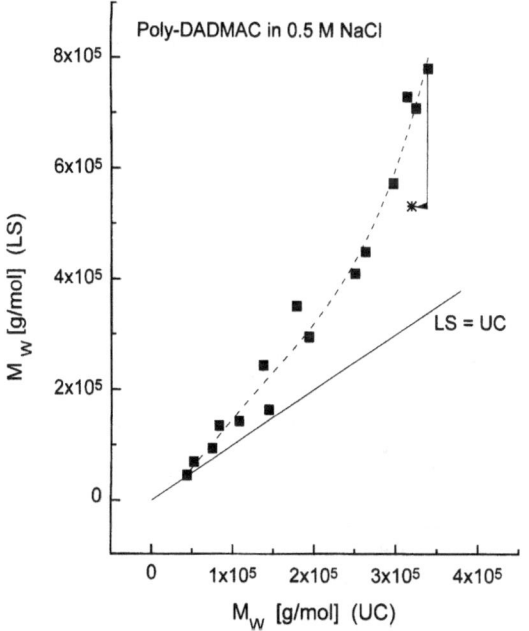

Fig. 9 Comparison between M_w obtained from sedimentation equilibrium and from light scattering for poly-DADMAC samples in 0.5 M NaCl (* – sample precentrifuged for 20 h at 12 000 rpm)

highest molecular weight sample the solution was centrifuged in a preparative tube under conditions comparable with those of the sedimentation equilibrium run (20 h at 12 000 rpm) and then reanalysed by light scattering and sedimentation equilibrium as well. By means of a density measurement it could be checked that the polymer concen-

tration in the solution was not decreased by this procedure within the experimental error of the concentration determination by density ($<4\%$). The results of the molecular mass determination are plotted as an additional point in the M_w(LS) vs. M_w(UC) plot of Fig. 9. The M_w-value of the precentrifuged sample obtained by light scattering is significantly decreased whereas the result from sedimentation equilibrium is nearly unaffected, but a discrepancy between both results still remains. It must be assumed that poly-DADMAC solutions (especially of the high molecular weight samples) contain small amounts of heavy aggregates which will be located at the cell bottom under the conditions of the equilibrium experiments and will not be considered in the evaluation of the equilibrium concentration profiles. Another hint of the existence of aggregates was already given by the synthetic boundary experiments. From the decrease of the upper plateau level (see Fig. 4) the mass fraction of aggregates could be estimated to approximately 8% for the high molecular weight poly-DADMAC's and was decreased by about the half for the precentrifuged sample.

It could be assumed that small amounts of heavy aggregates may be overestimated in light scattering experiments. To demonstrate the quality of the light scattering measurements a typical Zimm diagram of a high molecular weight sample is shown in Fig. 10. There are no noticeable indications for an anomalous behavior. Only a slight decrease of the concentration dependence with decreasing scattering angle gives a hint to the presence of small amounts of high molecular species. If they exist, they should have a more compact structure, because no

Fig. 10 Zimm diagram of a high molecular weight poly-DADMAC sample in 0.5 M NaCl: $M_w = 7.3 \times 10^5$ g/mol, $R_G = 61$ nm, $A_2 = 1.1 \times 10^{-3}$ ml mol/g^2

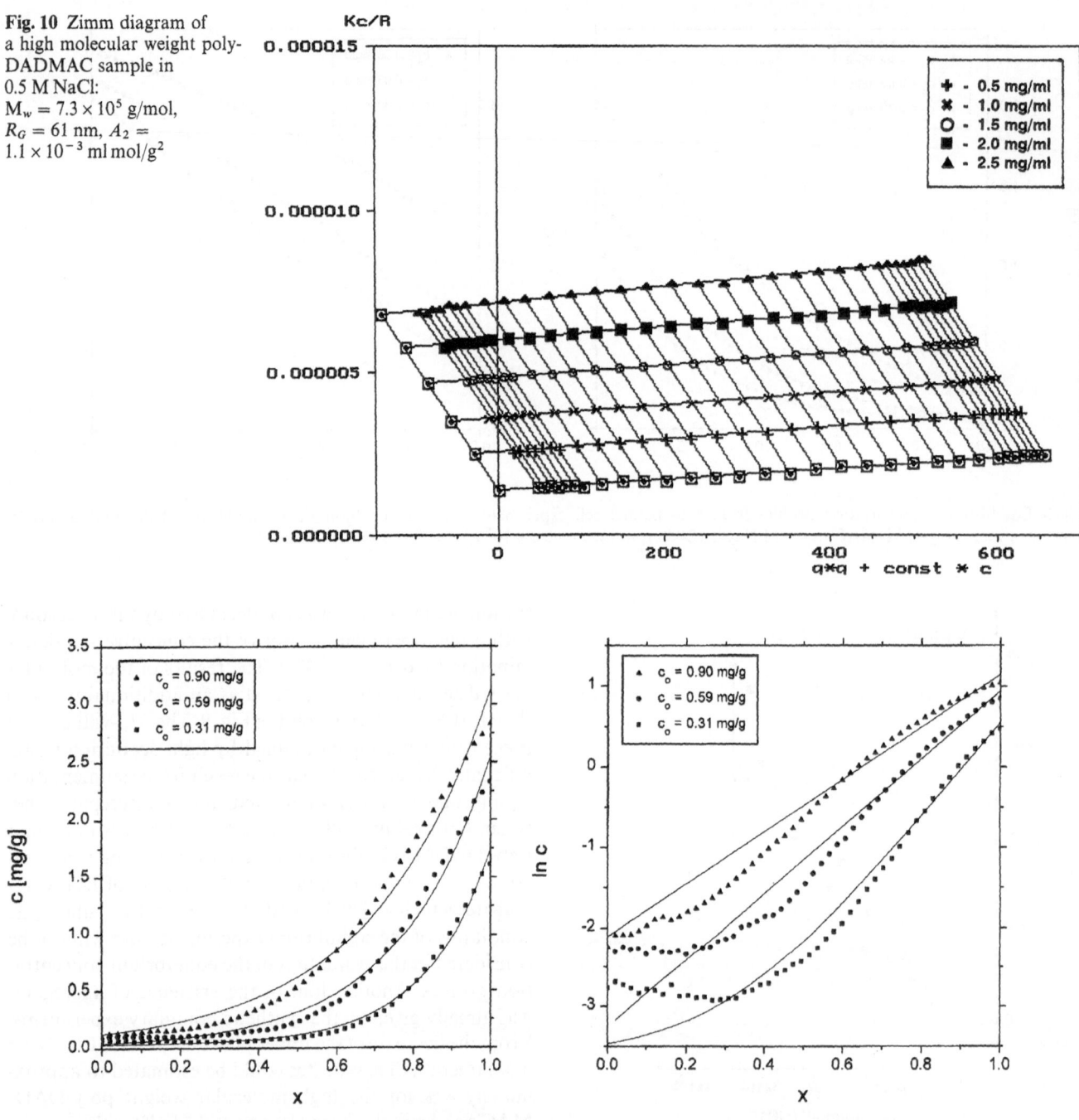

Fig. 11 Equilibrium concentration profiles from 6-channel cell experiment. Example for an extremely "non-ideal" equilibrium behavior: DADMAC-AAM copolymer (47% DADMAC) in 0.5 M NaCl, 10 000 rpm

stronger curvature of the scattering curves occurs. The Mark-Houwink-plot of log $[\eta]$ vs. log M_w (with M_w from light scattering) exhibits a curvature in the high molecular weight region. The same behavior was stated for poly-DADMAC fractions with $M_w > 2 \times 10^5$ g/mol in [12] and related to the presence of branched structures.

An extremely non-ideal behavior of the equilibrium concentration distributions was observed in the case of some of the DADMAC-AAM-copolymers (an example is given in Fig. 11). Here the non-ideal curve shape was detected even at the lowest concentration and an exact determination of M_w seems to be impossible. The virial coefficients were obtained somewhat lower than for DAD-MAC-homopolymers $(B = 0.8–2 \times 10^{-3}$ ml mol/g$^2)$, but these values may be incorrect because of the inexact determination of M_w.

Progr Colloid Polym Sci (1997) 107:127–135
© Steinkopff Verlag 1997

One reason for the non-ideal sedimentation equilibrium behavior may be that, in spite of the low initial cell loading concentrations, the concentration in the cell bottom region can reach the so-called critical or overlap concentration $c^* \approx 3M_w/(4\pi N_A R_G^3)$ of a network solution. Indeed, for high molecular weight poly-DADMAC ($M_w \approx 5 \times 10^5$ g/mol, $R_G \approx 60$ nm) one get overlap concentrations $c^* \approx 1$ mg/ml and for the DADMAC-AAM copolymers ($M_w \approx 10^6$ g/mol, $R_G \approx 100$ nm [7]) $c^* \approx 0.4$ mg/ml, which are in the order of the cell bottom concentrations of our experiments (Figs. 8 and 11). For these systems the experimental conditions for sedimentation equilibrium have to be improved minimizing the λ-value of Eq. (7) (i.e. rotation speed ω and filling height ($r_b - r_m$) have to be as low as possible) under the conditions of a sufficiently detectable fringe displacement. Generally, low cell loading concentrations c_0 are desired, but there is a limitation imposed by the sensitivity of interference detection at the given optical path length of the analytical cells.

References

1. Dautzenberg H, Jaeger W, Kötz J, Philipp B, Seidel Ch, Stscherbina D (1994) Polyelectrolytes: Formation, Characterization, Application. Karl Hanser Publ., Munich, Chap. 5.
2. Dautzenberg H, Jaeger W, Kötz J, Philipp B, Seidel Ch, Stscherbina D (1994) Polyelectrolytes: Formation, Characterization, Application. Karl Hanser Publ., Munich, Chap. 1, p. 11 ff and references therein
3. Hahn M, Jaeger W (1992) Ang Makromol Chem 198:165
4. Huang PC, Reichert K-H (1989) Ang Makromol Chem 162:19
5. Wandrey Ch, Görnitz E (1992) Acta Polymer 43:320
6. Brand F (1995) PhD Thesis, Tech Univ, Berlin
7. Brand F, Dautzenberg H, Jaeger W, Hahn M (1997) Ang Makromol Chem 248:41
8. Zimmermann A (1996) PhD Thesis, Tech Univ, Berlin
9. Fujita H (1975) Foundations of Ultracentrifugal Analysis. Wiley, New York
10. Lechner MD, Mächtle W (1992) Makromol Chem, Macromol Symp 61:165
11. Sticker M (1984) Ang Makromol Chem 123/124:85
12. Xia J, Dubin PL, Edwards S, Havel H (1995) J Polym Sci, B: Polym Phys 33:1117

Progr Colloid Polym Sci (1997) 107:136–147
© Steinkopff Verlag 1997

POLYMERS, COLLOIDS & SUPRAMOLECULAR SYSTEMS

H. Cölfen
T. Pauck
M. Antonietti

Investigation of quantum size colloids using the XL-I ultracentrifuge

Dr. H. Cölfen (✉) · T. Pauck · M. Antonietti
Max-Planck-Institut
für Kolloid-und Grenzflächenforschung
Kolloidchemie
Kantstraße 55
14153 Teltow
E-mail: coelfen@castor.mpikg-teltow.mpg.de

Abstract Quantum size colloids attract a steadily growing scientific and economical interest because of their size-dependent optical and electrical properties. It is shown that the fractionation in the analytical ultracentrifuge enables the characterization of individual species even in complicated mixtures. If an ultracentrifuge with combined UV-absorption and Rayleigh interference optics is used, dependence of the spectra as well as of extinction coefficients on the particle size can be examined. As it is also possible to monitor particle growth from its initial stages, the combined spectral and particle size information can lead to an understanding of the particle growth mechanism. This is demonstrated for model systems of aggregates of small individual particles and grown particles of the same size. Furthermore, sedimentation velocity experiments prove to be very useful for determining the effectiveness of reactions performed in "micellar nanoreactors" and for determining the extinction coefficient of one component in a complicated mixture.

Key words Quantum size effect – colloids, analytical ultracentrifugation – sedimentation velocity – particle growth

Introduction

Quantum size particles are a particularly interesting class of colloidal systems because their optical and electrical properties depend on the particle size. The probably most well-known example is the classical Faraday experiment with colloidal gold which exhibits different colors at different particle sizes. Size quantization arises when the particle radius is in the order of the exiton Bohr radius of the material of interest. If one considers only a single atom, the energy levels are well defined. For 2 atoms, these energy levels split up according to quantum mechanics. In the case of a colloidal system with n atoms, there are n energy levels resulting from the splitting of single-atom energy levels. These levels are rather similar in their energy. In the

extreme case, the differences between the energy levels smear completely, resulting in a semiconductor with a valence band and a conduction band. The energy between these bands depends on the particle size for particles being smaller than approximately 10 nm. In extreme cases, such particles can be tuned from metallic conductors (vanishing energy gap) to semi-conducters (small energy gap). Therefore, it is a challenge for colloid chemists to synthesize such particles with desired properties. In Fig. 1, the quantum size effect is demonstrated for ZnO of various particle size.

The different spectral properties become quite obvious. A decrease of the particle diameter from 6 to 1 nm shifts the onset of absorbance by about 20 nm. Thus, the quantum size effect could be exploited for the design of selective UV-absorbers.

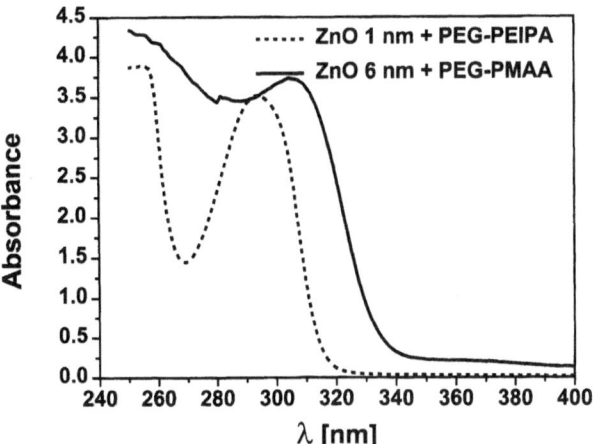

Fig. 1 UV-spectra of ZnO at different particle diameters

There are different approaches for the synthesis of such particles with defined size on the nm scale. One of them is the nanoreactor concept [1–3]. This approach applies monodisperse micelles formed by block copolymers in selective solvents. The block copolymers are designed such that the micelle core consists of strongly segregated functionalized polymer which interacts with metal cations. These metal cations can be added to such a micellar solution resulting in a metal-loaded micelle core. Now reducing the metal salt, one obtains the metal colloid with a defined size inside the micelle core. Alternatively, one can add appropriate anions which then leads to ceramic materials or oxidic semiconductors.

For all the systems described above, the characterization of the particle size as well as their spectral properties is very important. The complete analysis can be done using the absorption optics of an analytical ultracentrifuge. A combined use with Rayleigh interference optics provides the determination of extinction coefficients. In case of the block copolymer micelles as nanoreactors, a combined use of both optical systems yields a fast characterization of the efficiency of the reaction, e.g. the ratio of filled to empty micelles. This article will describe these capabilities of the analytical ultracentrifuge for selected examples.

Materials and methods

A Beckman Optima XL-I analytical ultracentrifuge (Beckman Instruments, Palo Alto, CA) equipped with Rayleigh interference and UV-absorption optics was used for all experiments. If not stated otherwise, all experiments were performed at 25 °C. The model colloid particle sizes were not determined by ultracentrifugation, because the density of most of the composite particles was not known. Instead, we used dynamic light scattering as all the particles were monodisperse. Dynamic light scattering was performed at 90° in a NICOMP particle sizer (NICOMP Model 370, Santa Barbara, CA). Transmission electron microscopy (TEM) was performed on an EM912 Omega electron microscope (Carl Zeiss, Jena). The solvents used were isopropanol for ZnO colloids, water for Au in microgels and Pt in PEO-PMAA, and toluene for Au and CdS in block copolymer micelles.

The ZnO colloids were prepared in isopropanolic solution via a sol-gel route [4]. Two-milliliters of a 20 mM NaOH dissolved in dried isopropanol were added to 2 ml of isopropanolic solution of a 0.2 mM $Zn(Ac)_2$. The mixture was then heated to 65 °C under reflux and vigorous stirring for 2 h to dehydrate the primarily formed $Zn(OH)_2$. The beginning of heating defines the origin of the time window in which the propagation of the reaction was monitored. All reagents were used as supplied from Aldrich, except for isopropanol which was dried over 0.4 nm molecular sieve.

The gold particles prepared in micelles and microgels have already been subject of other studies. Their synthesis is described in refs. [5, 6]. The synthesis of CdS in polystryrene-poly-4-vinylpyridine micelles is described in ref. [3]. The synthesis of ZnO in polystyrene-poly-methacrylic acid (PS-PMAA) micelles is subject of the same study [3]. Pt was prepared in polyethyleneoxide-block-polymethacrylic acid block copolymers (PEO-PMAA) as described in [7].

As these colloids only serve as model systems and their characterization has partly been performed in other studies (light scattering, electron microscopy) [3, 5, 6], we place the properties relevant to this study in the materials section.

The first model system is gold prepared in polystyrene-sulfonate (PSS) microgels [6]. Sample A is a microgel ($d = 110 \pm 27.5$ nm from light scattering) filled with large single crystalline gold particles (partly well exceeding 50 nm in diameter) derived by slow reduction of gold salt (see Fig. 2A). The spectrum of the gold in the visible range shows an increase in the absorbance between 500 and 600 nm and is well distinguished from the spectra of the other samples (see Fig. 2B). Sample A is a model system for a grown particle.

Sample B was prepared in the same microgel ($d = 110 \pm 27.5$ nm from light scattering) as sample A but by fast reduction of the Au-salt leading to much smaller gold particles (see Fig. 2A) with a diameter between 3 and 10 nm. Sample B is a model system for an aggregate of several small individual particles. The UV-spectrum (Fig. 2B) shows a plasmon band characteristic for gold particles of that size and differs much from that of sample A.

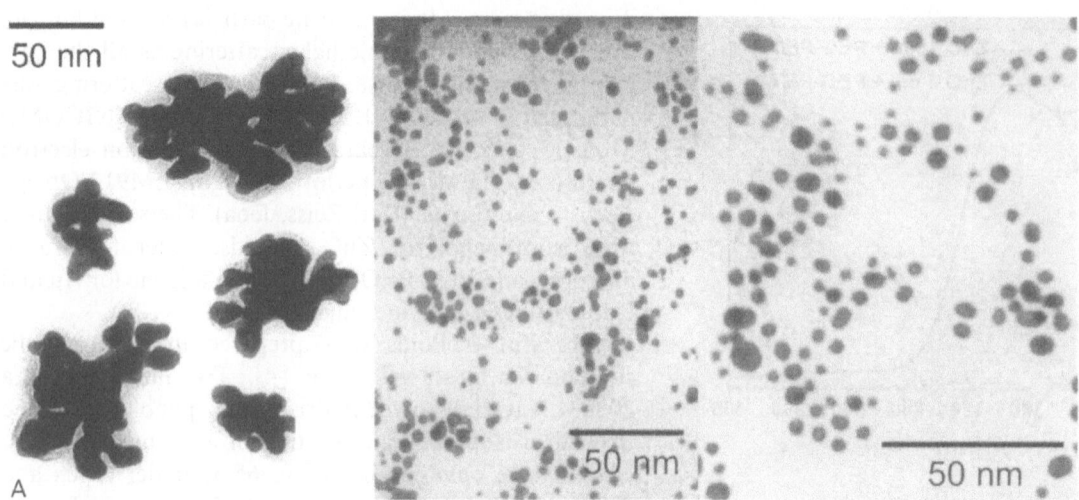

Fig. 2A TEM-photos of Au particles prepared in polystyrenesulfonate microgels. Samples from left to right: Sample A, sample B, sample C. In all cases, the contrast of the polymer was not sufficient for a visualization in electron microscopy

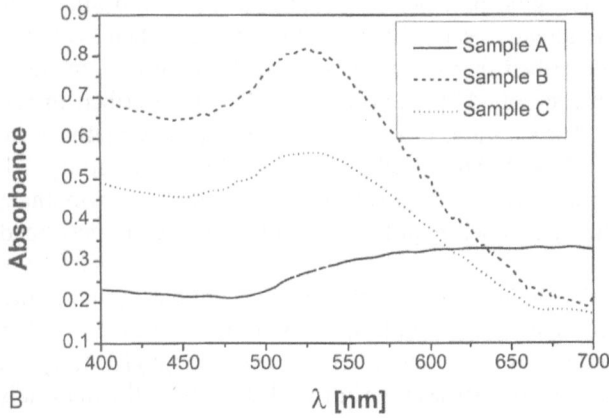

Fig. 2B UV-absorption spectra of Au particles prepared in PSS-microgels. The spectra correspond to the stock dispersions used for the later experiments and were recorded on the Optimal XL-I

Sample C was prepared in a smaller microgel ($d = 55.4 \pm 14.4$ nm) also by fast reduction of the gold salt leading to gold particles with diameters between 3 and 8 nm (see Fig. 2A). Thus, the UV-spectrum of this sample is identical to that of sample B. However, the ratio of metal to polymer was adjusted such that the sedimentation coefficients of samples B and C were almost identical. Sample C serves as a model for an aggregate of several small individual species but with only half the hydrodynamic radius as sample B.

Another class of model systems was prepared by the precipitation of CdS in polystyrene-block-poly-4-vinyl-pyridine (PS-P4VP) micelles [3]. The size of the CdS particles was controlled by the ratio of block copolymer to CdS. For a ratio of 50:1 (for polymer:CdS), small particles are formed (Fig. 3A). The particle size can be determined from the position of the band gap in the UV-spectrum to be 2 nm [8]. Light scattering yields the hydrodynamic diameter of the whole micelle to be 65.5 ± 9.8 nm. Sample A serves as a model for a grown particle.

If the ratio of polymer to Cd is decreased to 6:1, one obtains many small particles of 3 nm, as determined from the position of the band gap in the UV-spectrum [8] (see Figs. 3A and B). The diameter of the whole micelle is equal to that of sample A (68.9 ± 7.6 nm from light scattering). Sample B serves as a model for an aggregate of individual particles.

As a third model system, Au particles prepared in PS-P4VP micelles were chosen. If Au-salt in the micelle core is slowly reduced, Au colloids with a diameter of 6 nm are obtained (see Fig. 4A) [5]. This sample (sample A) shows a plasmon resonance in the UV-spectrum (see Fig. 4B). The hydrodynamic diameter of the whole micelle is 57.0 ± 6.2 nm as determined from light scattering. Sample A is a model for a grown particle.

If, on the other hand, the Au-salt in the micelle core is reduced fast (sample B), the gold particles in the micelle core are only as small as 2–3 nm (see Fig. 4A). These Au-particles show no plasmon resonance (Fig. 4B) and hence, can, again be distinguished from sample A via the absorption spectrum. The hydrodynamic diameter of sample B equals that of sample A (57.3 ± 5.5 nm). Sample B is a model for an aggregate.

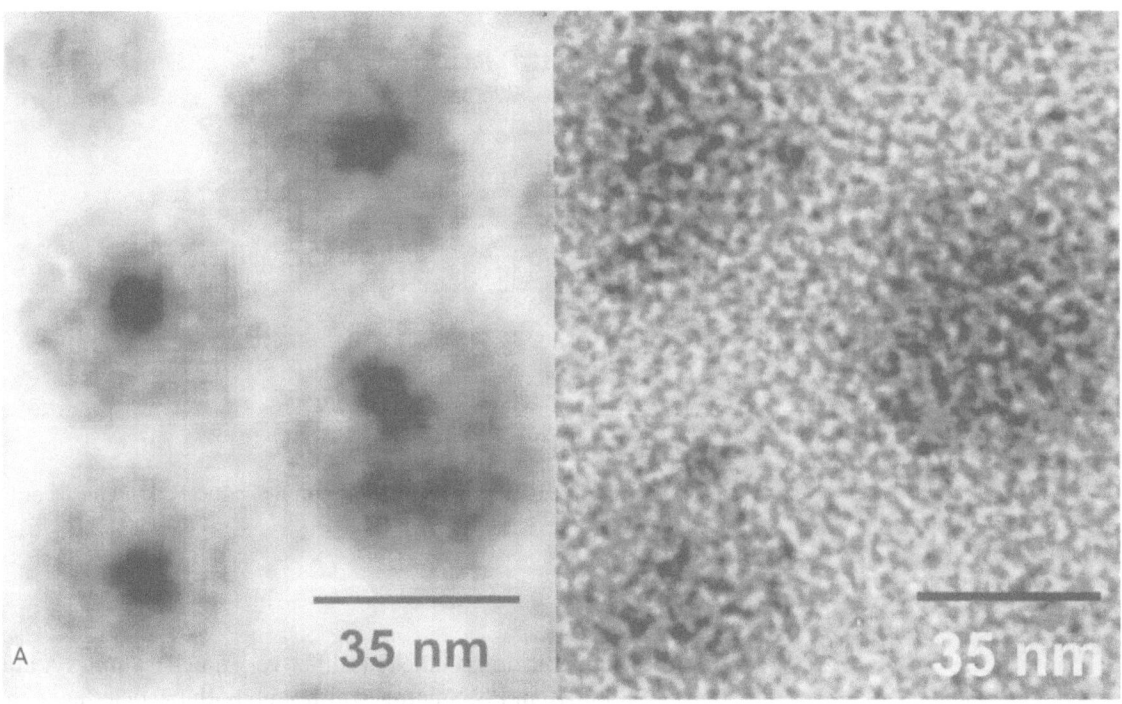

Fig. 3A TEM-photographs of CdS particles prepared inside the core of PS-P4VP micelles. Samples from left to right: sample A, sample B

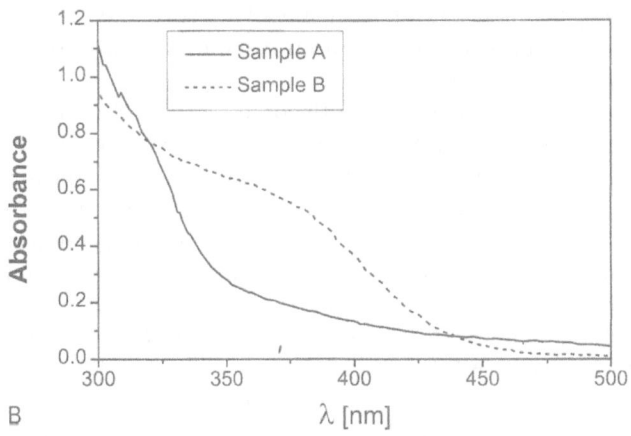

Fig. 3B UV-absorption spectra of CdS particles prepared inside PS-P4VP micelles. The spectra correspond to the stock dispersions used for the later experiments and were recorded in the Optima XL-I

From the samples mentioned above, 1:1 mixtures were prepared so that a model system for an aggregate was mixed with that of a grown particle of the same system where the particles have the same hydrodynamic radius (also see Table 1). The only exception is the mixture of samples B and C of the Au colloids prepared in microgels.

Results and discussion

Potential of the ultracentrifuge to distinguish between particle growth and aggregation

It is well known that the analytical ultracentrifuge is an excellent tool for determining particle size distributions [9–15]. In a recent study, it was shown that particle size distributions can be determined with a resolution of up to 1 Å [16]. This allows the detailed study of particle growth mechanisms [17]. However, a particle size distribution alone cannot give the information if the particle is solid or if it is an aggregate of smaller particles. These two models of particle growth are presented in Fig. 5.

It is desired to distinguish between these two models. Electron microscopy is one possibility to accomplish this. However, especially if the particles are very small or if they present only poor electron contrast, these two states may be hard to distinguish. Analytical ultracentrifugation in a density gradient – often a valuable tool to separate structures even with marginal differences – cannot be applied in most cases for inorganic colloids or polymeric hybrids because the particle density is too high to allow an appropriate banding by any known density gradient.

However, with quantum size particles sedimentation velocity analytical ultracentrifugation can be applied with

140

H. Cölfen et al.
Quantum size colloids in the XL-I

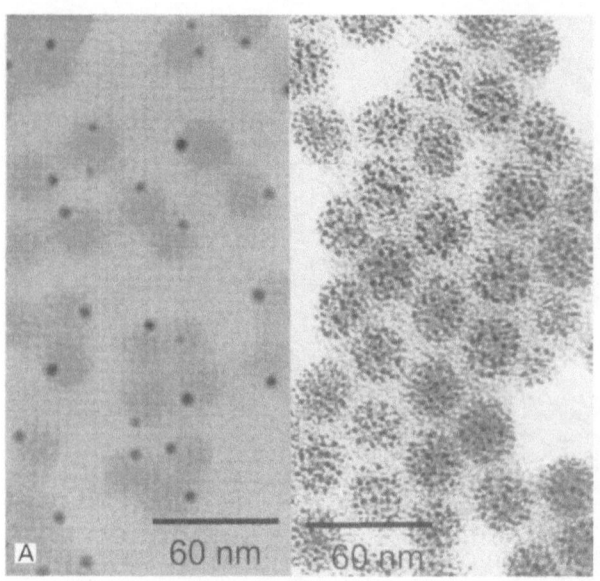

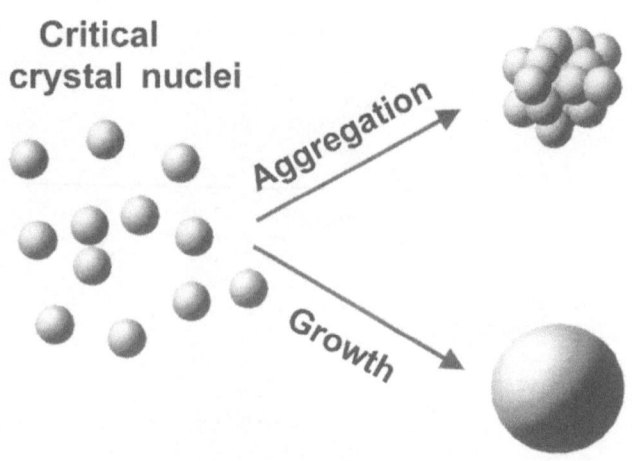

Fig. 5 Particle growth by aggregation of critical crystal nuclei or by growth of a single particle

Fig. 4A TEM-photos of Au particles prepared inside the core of PS-P4VP micelles. Samples from left to right: sample A, sample B

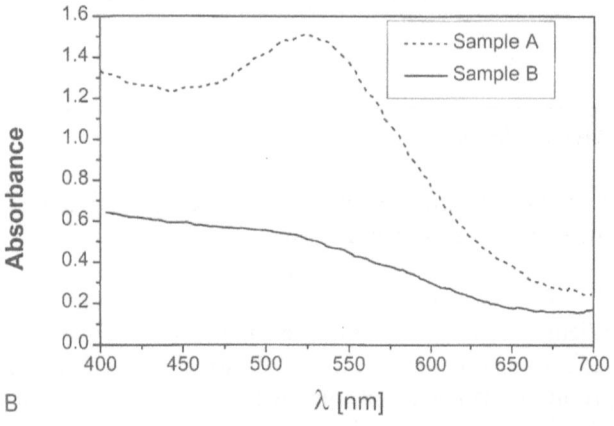

Fig. 4B UV-absorption spectra of Au particles prepared inside PS-P4VP micelles. The spectra correspond to the stock dispersions used for the later experiments and were recorded in the Optima XL-I

advantage because radial scans yield the particle size of the sedimenting particle, whereas wavelength scans yield the particle size via the band gap. In case of aggregates, these two particle sizes differ.

As mentioned above, we set up a system of model colloids and made diverse 1:1 mixtures of two colloids with equal external diameter. The four different investigated systems are summarized in Table 1.

The easiest case is present when aggregates and grown particles exhibit such a large difference in density that their

sedimentation coefficients differ very much (samples A and B for Au in PSS microgel) although both particles have the same hydrodynamic diameter. This makes them indistinguishable by all methods sensitive to the diffusion coefficient (dynamic light scattering, flow-field-flow-fractionation, etc.). If such mixtures are investigated by analytical ultracentrifugation, it is easily possible to detect both components applying radial scans at two different appropriate speeds in one experiment.

Figure 6A shows that both components in the mixture are separated completely. This can also be demonstrated by acquiring the absorption spectra at the radius indicated by a vertical line in Fig. 6A in a separate experiment. At 1100 rpm, the sum spectrum is derived for both mixture components. At 15 000 rpm, one obtains the spectrum for pure sample A (the slower sedimenting component), whereas the difference spectrum at 1100 rpm delivers the spectrum of sample B (see Fig. 6B).

These results indicate that it is possible to distinguish between aggregates and grown particles by analytical ultracentrifugation in case the grown particles have a larger density than the aggregates of individual smaller particles.

The next mixture of model colloids corresponds to the case of only slight density differences between the grown and the aggregated particles (CdS in PS-P4VP micelles, samples A and B). Radial scans during the sedimentation velocity experiment still indicate two components by the formation of a second step as the experiment proceeded, but the resolution achieved between the two components is still poor (see Fig. 7A).

A better possibility to distinguish between the aggregates and grown particles is to consider the absorption spectra acquired at the times and radii indicated in Fig. 7A.

Progr Colloid Polym Sci (1997) 107: 136–147
© Steinkopff Verlag 1997

Table 1 Model systems for 1:1 mixtures of aggregates and grown particles and the run conditions and detection with the analytical ultracentrifuge

System	Type of system	Detection
Au in PSS-microgel, equal hydrodynamic radius	Aggregates + particles with big density difference and different spectra	2 speeds Spectrum or radial scans
CdS in PS-P4VP micelles, equal hydrodynamic radius	Aggregates + particles with slightly different density and different spectra	1 speed Spectrum or radial scans
Au in PS-P4VP micelles, equal hydrodynamic radius	Aggregates + particles with almost equal density and slightly different spectra	1 speed Spectrum
Au in PSS-microgels, different hydrodynamic radius	Aggregates with different density but equal sedimentation coefficients and spectra	Not possible

For the model system CdS in PS-P4VP micelles, the "aggregates" have a slightly higher sedimentation coefficient than the "grown" particles, facilitated by the different amounts of inorganic materials in the hybrid colloid. This has the consequence that for the conditions outlined as No. 1 in Fig. 7A, the sum spectrum for both components is recorded (see Fig. 7B). In this spectrum, superposition of different band gaps results in an absorption spectrum with two steps corresponding to the band gaps of the two components in the mixture. At the radius and time No. 2 in Fig. 7A, only the spectrum of the slower-sedimenting component (sample A) is obtained. The difference spectrum with that from No. 1 is the spectrum of the faster sedimenting sample B. From such spectra, the particle sizes of the CdS can be derived using the wavelength of the band gap [8].

The next model system mixture represents the case of aggregates and grown particles which have an almost equal density but slightly different spectra. It is a mixture of samples A and B of Au prepared in PS-P4VP micelles. As the sedimentation coefficients of both samples in the mixture are almost equal, radial scans at a certain speed yield only one sedimenting boundary which could well be misinterpreted in a way that only one species is present (see Fig. 8A).

Nevertheless, it is still possible to detect both, aggregates and grown particles. If absorption spectra are acquired near the cell bottom depending on time, one can see that the spectra change from the sum spectrum of A and B showing the plasmon resonance of the grown particles A to the spectrum of the slightly slower sedimenting aggregates B. This is illustrated in Fig. 8B where the plasmon resonance in the dashed sum spectra vanishes more and more until the spectrum of the aggregates B is obtained (solid line). Hence, it is still possible to distinguish aggregates from grown particles even if the sedimentation velocity profile only shows one sedimenting component (see Fig. 8A). This is due to the steadily decreasing

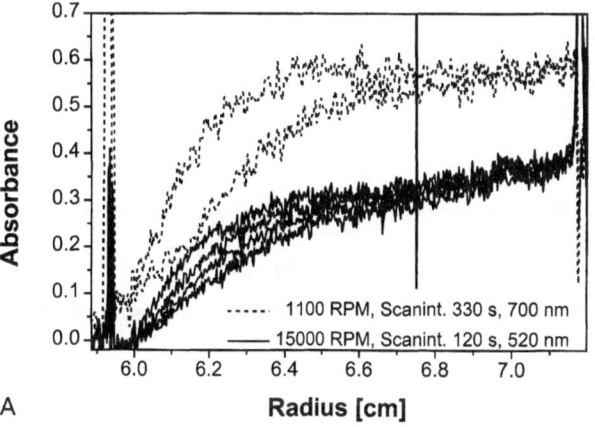

Fig. 6A Sedimentation velocity profiles for a 1:1 mixture of samples A and B of Au prepared in PSS microgels. The scans were performed in one experiment at the speeds and wavelengths indicated

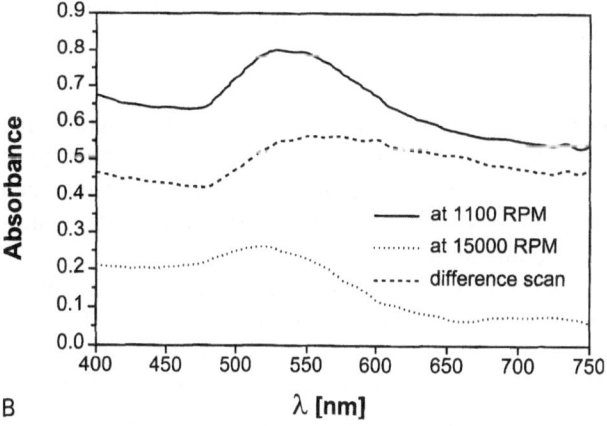

Fig. 6B Absorption scans for a mixture of samples A and B of Au in PSS microgels taken at the radius indicated as a vertical line in Fig 6a at different speeds of centrifugation

142

H. Cölfen et al.
Quantum size colloids in the XL-I

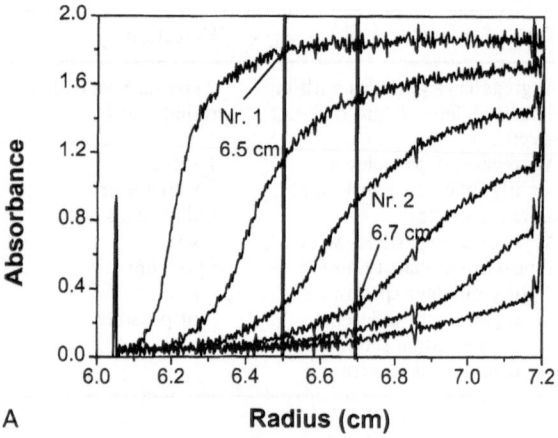

A   **Radius (cm)**

Fig. 7A Sedimentation velocity profiles for a 1:1 mixture of samples A & B of CdS prepared in PS-P4VP micelles. The scans were performed at 30 000 rpm and 300 nm. Scan interval = 2 min. The vertical lines No. 1 and 2 correspond to the radial positions and the time absorption spectra were acquired in a separate experiment (see Fig. 7b)

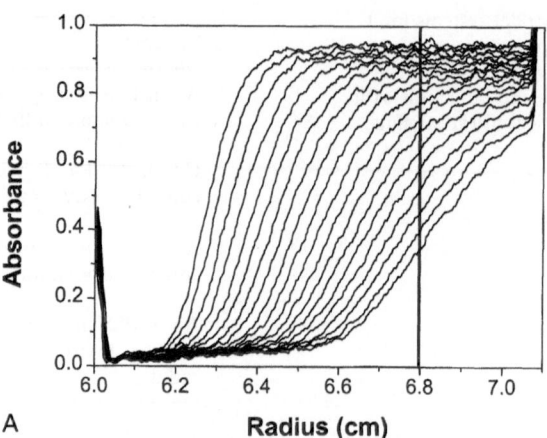

A   **Radius (cm)**

Fig. 8A Sedimentation velocity profiles for a 1:1 mixture of samples A and B of Au prepared in PS-P4VP micelles. The scans were performed at 10 000 rpm and 500 nm. Scan interval = 2 min. The vertical line corresponds to the radial position where absorption spectra were acquired in a separate experiment (see Fig. 8b).

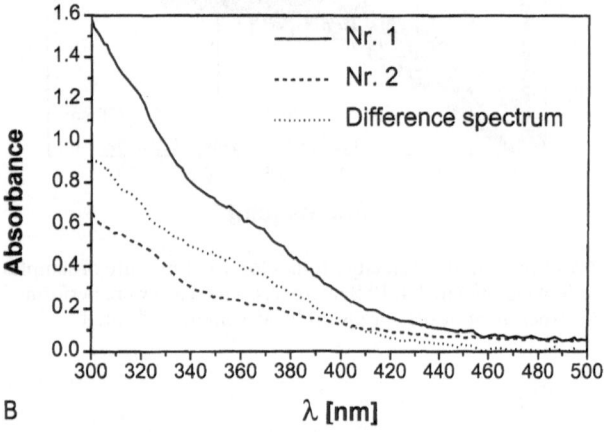

B λ [nm]

Fig. 7B Time-dependent absorption scans for a mixture of samples A and B of CdS in PS-P4VP micelles taken at the radii and times indicated as vertical lines No. 1 and 2 in Fig. 7a at 30 000 rpm

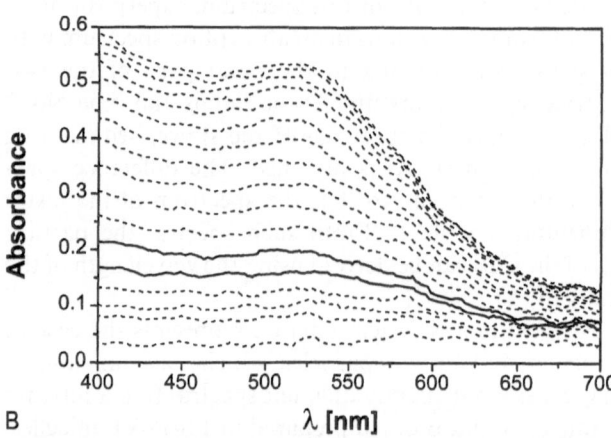

B λ [nm]

Fig. 8B Time-dependent absorption scans for a mixture of samples A and B of Au in PS-P4VP micelles taken at the radius indicated as a vertical line in Fig. 8a at 10 000 rpm and a scan interval of 140 s

concentration of the faster sedimenting component in the region of the cell bottom until it vanishes so that the pure slower-sedimenting component can be detected. Such analysis is only possible, if the spectra of the two components are distinguishable.

This leads to the case where the ultracentrifuge cannot distinguish between different particles anymore. If a mixture of samples B and C of Au prepared in PSS microgels is set up, the two components differ by a factor of 2 in their hydrodynamic radius but have the same sedimentation coefficients. This can be attributed to the appropriate counterbalancing of density difference and particle size. If this mixture is exerted to an ultracentrifugal field, no

separation of the mixture can be achieved either by radial scans or by absorption spectra as the spectra of both components are equal (see Figs. 9A and B).

The absorption spectra in Fig. 9B just decrease in intensity with time so that it is not possible to distinguish between the two components anymore. However, such a mixture can be separated well with techniques which are sensitive to the diffusion coefficient but not to particle density. One especially promising separation technique for such cases would be Field-Flow-Fractionation – in this particular case Flow-Field-Flow Fractionation.

The four model cases above have shown that it is well worth not only to look at radial scans from sedimentation

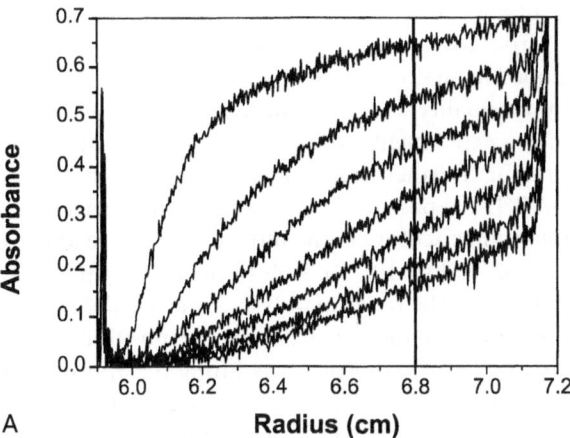

Fig. 9A Sedimentation velocity profiles for a 1:1 mixture of samples B and C of Au prepared in PSS microgels. The scans were performed at 35 000 rpm and 520 nm. Scan interval = 90 s. The vertical line corresponds to the radial position where absorption spectra were acquired in a separate experiment (see Fig. 9b)

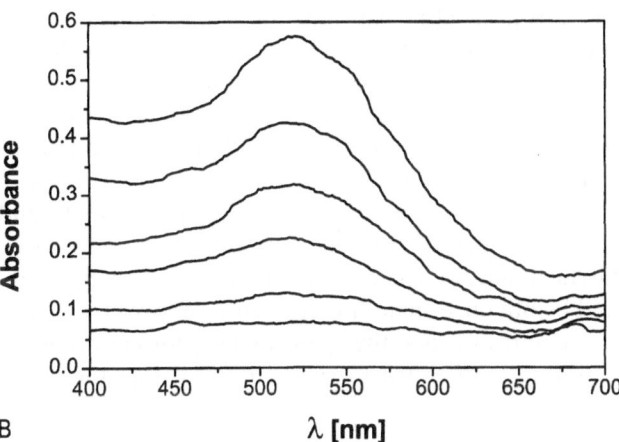

Fig. 9B Time dependent absorption scans for a mixture of samples B and C of Au in PSS microgels taken at the radius indicated as a vertical line in Fig. 9a at 35 000 rpm, scan interval 160 s

velocity experiments but also at spectra acquired at a constant radius but different times. Especially in the case of quantum size particles where spectral properties depend on the particle size, absorbance spectra are more likely to resolve the mixture components than radial scans.

Furthermore, the results from the measurements above show that even complicated cases of the characterization of organic–inorganic hybrid colloids can be addressed. It is principally possible to determine the particle size of the whole particle via the sedimentation coefficient distribution [9–14], whereas absorbance spectra can deliver the particle size of the inorganic component inside the hybrid

colloid if the position of the band gap is considered [8]. Therefore, the presented combination is a fast and effective alternative to electron microscopy which is commonly applied to characterize such particles.

Observation of the particle growth of quantum size ZnO

From an earlier study, it is known that particle growth can be observed via particle size distributions even with Ångström resolution [16]. Therefore, it is of interest, if the growth of a quantum size particle can be monitored via its particle size distribution and via the shift of the band gap in the absorbance spectrum. This would permit the correlation of particle sizes with the corresponding band gaps for different particle sizes in only one experiment and under exactly the same conditions. However, this requires a particle growth which is slow enough that the radial scans during the ultracentrifuge experiments represent snapshots of the particle size distribution at that time. In case of the ZnO formation, a decrease of the reaction speed could be achieved by a decrease of the water content in the reaction mixture. For such a reaction mixture, it is possible to take samples at different reaction times and to monitor the changes of the particle size distribution with time (see Fig. 10).

One can clearly see the particle growth in Fig. 10. The initial particle size observed is 3 nm. This certainly does not correspond to a critical crystal nucleus and it appears that the particle growth is fast in its initial stages. Particles with $d < 3$ nm are formed already in the first 10 min. After 45 min, the particles are still monodisperse but grown to 5.2 nm. After 120 min, one observes several monodisperse

Fig. 10 Time-dependent changes of the particle size distributions of growing ZnO in isopropanol as observed by analytical ultracentrifugation at a wavelength of 280 nm

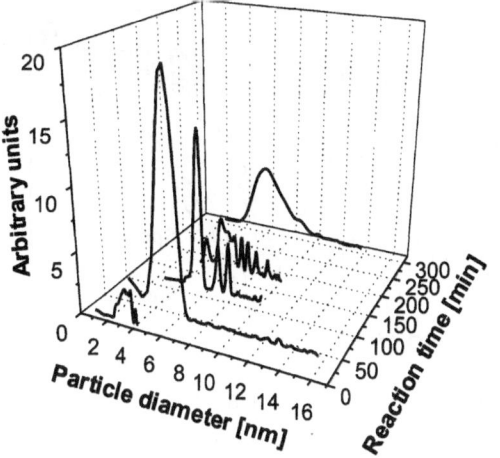

species, the smallest still being the 5.2 nm species already observed after 45 min. The two other species observed are 6.9 and 7.8 nm. After 180 min there is a smaller species at 4.1 nm, the species already observed at 5.5, 6.9 and 7.6 nm and two additional components at 8.2 and 9.0 nm. After 315 min, no more monodisperse species can be observed. Instead, a continuous particle size distribution is detected, which is nearly the envelope curve of the particle size distribution at 180 min. From such a result, one can at least qualitatively learn something about the growth of ZnO.

One explanation for the observations above is that up to 45 min, normal particle growth can be observed which is represented in the increase of the particle size of the detected species. Then, very defined particle sizes are favored. This can be attributed to Ostwald ripening where the structures being especially favorable from the energy point of view grow on cost of those structures which are less favorable. The continuous distribution after 315 min indicates another mechanism of growth again which we cannot explain yet.

With the results from Fig. 10, it is possible to perform a sedimentation velocity run on a mixture of several monodisperse ZnO species and to acquire spectra of the individual components once they are fractionated. A typical experiment is shown in Fig. 11. As soon as several components are fractionated, the radial scans are stopped and absorbance spectra are acquired in the appropriate wavelength range at a fixed radius in the range indicated as "spectra recorded" in Fig. 11. The spectrum of species 1 can be derived after the other 2 species are already sedimented. The sum spectrum of species 1 and 2 is derived after species 3 is already sedimented. To obtain the spec-

trum of species 2, it is necessary to subtract the spectrum of species 1 from the sum spectrum. The spectrum for species 3 is obtained in analogy. Such a procedure delivers spectra of only poor quality which is mainly caused by the already rather noisy experimental data obtained from scans without averaging. Furthermore, the wavelength resolution of the XL-A/XL-I is poor with ≤ 4 nm. As the particle sizes of the quantum size ZnO in the ultracentrifuge are already big (> 3 nm, see Fig. 10), the shift in the band gap is only very small and in the range of only 1–2 nm. We defined the maximum slope in the spectrum as the position of the band gap and derived this point by fitting the appropriate range of experimental data to a sigmoidal curve which could then be differentiated for determining the band gap position. The resulting band gaps, resp. the gap energy, should scale with the inverse squared particle radius due to the model of an electron in a box. The results of the ultracentrifuge experiments are compared with those from literature in Fig. 12.

It can be seen that the data from ultracentrifugation agree well with those of Weller et al. [8], although the band gap from ultracentrifugation is determined slightly higher compared to the literature reference. The origin of this deviation is caused by the determination of the band gap from the maximum slope of the absorption spectra and is also revealed in the determined bulk semiconductor band gap which is also higher than the reported value of 3.35 eV. The difference is nevertheless quite small, and if a larger range of particle sizes could be addressed by analytical ultracentrifugation, much more precise data could be obtained than those presented above. This, however, requires to slow the speed of ZnO formation in the reaction mixture down to such an extent that the initial

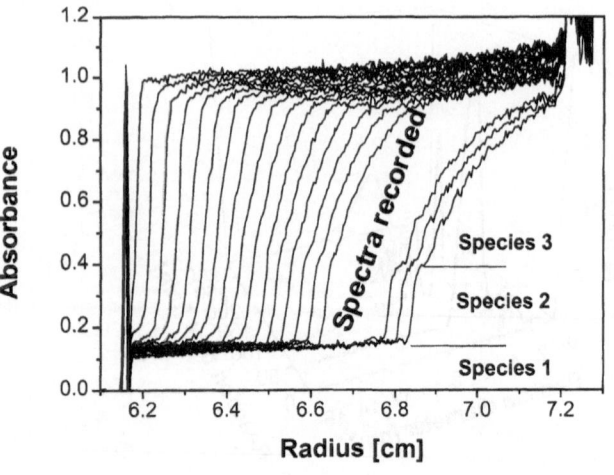

Fig. 11 Sedimentation velocity pattern of growing ZnO with different monodisperse components. Run parameters: 25 °C, 40 000 rpm, 280 nm, scan interval 2 min

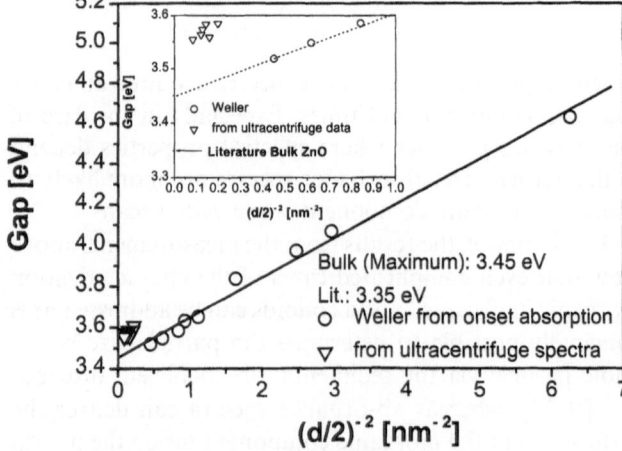

Fig. 12 Energy of the band gap in dependence of the reciprocal squared particle radius for quantum size ZnO from ultracentrifuge experiments compared with literature data [8]

Progr Colloid Polym Sci (1997) 107: 136–147
© Steinkopff Verlag 1997

stages of particle formation (even below 1 nm) could be detected as well.

Determination of extinction coefficients of quantum size particles

The extinction coefficient of a quantum size particle is another important quantity, both for basic research as well as for applications. The use of semiconductor colloids as highly selective UV-absorbers is especially promising, since tuning-in of the spectral characteristics by particle size can be varied in a considerable range. It is not always straightforward to measure the extinction coefficient of a defined quantum size particle, because it is often the case that these particles are mixed with unreacted and optically inactive material. Such questions can be addressed by analytical ultracentrifugation in a very advantageous manner. Sometimes, ultracentrifugation seems to be the only analytical technique as demonstrated in the following case.

ZnO was prepared in PS-PMAA micelles in THF as described in ref. [3]. The dispersion is a clear liquid without any precipitate. If an absorbance spectrum is acquired in a spectrophotometer, one can calculate an extinction coefficient at a certain wavelength from the determined absorption on the basis of the concentration of ZnO which should theoretically be formed. This coefficient is very low and in the range of only several hundred $l\,mol^{-1}\,cm^{-1}$. If this dispersion is investigated in a sedimentation velocity experiment using combined absorption and Rayleigh interference detection, one gets the patterns shown in Fig. 13A.

From Fig. 13A, three components can be detected from Rayleigh interference scans whereas absorption scans at 315 nm only detect ZnO. It becomes obvious that only the smallest part (0.18 fringes out of 2.8 fringes) is spectroscopically active ZnO if one compares the boundary positions from the absorption with those from Rayleigh interference scans. The boundary positions of the ZnO-filled micelles are not very well defined in Fig. 13A, but are seen more clearly in a zoomed plot. However, the largest part is a spectroscopically inactive component (1.88 out of 2.80 fringes) which is most likely Zn^{2+} attached to the COO^- groups of the PMAA block. In other words, this is the unreacted component. The third component is likely to be polymer which is not incorporated into the micelles (0.74 out of 2.80 fringes). This shows very clearly that most of the Zn^{2+} has not reacted to ZnO and thus the extinction coefficient which is derived from a spectrophotometer for such mixture can only be severely underestimated.

The fringe shifts for the ZnO^- filled micelles can be converted to concentration units by assuming a linear

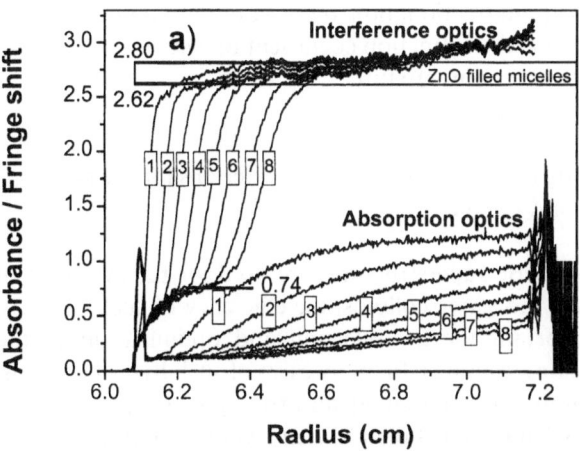

Fig. 13A Absorption (315 nm) and Rayleigh interference radial scans during a sedimentation velocity experiment with ZnO prepared in PS-PMAA micelles in THF. Run parameters: 20 000 rpm, scan interval 2 min. The numbers represent the corresponding interference and absorption scans

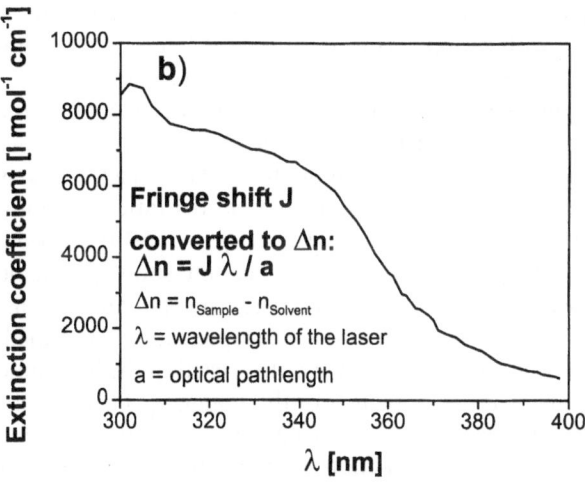

Fig. 13B Wavelength dependence of the extinction coefficient for ZnO prepared in PS-PMAA micelles. The particle diameter of ZnO is 3.2–6.7 nm from TEM. The diameter of the micelles is 64.4 ± 33.9 nm from light scattering

dependence between the refractive index of the solvent THF and that of pure ZnO. The slope of this line is taken as dn/dc of ZnO in THF. Then the absorption value of the upper plateau in Fig. 13A is addressed to be the one of ZnO. With this information, a spectrum for ZnO, the only light absorbing species in the centrifuge cell can be converted to a wavelength dependence of the extinction coefficient (Fig. 13B). It can be seen that the extinction coefficient is a magnitude higher than the value estimated from an absorption reading of the whole mixture and the amount of Zn.

146
H. Cölfen et al.
Quantum size colloids in the XL-I

Without fractionation, there would be no possibility to determine the extinction coefficient of ZnO is such a complicated reaction mixture.

Investigation of the efficiency of reactions for the synthesis of hybrid colloids

If polymeric superstructures are used for the synthesis of organic/inorganic colloids, the efficiency of such reaction is important for the selection and optimization of the template structures. In many cases of such hybrid colloids, the inorganic compound absorbs light (CdS, ZnO, Fe-oxides, TiO_2, nobel metals, etc.) – very often in the visible region – whereas the polymer does not absorb light or only in the UV-range. This makes the XL-I analytical ultracentrifuge a valuable tool to look at the concentration distributions of the two components during sedimentation velocity runs if Rayleigh interference and absorption optics are used simultaneously. The first sedimenting component is the inorganic particle formed either outside or inside the template superstructure. To distinguish if the template is loaded with cations prior to the reaction, a sedimentation velocity experiment can give appropriate information. If only the sedimenting polymer can be detected but the absorption scan for the cation shows a uniform distribution, no cation complexation has taken place. This is a most unlikely event due to the electrostatic attraction between the negatively charged template and the cation. If the absorption scans for the cation show the same sedimentation velocity as those for the polymeric template, interaction between these components is confirmed. Hence, it is most likely that after the reaction the inorganic component is still attached to the template. There might still be cations or polymers, which have not interacted due to deficiency of the reaction partner. This would result in the detection of a second component in the sedimentation diagrams of the polymer or in a baseline absorbance for the cation. For the following example, an interaction between the template polymer and Pt-cations was confirmed in a separate velocity experiment before the reduction to Pt was performed. Then the reduction to Pt was carried out [7]. The reaction mixture was then subjected to analytical ultracentrifugation. At 30 000 rpm, the sedimentation of the metallic Pt could well be detected with both optical detection systems (see Fig. 14A).

From the Rayleigh interference scans, the concentration of the polymer-Pt component can be determined to be 4.62 fringes. The second step in the Rayleigh interference fringes corresponds to the pure polymer or its superstructures because there is no equivalent step in the absorption scan which detects the Pt concentration. If the rotational speed is increased to 44 000 rpm, the polymer-Pt com-

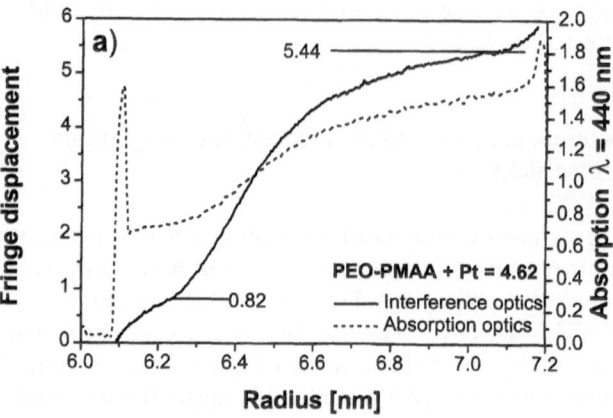

Fig. 14A Rayleigh interference and absorption radial scans during a sedimentation velocity experiment with Pt synthesized in PEO-PMAA aggregates in water at 30 000 rpm. For clarity of the figure, only one set of corresponding scans is shown

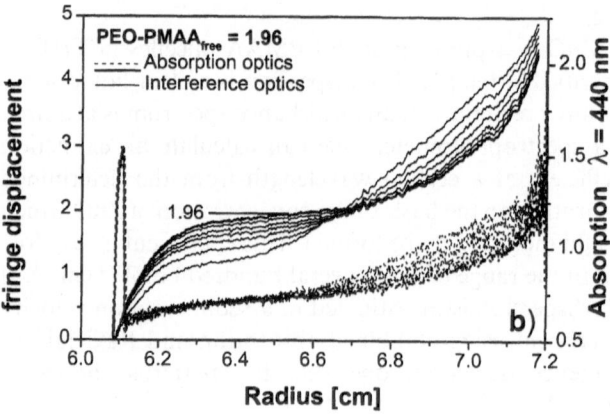

Fig. 14B Rayleigh interference and absorption radial scans during a sedimentation velocity experiment with Pt synthesized in PEO-PMAA aggregates in water at 44 000 rpm

pound sediments completely whereas the second compound starts to sediment. Due to the characteristics of the Fourier transformation of the raw fringe data, the first point in the Rayleigh interference data is always set to 0. This means that an apparent increase of the concentration is detected until a constant value of 1.96 fringes is observed for the polymer concentration without Pt (see Fig. 14B).

From the fringe shifts of the polymer with and without Pt (see Figs. 14A and B), the ratio of polymer attached to Pt to free polymer can be estimated under the assumption that the coordination of the metal does not change the dn/dc value of the polymer. For the example above, the ratio is 2.4 : 1 which means that all Pt is attached to polymer but there is still an excess of free polymer.

Conclusions

The ultracentrifuge is a mighty tool for the characterization of complex organic/inorganic hybrid colloids. If the inorganic component (absorption) can be detected separately from the polymeric component (Rayleigh interference), efficiency of colloid formation reactions, extinction coefficients of the inorganic component, the change of the spectral properties with the particle size and particle growth can be investigated. It is even possible to distinguish between aggregated and grown particles in the case of quantum size particles. This means that additional information beside the particle size distribution (from the sedimentation coefficient distribution) can be obtained from absorption spectra of the fractionated particles in the ultracentrifuge cell.

Acknowledgements We want to thank Michael Breulmann for the CdS in PS-P4VP micelles, Sascha Oestreich for the Au in PS-P4VP micelles, Franziska Gröhn for Au in PSS microgels and Olga Platonova for Pt in PEO-PMAA. The support of the Max-Planck-society is gratefully acknowledged.

References

1. Antonietti M, Förster S, Hartmann J, Oestreich S, Wenz E (1996) Nachr Chem Techn Lab 44:579
2. Spatz JP, Roescher A, Möller M (1996) Adv Mater 8:4
3. Förster S, Breulmann M, Wenz E, Antonietti M, Haase M, Schneider T, Weller H (1997) in preparation
4. Bahnemann DW, Kormann C, Hoffmann MR (1987) J Phys Chem 91:3789
5. Oestreich S (1997) Ph.D thesis, Potsdam
6. Antonietti M, Gröhn F (1997) Angew Chem, accepted
7. Bronstein L, Platonova O, Valetsky P, Sedlak M, Mayer A, Hartmann J, Breulmann M, Cölfen H, Antonietti M (1997) in preparation
8. Haase M, Weller H, Henglein A (1988) J Phys Chem 92:482
9. Mächtle W (1984) Makromol Chem 185:1025
10. Mächtle W (1988) Angew Makromol Chem 162:35
11. Müller HG (1989) Colloid Polym Sci 267:1113
12. Mächtle W (1992) Makromol Chem, Macromol Symp 61:131
13. Mächtle W (1992) In: Harding SE, Rowe AJ, Horton JC (eds), Analytical Ultracentrifugation in Biochemistry and Polymer Science, Chap 10. The Royal Society of Chemistry, Cambridge, UK p. 147
14. Müller HG, Herrmann F (1995) Progr Colloid Polym Sci 99:114
15. Rapoport DH, Vogel W, Cölfen H, Schlögl R, J Phys Chem B 101:4175
16. Cölfen H, Pauck T (1997) Colloid Polym Sci 275:175
17. Braun GA, Bönnemann H, Cölfen H, Antonietti M (1997) (in preparation)

Progr Colloid Polym Sci (1997) 107:148–153
© Steinkopff Verlag 1997

M.D. Lechner
W. Mächtle
U. Sedlack

Influence of pressure and solvent composition on the density gradient in the analytical ultracentrifuge I Extended *Hermans–Ende* equation for the equilibrium density gradient

Prof. Dr. M.D. Lechner (✉)· U. Sedlack
Physikalische Chemie
Universität Osnabrück
Barbarastraße 7
49069 Osnabrück, Germany

W. Mächtle
Kunststofflaboratorium
Polymerphysik
BASF AG
67056 Ludwigshafen, Germany

Abstract An extended *Hermans–Ende* equation for real solutions has been derived in order to calculate the density as a function of the radial distance in a density gradient mixture (light/heavy medium) inside an ultracentrifugal cell. The equation has been tested with equilibrium density gradient measurements of the system H_2O/methanol/metrizamide/polystyrene latex where metrizamide is the heavy medium. Additionally, density measurements of the systems with a vibrating tube densitometer at normal pressures and at pressures up to 400 bar have been performed in order to complete the measurements. The solvent compositions were 8–25% metrizamide, 20–100% water and 0–80% methanol. The latex concentration was 0.03%. It is demonstrated that the density gradient is highly influenced by nonideality effects with respect to the solvent composition and by the pressure gradient in the ultracentrifuge.

Keywords Analytical ultracentrifuge – density gradient – equilibrium – *Hermans–Ende* equation

Introduction

The Hermans–Ende equation for the calculation of equilibrium density profiles in an analytical ultracentrifuge is restricted to ideal solutions in which the solvent components have equal molar volumes and are mixed with zero volume change and zero heat of mixing [1]. Especially in the case of athermic and real solutions there are considerable discrepancies between the calculated and the experimentally determined densities as a function of the radial distance in the ultracentrifugal cell. Due to these discrepancies it would be worth extending the Hermans–Ende equation to real solutions.

There are two other methods for the determination of the equilibrium density profile. The method of Hearst and Vinograd [2] requires a detailed knowledge of the chemical potential of the solution and is up to now restricted to aqueous solutions of several heavy salts. The method of Munk [3,4] gives an empirical relationship with several constants which may be determined with marker polymers of known densities at several equilibrium runs with different speeds.

In this paper the extended Hermans–Ende equation will be derived and compared with experimental results of polystyrene latex dispersed in water/methanol/metrizamide density gradient mixtures.

Theory

Consider a mixture of two solvents with indices $k = 0$ and 1. The basic equation for the sedimentation equilibrium in the analytical ultracentrifuge is then for component 1 [1],

$$d\Delta\mu_1 = M_1\omega^2 r\, dr(1 - \upsilon_1\rho)\,, \tag{1}$$

where $\Delta\mu_1$ is the chemical potential, M_1 the molar mass, υ_1 the partial specific volume of component 1, ω the angular velocity, r the distance from center of rotation, and ρ the density of the solution at the distance r. The relationship between the density ρ and the volume fractions

$\varphi_k(k = 0, 1)$ is given by

$$\rho = \varphi_0/\upsilon_0 + \varphi_1/\upsilon_1 = \varphi_0\rho_0 + \varphi_1\rho_1\,, \tag{2}$$

where $\varphi_0 + \varphi_1 = 1$, $\varphi_k = n_k V_k^\circ/(n_0 V_0^\circ + n_1 V_1^\circ)$, $v_k = (\partial V/\partial m_k)_{p,T,m_{j\neq k}}$, v_k is the partial specific volume, V_k° the molar volume, $\rho_k = 1/\upsilon_k$ the partial density, n_k the number of moles, and m_k the mass of the component $k(k = 0, 1)$.

The calculation of the density ρ as a function of the radial distance r requires an expression for the chemical potential $\Delta\mu_1$. A suitable equation for the free enthalpy of mixing ΔG_m is the extended Flory–Huggins relationship [5]

$$\Delta G_m = RT(n_0 \ln\varphi_0 + n_1 \ln\varphi_1 + \chi_{01}n_0\varphi_1) \tag{3}$$

with χ_{01} being the Flory–Huggins parameter. Differentiation with respect to n_1 yields

$$\Delta\mu_1 = (\partial\Delta G_m/\partial n_1)_{p,T,n_0} = RT\,[\ln\varphi_1 + (1 - V_1^\circ/V_0^\circ)\varphi_0$$
$$+ \chi_{01}(V_1^\circ/V_0^\circ)\varphi_0^2]\,. \tag{4}$$

Replacing φ_0 by $1 - \varphi_1$ and differentiating Eq. (4) with respect to φ_1 yields

$$d\Delta\mu_1 = d\varphi_1\, RT\,[1/\varphi_1 + 2\chi_{01}(V_1^\circ/V_0^\circ)\varphi_1 - (1 - V_1^\circ/V_0^\circ)$$
$$- 2\chi_{01}(V_1^\circ/V_0^\circ)]\,. \tag{5}$$

Using Eq. (2), Eq. (1) and (5) yield

$$d\ln\varphi_1 + 2\chi_{01}(V_1^\circ/V_0^\circ)\varphi_1\, d\varphi_1$$
$$- [1 - V_1^\circ/V_0^\circ + 2\chi_{01}(V_1^\circ/V_0^\circ)]\, d\varphi_1 = \beta 2r\, dr\varphi_0 \tag{6}$$

with $\beta = \omega^2(M_1/\rho_1)\,(\rho_1 - \rho_0)\,/(2RT)$. ρ_0 and ρ_1 are assumed as constants, i.e. they do not depend on the volume fractions φ_0 and φ_1. Integration of Eq. (6) gives the extended Hermans–Ende equation

$$\frac{\varphi_1\exp[-2\chi_{01}(V_1^\circ/V_0^\circ)\varphi_1]}{(1 - \varphi_1)^{V_1^\circ/V_0^\circ}} = \alpha\exp(\beta r^2)\,. \tag{7}$$

The integration constant α may be determined by the condition of mass conservation:

$$\alpha = \frac{\beta\varphi_1^{in}(r_b^2 - r_m^2)}{(1 - \varphi_1^{in})^{V_i/V_0^\circ}\exp[-2\chi_{01}(V_1^\circ/V_0^\circ)\varphi_1^{in}]\,[\exp(\beta r_b^2) - \exp(\beta r_m^2)]}\,, \tag{8}$$

where φ_1^{in} is the initial volume fraction of component 1 and r_m and r_b the radial distances of the meniscus and the bottom of the cell, respectively.

It has been pointed out in the literature [11] that χ_{01} is not a constant, as has been postulated by the lattice theory. χ_{01} depends, in strong nonideal systems, on concentration and pressure. Especially with respect to large density profiles which may arise at high rotor speeds and high concentrations of the heavy solvent component, one has to consider the concentration dependence of the Flory–Huggins parameter χ_{01}:

$$\chi_{01} = \chi_{01}' + a_2\varphi_1^{in} + a_3(\varphi_1^{in})^2 + \cdots\,. \tag{9}$$

Introducing Eq. (9) into Eqs. (7) and (8) yields:

$$\frac{\varphi_1\exp[-2(\chi_{01}' + a_2\varphi_1^{in} + a_3(\varphi_1^{in})^2 + \cdots)(V_1^\circ/V_0^\circ\varphi_1]}{(1 - \varphi_1)^{V_i/V_0^\circ}} = \alpha\exp(\beta r^2)\,, \tag{10}$$

$$\alpha = \frac{\beta\varphi_1^{in}(r_b^2 - r_m^2)}{(1 - \varphi_1^{in})^{V_i/V_0^\circ}\exp[-2(\chi_{01}' + a_2\varphi_1^{in} + a_3(\varphi_1^{in})^2 + \cdots)(V_1^\circ/V_0^\circ)\varphi_1^{in}]\,[\exp(\beta r_b^2) - \exp(\beta r_m^2)]}\,. \tag{11}$$

Note that the concentration dependence of χ_{01}, Eqs. (9)–(11), is connected to the volume fraction of the initial solution of solvent 1 φ_1^{in}. There are two simplifications of Eq. (7):

Ideal solution: $V_0^\circ = V_1^\circ$; $\chi_{01} = 0$:

$$\frac{\varphi_1}{1 - \varphi_1} = \alpha\exp(\beta r^2)\,, \quad \varphi_1 = 1 - 1/[\alpha\exp(\beta r^2) + 1]\,. \tag{12}$$

This is the well-known equation derived by Hermans and Ende [1].

Athermic solution: $V_0^\circ \neq V_1^\circ$; $\chi_{01} = 0$:

$$\frac{\varphi_1}{(1 - \varphi_1)^{V_i/V_0^\circ}} = \alpha\exp(\beta r^2)\,. \tag{13}$$

The integration constant α, Eqs. (8) and (11), may be simplified with respect to Eqs. (12) and (13).

150

M.D. Lechner et al.
Extended *Hermans-Ende* equation for the equilibrium density gradient

Equations (7), (10) and (13) are implicit functions; they cannot be resolved to $\varphi_1 = f(r^2)$. Therefore one has to calculate φ_1 by regression procedures.

The influences of the composition and the pressure on the density profile may be assumed to be independent from each other. Therefore, it is convenient to extract the pressure dependence from Eqs. (7), (10), (12) and (13) and to determine the pressure dependence of the density gradient solution separately:

$$\rho(r) = \rho_{p=1}(r)/(1 - \kappa p), \quad p = (\rho_{p=1}\omega^2/2)(r^2 - r_m^2) \quad (14)$$

where $\rho(r)$ is the density at radial distance r and pressure p, $\rho_{p=1}(r)$ the density at radial distance r and pressure $p = 1$ bar, and κ the compressibility of the solvent mixture.

The calculated volume fraction φ_1 in Eqs. (7), (10), (12), and (13) has to be transformed into the density of the solution. In some binary solvent mixtures (e.g., water/metrizamide) the density behaves nearly ideally,

$$\rho = \varphi_0/\upsilon_0^\circ + \varphi_1/\upsilon_1^\circ = \varphi_0\rho_0^\circ + \varphi_1\rho_1^\circ = \rho_0^\circ + (\rho_1^\circ - \rho_0^\circ)\varphi_1 ,$$

$$(15a)$$

where υ_k° and ρ_k° ($k = 0, 1$) are the specific volumes and the densities of the pure solvents. In case of nonideality of the solution a $\rho = f(\varphi_1)$ relationship has to be used

$$\rho = a + b\varphi_1 + c\varphi_1^2 + \cdots , \quad (15b)$$

where the constants $a, b, c, \ldots$ may be determined by density measurements.

In the case of ternary systems, the given equations have to be extended by a third solvent 2. These equations are unwieldy and complicated to handle. For example, there are three Flory–Huggins parameters χ_{01}, χ_{02} and χ_{12} instead of one parameter χ_{01}. Nevertheless, in many cases (e.g., water/methanol/metrizamide) it would be possible to consider one solvent pair (e.g., water/methanol) as solvent 0 and the other component (e.g., metrizamide) as solvent 1. In this way, Eqs. (7)–(13) are used for evaluating three-component solvent systems. Normally, it does not matter that "solvent 0" (water/methanol) is highly nonideal. One has to insert for the density of solvent 0, ρ_0°, the experimentally determined density of the water/methanol mixture and for the molar volume V_0°,

$$V_0^\circ = M_0\upsilon_0^\circ = M_0/\rho_0^\circ , \quad M_0 = x_{H_2O}M_{H_2O} + x_{CH_4O}M_{CH_4O} ,$$

$$(16)$$

where x_{H_2O}, x_{CH_4O}, M_{H_2O}, and M_{CH_4O} are the mole fractions and the molar masses of water and methanol, respectively.

Experimental part

The following materials were used: double-distilled water, p.a. methanol, alkylsulfonate (BASF AG, Ludwigshafen, Germany), metrizamide (2-(3-acetamido-5-N-methyl-acetamido-2,4,6-tri-iodobenzamido)-2-deoxy-D-glucose, Nyegaard, Oslo, Norway) and polystyrene latex (BASF AG). The physical constants are: water/alkylsulfonate (1%), $M = 18.0$ g/mol, $\rho^0 = 0.9987$ g/cm^3, and $V^0 = 18.02$ cm^3/mol; methanol, $M = 32.0$ g/mol, $\rho^0 = 0.7865$ g/cm^3, and $V^0 = 40.69$ cm^3/mol; metrizamide, $M = 789.1$ g/mol; $\rho^0 = 2.1552$ g/cm^3, and $V^0 = 370.3$ cm^3/mol. The properties of the polystyrene latex were: density $\rho_{PM} = 1.059_5$ g/cm^3 and mass average diameter $d = 190$ nm. The polystyrene latex concentration was 0.03 mass%. The density of the polystyrene latex was measured independently by dynamic density gradient measurements and particle density analysis [7]. All values are given at 25 °C.

The measurements were done with the following machines: Kratky-type vibrating tube densitometer with glass capillary and steel capillary for measurements at elevated pressures (A. Paar, Graz, Austria) and analytical ultracentrifuge Spinco Model E (Beckman, Palo Alto, CA, USA). The ultracentrifuge was equipped with a 8-hole rotor and Schlieren optics with multiplexer [8]. All measurements were done at 25 °C. The evaluation of the Schlieren photos was done with a graphics tablet and a computer. The data were transposed into an ASCII file which was accessible to a PC. The programs are available on request.

Results and discussion

The evaluation of the measurements needs a precise determination of the initial density of the solvents as function of the composition. In addition, pressure effects according to Eq. (14) have to be taken into account. Figure 1 demonstrates the density of water/methanol mixtures as a function of the methanol content. Water/methanol is a strong nonideal system with densities which are remarkably higher than that for ideal mixtures. Figure 2 shows the density of water/methanol/metrizamide mixtures as a function of the metrizamide content. The indicated values on both figures are our own measurements and data taken from literature [9]. Water/metrizamide is a nearly ideal system with respect to the density. The system water/methanol/metrizamide can also be regarded as a nearly ideal system if the solvent pair water/methanol is considered as one solvent.

Under the assumption that the contribution of metrizamide to the compressibility of the solvent is negligible,

Progr Colloid Polym Sci (1997) 107:148–153
© Steinkopff Verlag 1997

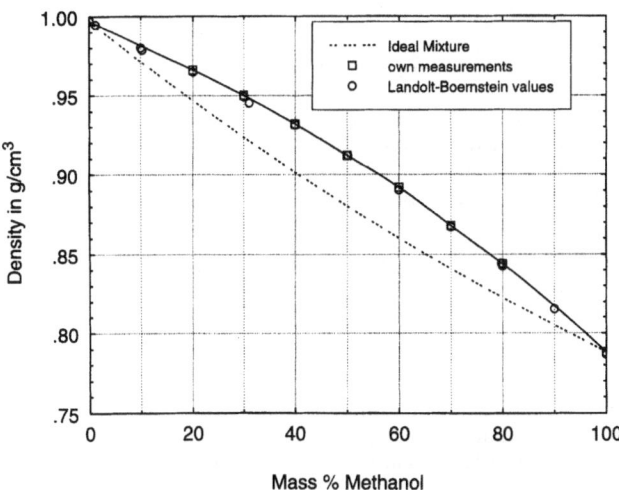

Fig. 1 Density of water/methanol mixtures as a function of the methanol content at 25 °C

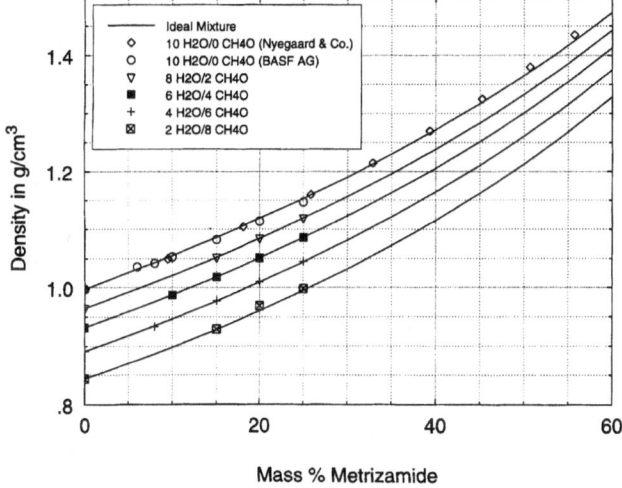

Fig. 2 Density of water/methanol/metrizamide mixtures as a function of the metrizamide content at different ratios m_{H_2O}/m_{CH_4O} at 25 °C

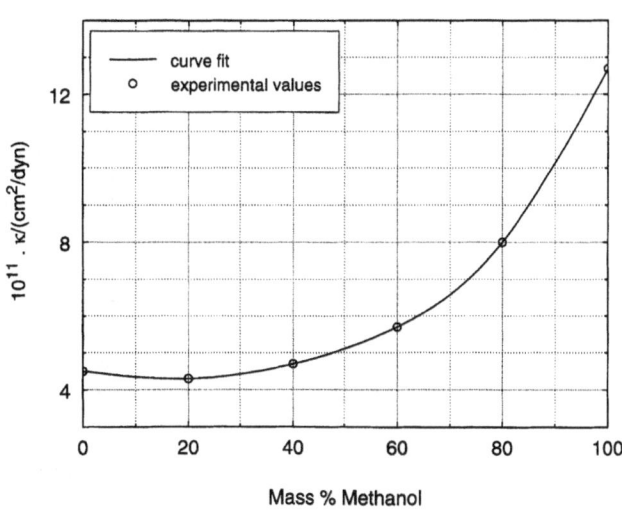

Fig. 3 Compressibility κ of water/methanol mixtures as a function of the methanol content at 25 °C

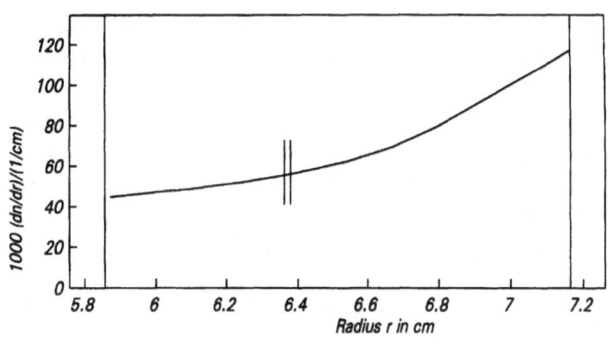

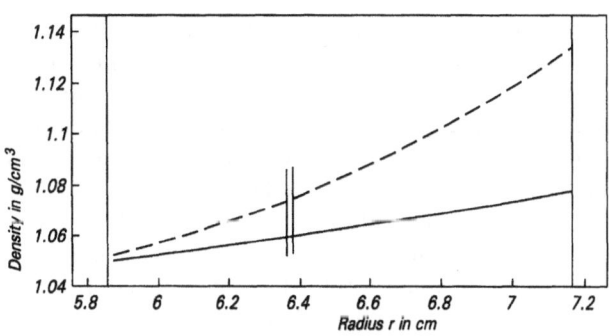

Fig. 4 Top: Schlieren plot of run A4 (see Tables 1 and 2). **Bottom**: Calculated density profile. ------ = ideal solution, Eqs. (12), (14), (15), original Hermans–Ende equation ———— = real solution, Eqs. (10), (14), (15), new extended Hermans–Ende equation

the compressibility of water and water/methanol mixtures is required. Measurements were performed with a special vibrating tube densitometer with a steel capillary. Experimental details are given elsewhere [10]. The measurements together with literature values are summarized in Fig. 3. The nonideality of this system with respect to the compressibility is remarkable.

Table 1 shows the conditions of the analytical ultracentrifuge experiments. The mass content of metrizamide was 8–25%. The polymer solution includes 1% alkylsulfonate as emulsifying agent. As the extended Hermans–Ende equation strictly refers to equilibrium conditions, it is absolutely necessary to achieve the equilibrium state. We

are sure that the equilibrium state was achieved after 40 h; no changes were observed up to 70 h.

The calculation of the parameters χ'_{01}, a_2, and a_3 was done with the help of Eqs. (10), (14), and (15), using optimization procedures. At least two runs (i.e., runs A1 and A2)

152

M.D. Lechner et al.
Extended *Hermans-Ende* equation for the equilibrium density gradient

Table 1 Conditions of the equilibrium density gradient runs

Run	t [h]	N [min]	C_{PM} [g/l]	w_1	$\dfrac{m_{H_2O}}{m_{CH_4O}}$	ρ_0° [g/cm^3]	ρ^{in} [g/cm^3]	$10^{11}\kappa$ [cm^2/dyn]
A1	46	24930	0.288	0.08	10:0	0.9970	1.043	4.5
A2	46	24930	0.284	0.10	10:0	0.9970	1.055	4.5
A3	46	24930	0.275	0.12	10:0	0.9970	1.067	4.5
A4	46	24930	0.272	0.15	10:0	0.9970	1.086	4.5
B1	47	24660	0.299	0.15	6:4	0.9317	1.018	4.7
B2	47	24660	0.294	0.20	6:4	0.9317	1.051	4.7
B3	47	24660	0.282	0.25	6:4	0.9317	1.086	4.7
C1	114	17810	0.304	0.25	2:8	0.8437	0.995	8.0

Note: t = run time; N = rotor speed; C_{PM} = concentration, latex particle w_1 mass fraction, solvent 1 (metrizamide); m_{H_2O}/m_{CH_4O} = mass ratio, water/methanol ρ_0 = density, solvent 0; ρ^{in} = initial density, solvent mixture, p = 1 bar; κ = compressibility

were needed to calculate χ'_{01} and $a_2 (a_3 = 0)$, and at least one more (i.e., runs A1, A2, and A3) for χ'_{01}, a_2, and a_3. The optimization procedure includes the pressure effect, Eq. (14). Compared with the composition effect, Eq. (10), the pressure effect for the system water/methanol/metrizamide is relatively small $(\Delta\rho = 0.001–0.003 \text{ g/cm}^3)$; it increases with increasing methanol content.

Figure 4 demonstrates as an example the original Schlieren plot of run A4 (top diagram with the narrow turbidity band of the polystyrene latex particles in the middle) and the calculated density profiles (bottom diagram) with respect to an ideal solution, Eqs. (12), (14) and (15) (= original Hermans–Ende equation) and to a real solution Eqs. (10), (14) and (15) (= new extended Hermans–Ende equation). The differences are remarkable. Normally, the partial molar volume of solvent 1, (metrizamide) V_1 is larger than that of solvent 0 (water or water/methanol) V_0. Equations (10) and (15) demonstrate that the density gradient becomes smaller with increasing values of V_1/V_0. Figure 5 shows an original Schlieren plot of run B2 and the calculated density profile for a real solution, Eqs. (10), (14) and (15). The runs B1–B3 and C1 demonstrate that it would be possible to handle the ternary system water/methanol/metrizamide as a binary system with water/methanol as light solvent and metrizamide as heavy solvent according to Eq. (16).

Table 2 summarizes the results concerning the calculated particle densities from the radial turbidity band position of the polystyrene particle marker position inside the density gradient. The differences between the original Hermans–Ende equation, (12) and the extended Hermans–Ende equation (10), are remarkable with respect to the density. The discrepancies between calculated and experimentally determined densities would be overcome if the extended Hermans–Ende equation (10), is used with an experimentally determined value of V_1/V_0 and calculated values of χ_{01} or χ'_{01}, a_2 and a_3. Table 2 demonstrates that

Fig. 5 Top: Schlieren plot of run B2 (see Tables 1 and 2). **Middle**: Calculated volume fraction profile. **Bottom**: Calculated density profile. Real solution, Eqs. (10), (14), (15)

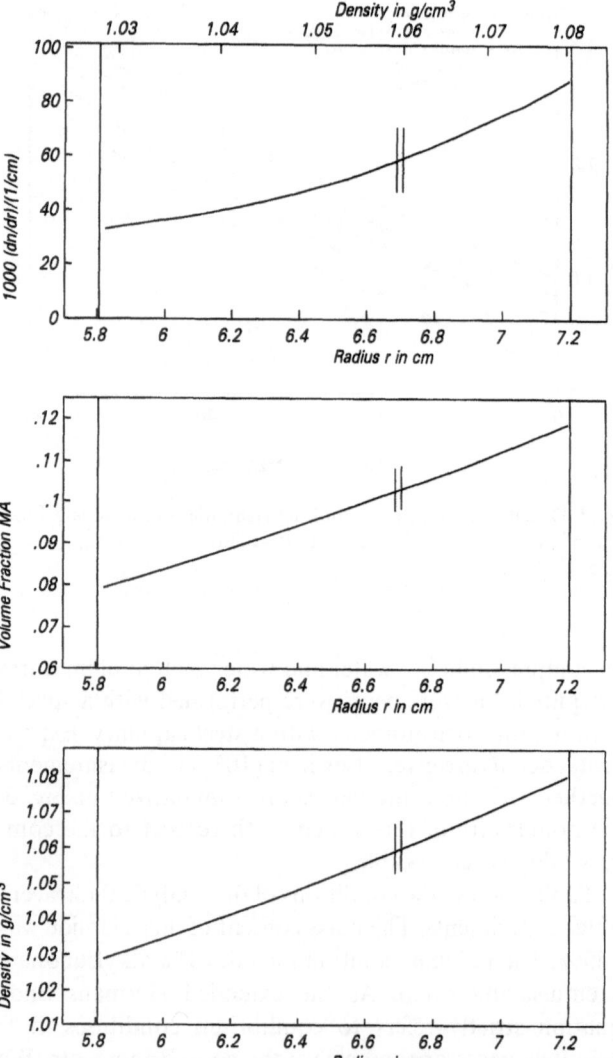

Table 2 Results of the equilibrium density gradient runs; calculated polystyrene particle densities from the radial turbidity band position r_{PM} inside the density gradient. $\Delta\rho_{PM}$ (p effect) = pressure effect, Eq. (14)

Run	r_m[cm]	r_{PM} [cm]	r_b [cm]	ρ_{PM} [g/cm³]	ρ_{PM} [g/cm³]	ρ_{PM} [g/cm³]	p [bar]	$\Delta\rho_{PM}$ [g/cm³]
				$V_1^\circ = V_0^\circ$ $\chi_{01} = 0$	$V_1^\circ/V_0^\circ = 20.5$ $\chi'_{01} = 0.32$ $a_2 = -6.18$ $a_3 = 0$	$V_1^\circ/V_0^\circ = 20.5$ $\chi'_{01} = 0.51$ $a_2 = -13.2$ $a_3 = 61.2$	At r_{PM}	p effect
A1	5.854	6.910	7.198	1.057	1.059_5	1.060_5	48.9	0.002
A2	5.854	6.713	7.196	1.061	1.061	1.060	39.8	0.001_5
A3	5.884	6.562	7.197	1.066	1.061	1.059_5	31.7	0.001_5
A4	5.856	6.370	7.164	1.074	1.059	1.060	24.2	0.001_5
				$V_1^\circ = V_0^\circ$ $\chi_{01} = 0$	$V_1^\circ/V_0^\circ = 15.6$ $\chi'_{01} = 0.33$ $a_2 = -3.18$ $a_3 = 0$	$V_1^\circ/V_0^\circ = 15.6$ $\chi'_{01} = 0.50$ $a_2 = -6.71$ $a_3 = 17.9$	At r_{PM}	p effect
B1	5.780	6.986	7.184	1.052	1.059	1.060_5	53.2	0.002_5
B2	5.805	6.696	7.201	1.061	1.062	1.059_5	40.0	0.002
B3	5.837	6.436	7.180	1.070	1.058	1.059_5	27.5	0.001_5
				$V_1^\circ = V_0^\circ$ $\chi_{01} = 0$	$V_1^\circ/V_0^\circ = 11.1$ $\chi'_{01} = .11$		At r_{PM}	p effect
C1	5.882	7.163	7.169	1.037	1.060		53.2	0.002_5

the calculated densities of the polystyrene latex are closer to the known value of $\rho_{PM} = 1.059_5$ g/cm³ if the parameter set of the fifth column with $\chi'_{01} \neq 0$, $a_2 \neq 0$, and $a_3 \neq 0$ is used. On the other hand it would be sufficient in some cases to use the parameter set of the fourth column with $\chi'_{01} \neq 0$, $a_2 \neq 0$, and $a_3 \neq 0$.

The values χ_{01}, χ'_{01}, a_2 and a_3 are constants for a given system, e.g., water/metrizamide with different initial compositions. For other sytems, these values have to be recalculated. Ternary systems such as water/methanol/metrizamide with different mass ratios of water/methanol have to be considered as different systems with different constants.

Acknowledgements Support of this work by *BASF AG*, Ludwigshafen and by the *Fonds der Chemischen Industrie* is kindly acknowledged. The measurements with the analytical ultracentrifuge were performed by *M. Kaiser*, Kunststofflaboratorium, *BASF AG*, Ludwigshafen.

References

1. Hermans JJ, Ende HA (1963) J Polym Sci C1:161
2. Hearst JE, Vinograd J (1961) Proc Nat Acad Sci USA 47:999, 1005
3. Munk P (1982) Macromolecules 15: 500
4. Härzschel R, Meyerhoff G (1990) Makromol Chem 191:3139
5. Lechner MD, Gehrke K, Nordmeier EH (1996) Makromolekulare Chemie. Birkhäuser, Basel
6. Fujita H (1975) Foundations of Ultracentrifugal Analysis. Wiley, New York
7. Mächtle W, Lechner MD, unpublished results
8. Mächtle W, Klodwig U (1979) Makromol Chem 180:2507
9. Landolt-Börnstein (1971), 6. Aufl., Bd. II/1. Springer, Berlin
   Rickwood D (1979) Metrizamide. Nyegaard & Co., Oslo, Norway
10. Vennemann N, Lechner MD, Oberthür RC (1987) Polymer 28:1738;
    Busch B, Lechner MD, Kleintjens LA (1992) Angew Makromol Chem 194:35
11. Elias HG (1990) Makromoleküle. Hüthig & Wepf, Basel

Progr Colloid Polym Sci (1997) 107:154–158
© Steinkopff Verlag 1997

POLYMERS, COLLOIDS & SUPRAMOLECULAR SYSTEMS

M.D. Lechner
W. Mächtle
U. Sedlack

Influence of pressure and solvent composition on the density gradient in the analytical ultracentrifuge. II Direct refractometric determination of the equilibrium and non-equilibrium density gradient

Prof. Dr. M.D. Lechner (✉) · U. Sedlack
Physikalische Chemie
Universität Osnabrück
Barbarastraße 7
49069 Osnabrück, Germany

W. Mächtle
Kunststofflaboratorium
Polymerphysik
BASF AG
67056 Ludwigshafen, Germany

Abstract A method for the direct refractometric determination of the density gradient in an analytical ultracentrifuge has been described. It is based on two assumptions: 1) the *Schlieren* optics measure the refractive index gradient, the interference optics the refractive index, and the absorption optics the absorption of the density gradient solution (consisting of a light and a heavy solvent component) and 2) there is a relationship between the refractive index or the absorption, on the one hand and the density of the density gradient solution, on the other. The method has been demonstrated on water/metrizamide/ polystyrene latex solutions with different concentrations of metrizamide (a heavy sugar). Advantages of this method are: first, the pressure and the composition effect are included in this method and must not be corrected as usual and second, it is not necessary to measure at sedimentation equilibrium; it would be possible to measure at times before the equilibrium is achieved. That means the method represents a quasi-dynamic density gradient method.

Key words Analytical ultracentrifuge – density gradient – refractometric determination

Introduction

Up to now there exist two possibilities for the determination of the static density gradient in an analytical ultracentrifuge (i.e. the density $\rho(r)$ as a function of the rotor radius r): 1) calculation of the density gradient according to Hermans and Ende [1] or Hearst and Vinograd [2] and 2) determination of the density gradient by using tracer materials with known density and an empirical relationship [3]. In this paper we will demonstrate that it is also possible to determine directly the density gradient from dn/dr values (*Schlieren* optics), or the number of fringes (interference optics), or the absorption (absorption optics). Advantages of this method are: pressure and composition effects of the density gradient are included in this method, and the method is also applicable for conditions under which sedimentation equilibrium has not been reached yet (i.e. "dynamic" density gradient).

Theory

The analytical ultracentrifuge uses frequently the interference optics, the *Schlieren* optics and the absorption optics as detectors for the determination of the composition of a two-component solution (i.e. a light and a heavy component in a density gradient) as a function of the radial distance r. The interference optics measure the refractive index difference $\Delta n(r)$, whereas the Schlieren optics measure the refractive index gradient dn/dr. The absolute refractive index $n(r)$ in the cell is

$$n(r) = n_\mathrm{m} + \Delta n(r) = n_\mathrm{m} + \int_{n_\mathrm{m}}^{n'} dn = n_\mathrm{m} + \int_{r_\mathrm{m}}^{r'} (dn/dr)\,dr , \quad (1)$$

where n_m and r_m are the refractive index and the radial distance at the meniscus, respectively. n_m may be calculated by the condition of mass conservation:

$$n_0(r_b^2 - r_m^2) = n_m(r_b^2 - r_m^2) + \int_{r_m}^{r_b} \Delta n(r)\, 2r\, dr \,, \tag{2}$$

$$n_m = n_0 - [1/(r_b^2 - r_m^2)] \int_{r_m}^{r_b} \Delta n(r)\, 2r\, dr$$

interference optics , (3)

$$n_m = n_0 - [1/(r_b^2 - r_m^2)] \left[r_b^2 \int_{r_m}^{r_b} (dn/dr)\, dr - \int_{r_m}^{r_b} r^2(dn/dr)\, dr \right]$$

Schlieren optics . (4)

r_b is the radial distance at the bottom of the sector shaped cell, and n_0 is the refractive index of the initial binary solution. $\Delta n(r)$ and (dn/dr) may be calculated by

$$\Delta n(r) = \Delta j(r)\, \lambda/l \quad \text{interference optics,} \tag{5}$$

$$dn/dr = y \tan \theta/(l m_x L' m_y) \quad \textit{Schlieren optics} \tag{6}$$

with $\Delta j(r)$ being the fringe shift at distance r, λ the wavelength of the light source, l the cell length, y the y-value of the Schlieren optics, θ the Philpot angle, m_x the magnification in the x direction, m_y the magnification in the y direction, and L' the optical lever.

Equations (1)–(6) enable to calculate the refractive index as a function of the radial distance in the ultracentrifuge cell. The determination of the density $\rho(r)$ requires a relationship between n and $\rho = f(n)$:

$$\rho = a + bn + cn^2 + \cdots \,. \tag{7}$$

This may be easily done by preparing mixtures of the two density gradient (light/heavy) components with compositions within the interesting region and determining the refractive index n with a refractometer and the density ρ with a densitometer (i.e., vibrating tube densitometer). Another possibility is to calculate the $\rho = f(n)$ relationship with the Lorentz–Lorenz equation and the mixing rule:

$$R_i = [(n_i^2 - 1)/(n_i^2 + 2)](1/\rho_i) \,,$$

$$R = [(n^2 - 1)/(n^2 + 2)](1/\rho) \,, \tag{8}$$

$$R = \sum_{i=1}^{k} w_i R_i \,. \tag{9}$$

The absorption optics allow a direct determination of the concentration in the cell via the Lambert–Beer law:

$$A = \lg(I_0/I) = \varepsilon Cl \,, \tag{10}$$

where ε is the specific decadic absorption coefficient of the heavy component, C the volume concentration of the heavy component, and l the cell length. The determination of the density from the concentration in the cell needs a calibration with mixtures of known densities and concentrations:

$$\rho = a_1 + a_2 C + a_3 C^2 + \cdots \,. \tag{11}$$

In case were ε depends on the concentration C one has to calibrate the absorption optics with mixtures of known densities:

$$\rho = a_1 + a_2 A' + a_3 A'^2 + \cdots \,, \quad A' = A/(\varepsilon_0 l) \tag{12}$$

with ε_0 being the absorption coefficient at zero concentration.

An advantage of the present method is that the pressure effect and the composition effect [2] of the density gradient are included in this method and must not be corrected as usual. We will show in this paper that it is possible to measure directly the density ρ in the ultracentrifuge cell via a refractive index or absorption measurement as a function of the radial distance r, instead of calculating $\rho(r)$ theoretically with an equilibrium equation like that of Hermans and Ende [1] or Hearst and Vinograd [2]. The proposed new direct density gradient method has the advantage that we do not really need equilibrium. It would be possible to measure at times before the equilibrium has been reached. This means one has a new dynamic density gradient method additionally to the formerly described two dynamic density gradient methods [8], based on the light/heavy components H_2O/D_2O and $H_2O/Percoll$.

Experimental part

The following materials were used: double-distilled water, alkylsulfonate (BASF AG, Ludwigshafen, Germany), metrizamide (2-(3-acetamido-5-N-methylacetamido-2,4,6-triiodobenz-amido)-2-deoxy-D-glucose, Nyegaard, Oslo, Norway), and polystyrene latex (BASF AG). The physical constants of metrizamide are $M = 789.1$ g/mol; $\rho^0 = 2.1552$ g/cm³ (25°C). The density of the polystyrene latex was $\rho - 1.059_5$ g/cm³ (25 °C) and its mass average diameter was $d = 190$ nm. The density of the polystyrene latex was measured independently with dynamic density gradient measurements and particle density analysis [4].

All measurements were done with the following machines: Kratky-type vibrating tube densitometer with a glass capillary (A. Paar, Graz, Austria), Abbe refractometer (Carl Zeiss, Oberkochen, Germany), and the analytical ultracentrifuge Spinco Model E (Beckman, Palo Alto, CA, USA). The ultracentrifuge was equipped with a 8-hole rotor and Schlieren optics with multiplexer [7]. The rotor was equipped with double-sector cells containing water in one sector and water/metrizamide/polystyrene latex in the other. The polystyrene latex concentration was 0.03 mass%. Evaluation of the Schlieren photos was done with a graphics tablet and a computer. The data were transposed into an AC SII file which was accessible to a PC. The programs are available on request.

Results and discussion

The evaluation of the measurements needs a precise knowledge of the refractive index, n_0, of the initial water/metrizamide solution as a function of the metrizamide content. Figure 1 demonstrates this function [5]. The experimental values were controlled and the function recalculated for 25 °C and 546 nm [4]:

$$n_{546} = 1.3343 + 1.4260 \times 10^{-3} w_{MA} + 1.5728 \times 10^{-5} w_{MA}^2,$$

$$(25\,°C)\,. \tag{13}$$

n_{546} is the refractive index at $\lambda = 546$ nm and w_{MA} the mass fraction of metrizamide.

Furthermore, a precise relationship between the density and the refractive index of the water/metrizamide solution is necessary. Figure 2 shows that this is a linear function [5]:

$$\rho = -3.5232 + 3.3889 n_{546} \quad (25\,°C) \tag{14}$$

Equation (10) was controlled by experimental measurements [4] and recalculated for 25 °C and 546 nm.

Sedimentation equilibrium measurements need the baseline to be predetermined because $\Delta n(r)$ (interference optics) and dn/dr (Schlieren optics) have to be absolute. This could be achieved with double-sector cells, containing the light solvent in one sector and the whole two-component density gradient solution in the other.

Figure 3 shows seven original Schlieren photos together with their digitized version. This is a demonstration of the evolution of the density gradient from strong

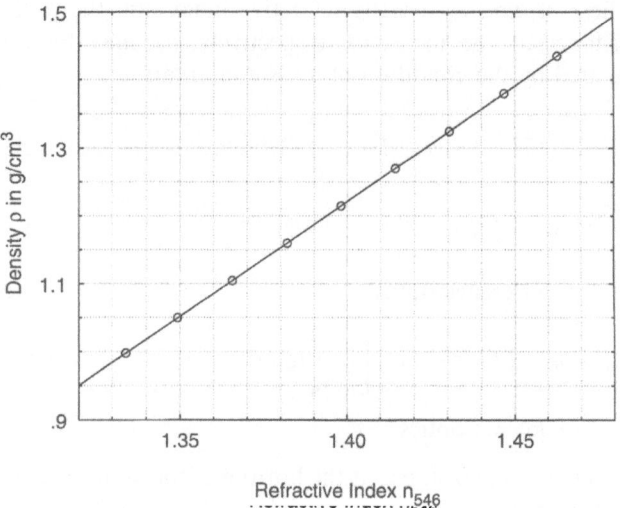

Fig. 2 Density of water/metrizamide mixtures as a function of the refractive index at 25 °C and $\lambda = 546$ nm

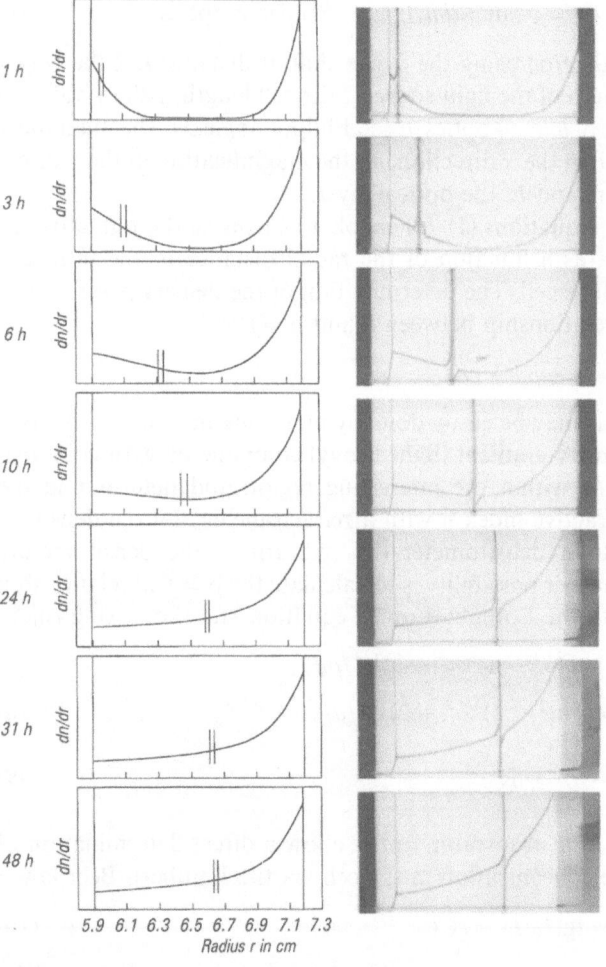

Fig. 3 Evolution of the density gradient of water/metrizamide/polystyrene latex at different times of the ultracentrifuge run; mass concentration of metrizamide = 12% and of polystyrene latex = 0.03%. **Right**: original Schlieren photos; **Left**: digitized Schlieren photos

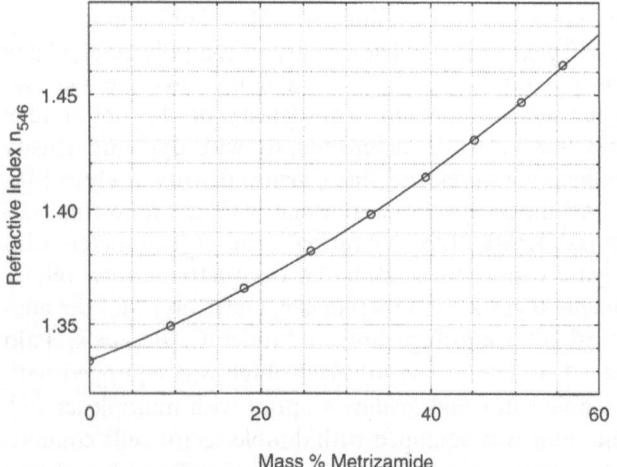

Fig. 1 Refractive index of water/metrizamide mixtures as a function of the metrizamide content at 25 °C and $\lambda = 546$ nm

nonequilibrium to equilibrium during different run times. The system is water/metrizamide/polystyrene latex with a metrizamide concentration of 12 mass%. It is assumed that the fast latex particles move within very short times to the appropriate density of the water/metrizamide density gradient, where they are gathered in a narrow turbidity band. At 24 h the equilibrium has been achieved.

Figures 4 and 6 demonstrate as examples the systems water/metrizamide/polystyrene latex at run times of 31 and 93 h with concentrations of 15% and 18% metrizamide. The experimental conditions are indicated in the

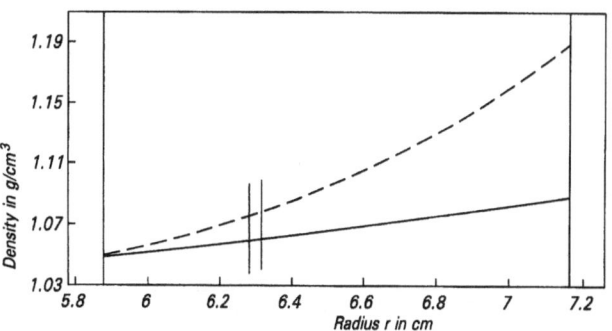

Fig. 5 Same run as Fig. 3: --- original Hermans–Ende equation; —— extended Hermans–Ende equation

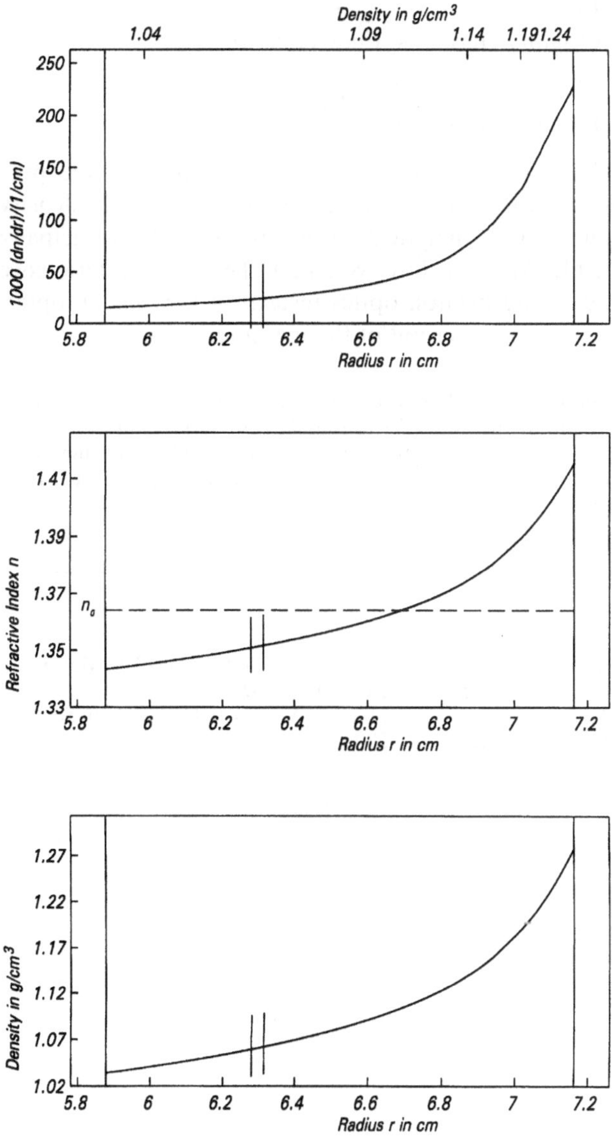

Fig. 4 Sedimentation equilibrium run of water/metrizamide/polystyrene latex; rotor speed $N = 30\,100\ \mathrm{min}^{-1}$; time of the run $t = 31$ h; path length of the cell $l = 0.3$ cm; mass fraction of metrizamide $w_{\mathrm{MA}} = 0.18$; concentration of polystyrene latex $C_{\mathrm{PM}} = 0.274$ g/l. **Top:** original digitized *Schlieren* photo; **Middle:** calculated refractive index profile, Eq. (4); **Bottom:** calculated density profile, Eq. (14)

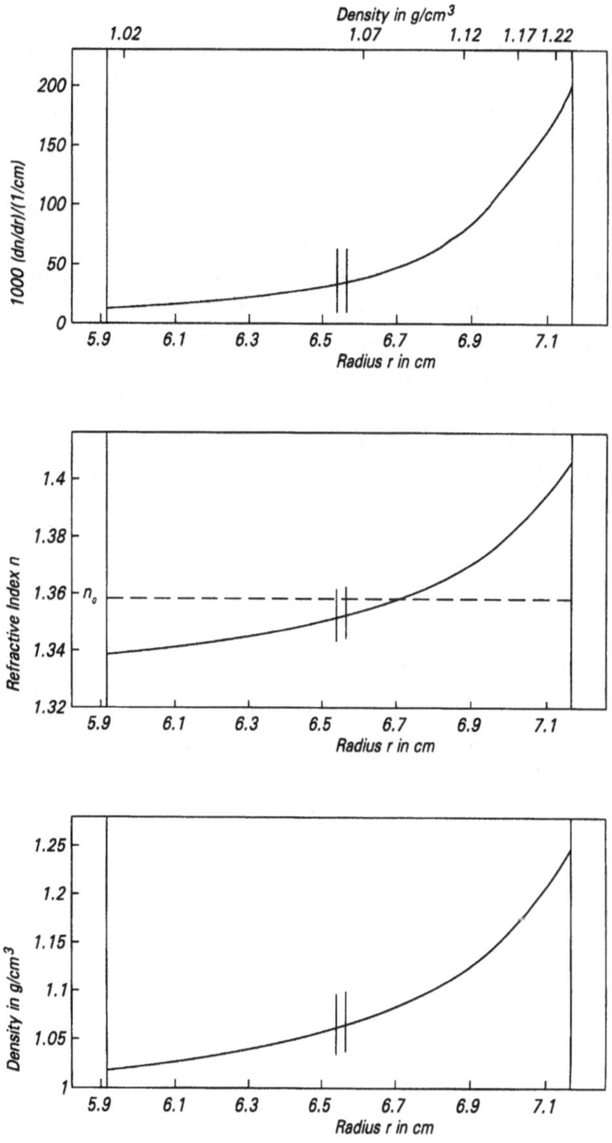

Fig. 6 Sedimentation equilibrium run of water/metrizamide/polystyrene latex; rotor speed $N = 30\,100\ \mathrm{min}^{-1}$; time of the run $t = 93$ h; path length of the cell $l = 0.3$ cm; mass fraction of metrizamide $w_{\mathrm{MA}} = 0.15$; concentration of polystyrene latex $C_{\mathrm{PM}} = 0.269$ g/l. **Top:** original digitized *Schlieren* photo; **Middle:** calculated refractive index profile, Eq. (4); **Bottom:** calculated density profile, Eq. (14)

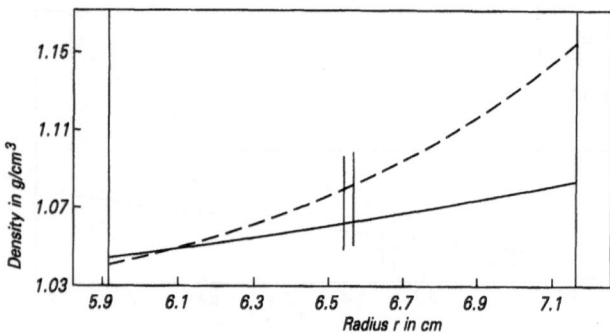

Fig. 7 Same run as Fig. 5: --- original Hermans–Ende equation; ——— extended Hermans–Ende equation

legends. Figures 4 and 6 (top) show the digitized original *Schlieren* curve, the same figures (middle) demonstrate the calculation of 1) the refractive index at the meniscus, n_m, according to the condition of mass conservation, Eq. (4), and the whole refractive index profile $n(r)$ from meniscus to bottom according to Eq. (1) and (2). The same figures (bottom) show the calculation of the density profile according to Eq. (14). For these calculations the measured (digitized) *Schlieren* curves have to be used in both cases.

Figures 5 and 7 compare the results of the new (quasi dynamic) method with those obtained by the Hermans–Ende equation [1] and the extended Hermans–Ende equation [6]. The differences are remarkable with respect to the known density of the polystyrene latex, but the values of the extended Hermans–Ende equation are closer to the

known value of $\rho_{PM} = 1.059_5$ g/cm^3 than those of the original Hermans–Ende equation.

Fig. 3: $\rho_{PM} = 1.060_5$ g/cm^3
Direct determination of the density gradient

Fig. 4: $\rho_{PM} = 1.075$ g/cm^3
Original Hermans–Ende equation

Fig. 4: $\rho_{PM} = 1.060$ g/cm^3
Extended Hermans–Ende equation

Fig. 5: $\rho_{PM} = 1.063$ g/cm^3
Direct determination of the density gradient

Fig. 6: $\rho_{PM} = 1.078_5$ g/cm^3
Original Hermans–Ende equation

Fig. 7: $\rho_{PM} = 1.062$ g/cm^3
Extended Hermans–Ende equation.

Our examples (Figs. 3–7) demonstrate that the proposed new method of a direct determination of the density gradient in an analytical ultracentrifuge cell is principally possible. We feel, however, that the use of interference optics or adsorption optics instead of a *Schlieren* optics increases the precision of the measurements.

Acknowledgements Support of this work by *BASF AG*, Ludwigshafen and by the *Fonds der Chemischen Industrie* is kindly acknowledged. The measurements with the analytical ultracentrifuge were performed by *M. Kaiser*, Kunststofflaboratorium, *BASF AG*, Ludwigshafen.

References

1. Hermans JJ, Ende HA (1963) J Polym Sci C1:161
2. Hearst JE, Vinograd J (1961) Proc Nat Acad Sci US 47:999, 1015
3. Munk P (1982) Macromolecules, 15:500
4. Lechner MD, Mächtle W, unpublished
5. Rickwood D (1979) Metrizamide. Nyegaard & Co., Oslo, Norway
6. Lechner MD, Mächtle W, Sedlack U (1997) this volume
7. Mächtle W, Klodwig U (1979) Makromol Chem 180:2507
8. Mächtle W (1984) Coll Polym Sci 262: 270

Progr Colloid Polym Sci (1997) 107:159–165
© Steinkopff Verlag 1997

POLYMERS, COLLOIDS & SUPRAMOLECULAR SYSTEMS

P. Rossmanith
W. Mächtle

First experiences with the new XL-I AUC: Applications in polymer and colloid science

P. Rossmanith (✉) · W. Mächtle
BASF Aktiengesellschaft
Polymer Physics, Solid State Physics
ZKM-G201
67056 Ludwigshafen
Germany

Abstract In our lab we have transformed all the application of the analytical ultracentrifuge (AUC) in colloid and polymer science from the Schlieren optical system of the Beckman model-E to the interference optics of the Beckman XL-I. Two basic experiments, the sedimentation velocity run and the density gradient run are described in this paper. The advantages of the interference optics together with the sedimentation velocity run are a much higher data density that gives together with the "dc/dt"-analysis method of W. Stafford very accurate results. In the field of colloid science this means that very precise particle size distributions are available, with an especially high resolution for small particles. Density gradient experiments were carried out and gave very good interference pictures. These pictures were analyzed after the formation of the density gradient was complete and also during its formation ("dynamic density gradient"). The method of analysis described in the paper is the same for both cases resulting in very good values for the particle density of the sample. The method is based on the determination of the absolute refractive index of the gradient forming medium and the transformation of the refractive index into a radial density distribution within the measuring cell.

Key words Analytical ultracentrifugation – XL-I, interference optics – sedimentation velocity run – density gradient run

Introduction

The analytical ultracentrifuge (AUC) is a versatile and well-known instrument in biochemistry (see for example [1]). Besides the analytical ultracentrifuge is extremely useful for the analysis of dispersions and colloid systems and in polymer science. The model E from Beckman has proven this statement for the last decades [2]. One and a half year ago Beckman came up with the XL-I. This analytical device on the basis of a preparative XL centrifuge is equipped with both an UV/VIS-detector and a re-fractive index detector which is an interference optical system with a CCD-camera. As the XL-I is a new machine, a computer controls both the centrifuge and the detectors. Especially for the detection, this is a very big advantage compared to the Schlieren optics, where photos have to be taken, manually developed and digitized for computer analysis. This time-consuming procedure is no longer necessary by using the completely digital XL-I. In polymer and colloid science the refractive index detector is more useful than the absorption detector, for most of the polymers have no good UV or visible absorption. We have shown in our lab that all the experiments necessary in colloid science can be adapted from the model E and the Schlieren optics to the XL-I and the interference optics. Out of the four experiments – the sedimentation velocity run, the synthetic boundary run, the sedimentation equilibrium run, and the density gradient run – this paper will show applications and the analysis of the sedimentation velocity run and the density gradient run. In the examples discussed, the samples will be polystyrene latices in water.

The Sedimentation velocity run

The biggest advantage of the interference optics of the XL-I is the ability of automatically taking and analyzing interference pictures of the whole cell. By using a four-hole rotor with three samples, each cell can be scanned every 20 s. This high repetition rate is extremely useful for sedimentation velocity runs, where you want to see the moving of a sedimentation boundary. In the inset of Fig. 1 the raw data of a sedimentation velocity experiment can be seen. The sample is a 65 nm polystyrene latex in water. The figure shows every 10th scan, so there are about 150 scans for the whole run. The analysis of this huge amount of data is not possible with a program that analyzes only a single scan. Therefore the "dc/dt"-program of W. Stafford was used [3]. With the help of this program several files can be

analyzed simultaneously by pairwise subtraction and averaging. For this experiment the analysis was made with 20 files at different times. For the files chosen the boundary had completely moved away from the meniscus but did not touch the bottom of the cell. The result of a "dc/dt"-analysis is the distribution of the apparent sedimentation coefficients, s^*, in units of fringe displacement $g(s^*)$. When dealing with polymers with a remarkable diffusion on the time scale of the experiment, the curves are broader for the analyzed data taken later during the run. The $g(s^*)$-curves will represent the distribution in units of apparent sedimentation coefficient s^*. To get the real distribution of the s-values, the curves have to be extrapolated to infinite time. For the experiment described the resulting $g(s^*)$-curves all represent the same line. The lack of broadening for the sample under discussion shows that there is no remarkable diffusion for these latex particles. Therefore, each curve gives the distribution of the real sedimentation coefficients, s, and no extrapolation is necessary.

The important value in colloid science, however, is not the sedimentation coefficient, s, of a latex particle but its diameter, D. Fortunately, the diameter of a latex particle can be determined out of the sedimentation coefficient with the help of the Stokes' law:

$$D = \sqrt{\frac{18\eta_{DM}\,s}{\rho_{PM} - \rho_{DM}}} \tag{1}$$

Fig. 1 Differential and integral particle size distribution of a 65 nm polystyrene latex. The axis on the left side refers to the differential particle size distribution $g(D)$, the axis on the right side to the integral particle size distribution $G(D)$. The concentration of the latex was $c = 2.1$ g/l, the machine ran at 15 000 rpm

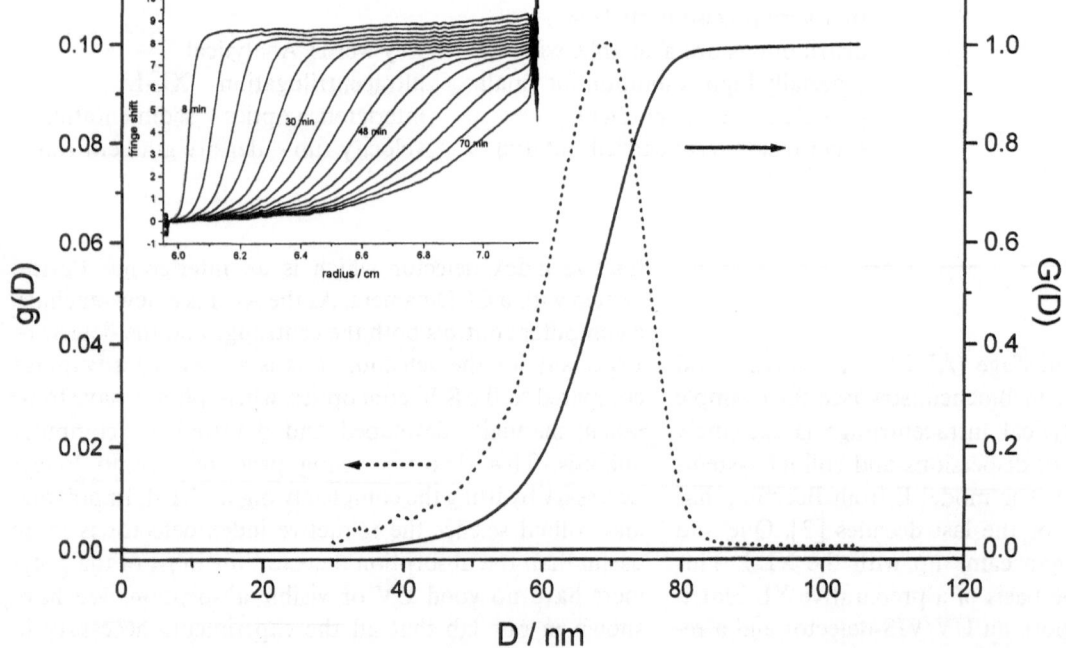

Progr Colloid Polym Sci (1997) 107: 159–165
© Steinkopff Verlag 1997

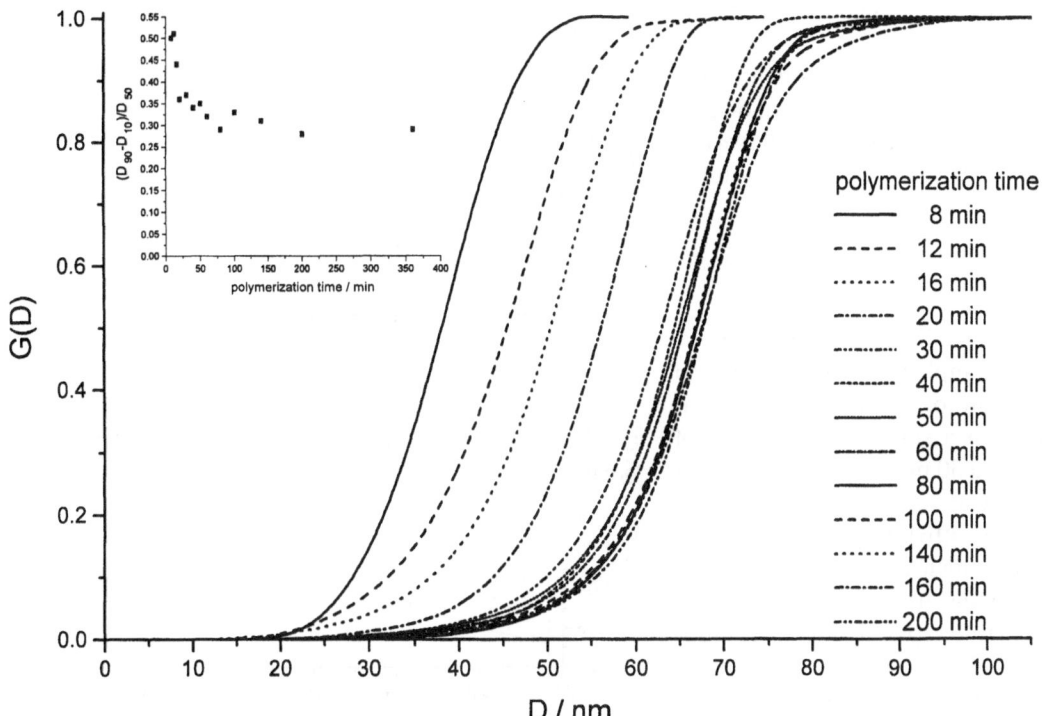

Fig. 2 Integral particle size distributions of samples taken at different polymerization times during an emulsion polymerization experiment. The inset shows the polydispersity $((D_{90} - D_{10})/D_{50})$ of the diameters

with

η_{DM}:   viscosity of the dispersing medium,
s: sedimentation coefficient of the polymer,
ρ_{DM}/ρ_{PM}: density of the dispersing medium/polymer.

By using this equation the sedimentation coefficient, s, of the $g(s)$-curve is transformed to the diameter, D. The y-axis of a plot of the $g(s)$-curve just changes the label from $g(s)$ to $g(D)$ without any transformation. In Fig. 1 both the differential and the integral particle size distribution can be seen. The distribution is very reliable for its shape, as proven by the sedimentation coefficient distribution at different times. To get this curve with its very low noise level no smoothing was done at any step during the analysis. The mean diameter according to capillary hydrodynamic fractionation is 65 nm, which is in excellent agreement with the AUC value.

With the help of this method even very small amounts of different particles can be detected. On the one hand, there is a very good signal-to-noise ratio due to averaging over a couple of data sets. On the other, the analysis at different times gives the chance to distinguish between the remaining noise and tiny signals. As the particle diameter is a very important value, the ultracentrifuge with interference optics is a very precise tool for analysis in colloid science.

The example shown was just one sample out of a whole series of measurements. These samples were taken at different times during 6 h after starting an emulsion polymerization. The development of the particle size distribution during this period of time can be seen in Fig. 2. The figure shows, that the growth of particles is much faster in the beginning of the experiment than at the end. This statement is not very exciting and has been expected in this way. What is more interesting is the shape of the distribution. The width of the distribution is dropping dramatically at the beginning of the experiment (see inset of Fig. 2). That means that the particle formation takes place only at the beginning. Then the existing particles are just growing. There is no new formation of particles at the end of the polymerization.

Density gradient

The last section has shown that the ultracentrifuge is a tool for the determination of particle sizes. The resulting particle size distribution is a very important value for the properties of a dispersion. Another value that is as important as the particle diameter is the particle density, i.e. the chemical composition. With the help of the density gradient experiment of the ultracentrifuge we are able to determine the density of particles [4].

This section deals with the analysis of density gradient data. With the help of a polystyrene latex of known density the method of analysis is sketched out, and this allows one to analyze density gradient experiments both at equilibrium and at nonequilibrium.

Equilibrium analysis

The density gradient in the example that will be discussed here was built up by a mixture of 92 wt% of water and 8 wt% of metrizamide which is a sugar with a density of $\rho = 2.17$ g/cm³. The sample is a 30 nm polystyrene latex at a concentration of 0.25 g/l. The signal at equilibrium is plotted in Fig. 3(a). This signal is obviously a superposition of the signal of the sample, which is the peak at about 6.65 cm radial position, and the underlying gradient of water/metrizamide. The goal is to separate the two components of the signal and to calculate the density distribution in the cell out of the signal of the gradient. As a first step, that part of the signal which is originated by the sample is cut. The resulting gradient is fit with a fifth-order polynomial function. The fit curve is extrapolated to meniscus and bottom. We now have the signal of the pure density gradient in units of fringe displacement. The fringe displacement, $\Delta\Phi$, is converted in units of refractive index relative to the refractive index of the first point at the meniscus, Δn:

$$\Delta n(r) = \Delta\Phi(r)\lambda/l \ . \tag{2}$$

λ is the wavelength of the laser and l the thickness of the centerpiece. To get the density at each position, however, it is necessary to know the absolute rather than the relative refractive index. The absolute refractive index, n, is calculated with the help of mass conservation: Assuming that the refractive index at the meniscus is the refractive index of water, it is possible to calculate the concentration of metrizamide at each position in the cell, and it is then converted to mass.

$$m_{MA}(r_i) = \{-8.236 + 6.181(\Delta n(r_i) + n_{H_2O})\}$$

$$\times l\frac{\phi\pi}{360}(r_i^2 - r_{i-1}^2) \tag{3}$$

where $m_{MA}(r_i)$ is the mass of metrizamide at ith radial position, n_{H_2O} is the refractive index of water, ϕ is the angle of the sector of the cell, and r_i^2 and r_{i-1}^2 are the ith and $(i-1)$st radial position, respectively. Integration results in the mass of metrizamide under the curve. By subtracting this value from the total mass of metrizamide a concentration of metrizamide at the meniscus and the refractive

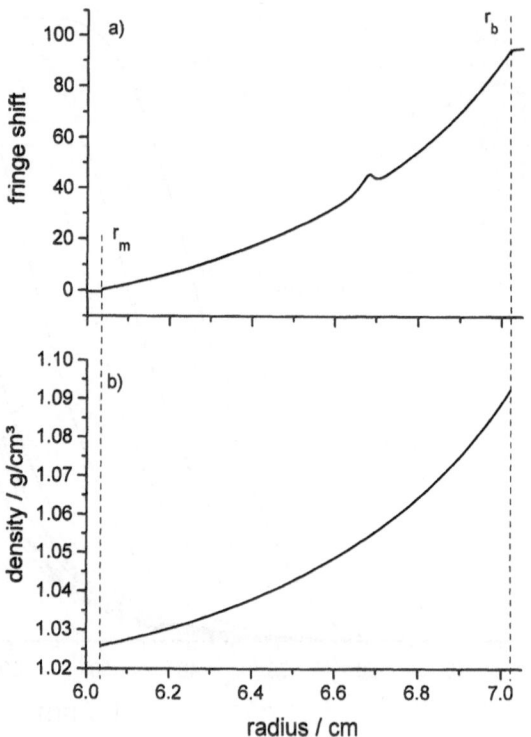

Fig. 3 (a) The signal of a density gradient experiment at equilibrium. The gradient was formed by 92 wt% of water and 8 wt% of metrizamide. The sample has been a 30 nm polystyrene latex with a concentration of 0.25 g/l. The centrifuge ran at 40 000 rpm for 92 h. (b) The density distribution in the cell can be calculated from the distribution of the absolute refractive index

index ($n = 1.332 + 0.162 c_m$; c_m is the metrizamide concentration at the meniscus) are calculated. With this value a better estimate of the metrizamide mass at each position is determined (Eq. (3)). After passing this loop for 3 to 4 times the refractive index at the meniscus is constant and the distribution of the absolute refractive index is determined.[1] Now, the radial density distribution in the cell is calculated by an equation that gives a relation between the refractive index and the density for the system metrizamide/water [5]:

$$\rho = n * 3.350 - 3.462 \text{ g/cm}^3 \ . \tag{4}$$

The resulting radial density distribution is plotted in Fig. 3b. To get the density distribution of the sample, the signal of the gradient is subtracted from the original data to give the signal of the sample. By converting the x-axis from radial position to density, the density distribution is plotted as a curve of density vs fringe displacement as a measure of concentration, as shown in Fig. 4. The analysis of this peak offers two informations: The peak

---

[1] The equations above were all obtained by linearizing the tabulated values from [5].

Progr Colloid Polym Sci (1997) 107:159–165
© Steinkopff Verlag 1997

Fig. 4 The "density distribution" of the sample in the cell (solid line). The result from the nonequilibrium analysis is also plotted as the dashed line

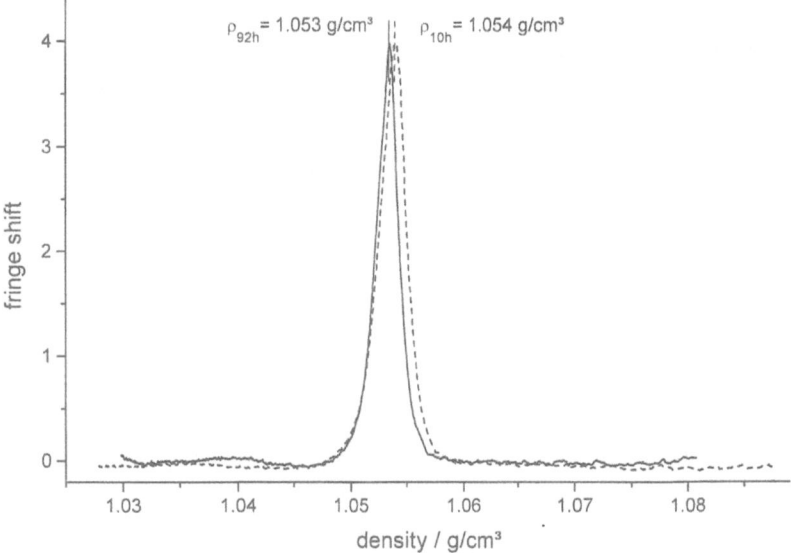

Fig. 5 A 30 nm polystyrene latex in different density gradients. The gradient forming media vary between 94 wt% water (H_2O) and 6 wt% metrizamide (MA) and 85 wt% water and 15 wt% metrizamide as denoted at the interference pictures. The determined densities are written under the peak in the interference picture

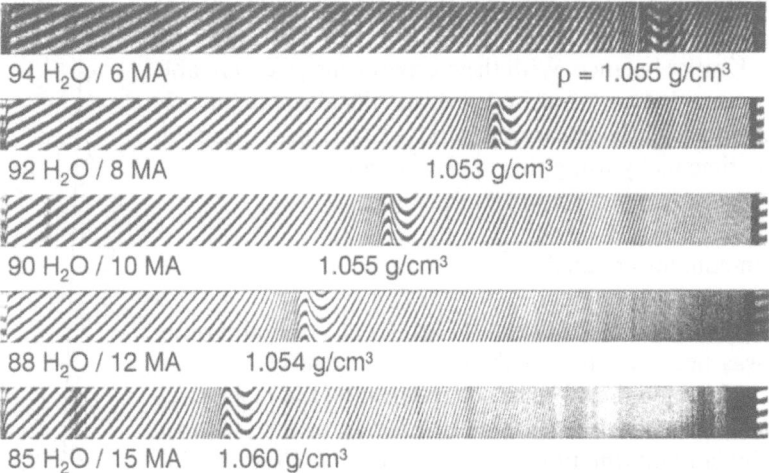

maximum is the particle density of the sample. If the sample is chemically inhomogeneous there would be a couple of peaks for a sample with some distinct densities or a broad and most of the time asymmetric peak for a continuous chemical distribution of the sample. The width of the peak is a result of diffusion broadening in case of a chemically homogeneous sample, which gives a clue to the molar mass of the sample. For the example in Fig. 4 only the peak maximum is of interest. The value for the polystyrene latex was obtained with $\rho = 1.053$ g/cm³. This is in excellent agreement with the value we got from the Kratky balance, which is $\rho = 1.055$ g/cm³.

To prove the reliability of this method of density analysis, several different density gradients with different amounts of metrizamide were run with the same polysty-rene latex as the sample. The interference pictures and the determined densities of these runs are shown in Fig. 5. The results for all experiments are in very good agreement with the density measured by the Kratky balance. The only experiment that does not give the right density was obtained with the steepest gradient with an amount of 15 wt% metrizamide. The reason for this mismatch can easily be seen in Fig. 5. The gradient at the bottom of the cell is so steep, that the CCD-camera hits its resolution limit. The interference lines are no longer resolved and it is therefore difficult to determine the right position of the bottom of the cell. On the other hand, the position of the cell bottom is extremely important for the calculation of the absolute refractive index. This calculation is based on mass conservation and since the concentration at the

Fig. 6 Signal of the cell from
a density gradient experiment
at different times while the
gradient is built up

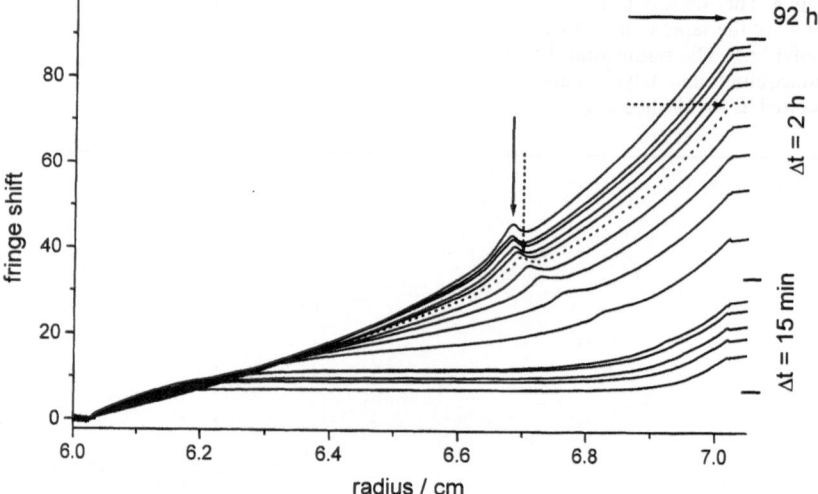

bottom is very high, the precision of the result will be
determined by the precision of the bottom position. All the
other measured densities are within 0.1% of the value from
the Kratky balance. With these experiments, we were able
to prove that very precise values for the absolute particle
densities can be measured with AUC density gradient
experiments by using this method of analysis.

Non-equilibrium analysis

To get the density gradients of Figs. 4 and 5 to equilibrium,
it was necessary to run the centrifuge at 40 000 rpm for
92 h. It would therefore be advantageous if the experiment
could be analyzed after a shorter period of time without
losing any information. For the analysis as shown in the
last section, no specific purpose of the equilibrium was
used to get the desired information. It should therefore be
principally possible to get this information also at a nonequilibrium state. The only requirements are that the
sample has moved to the position of the proper density
and the width of the peak is the same as at equilibrium.
Figure 6 shows the refractive index distribution in the cell
at different times during the formation of the gradient.
Obviously, the requirements for a proper density analysis
are given after a much shorter running time. The dotted line
in the figure is the signal of the cell after 10 h. The comparison of the solid and the dotted arrows shows the difference of the 2 signals. Both the height of the gradient
and the radial position of the peak in the cell are different.
The gradient and the sample are the same as in the
example for the equilibrium analysis: 92 wt% water with
8 wt% metrizamide as the gradient–forming system and
a 30 nm polystyrene latex as the sample. So the analysis

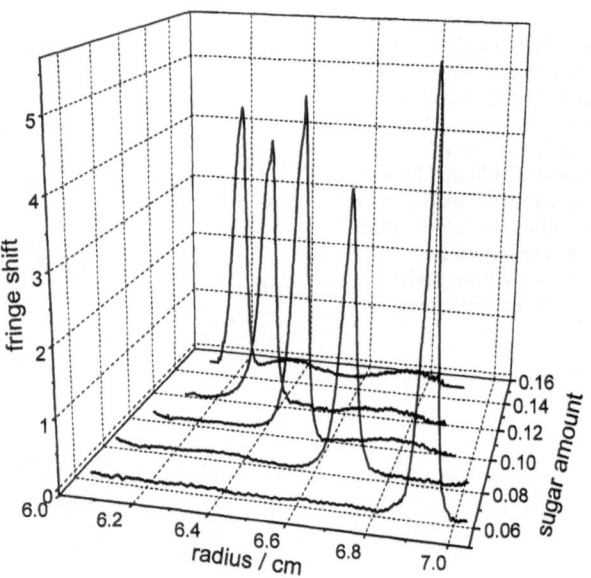

Fig. 7 The pure signal of the sample of all the density gradient
experiments

after 10 h running time should give the same result as
after 92 h.

The procedure for the analysis is the same as for the
analysis at equilibrium. The part of this signal that is
influenced by the sample is cut out and the underlying
gradient fitted with a fifth order polynomial function. After
extrapolating the curve to meniscus and bottom the absolute refractive index of the gradient is calculated by converting the fringe shift into a relative refractive index and
taking mass conservation into account. With the help of
the relation between the absolute refractive index and
the density, the radial density distribution in the cell is

Progr Colloid Polym Sci (1997) 107:159–165
© Steinkopff Verlag 1997

Table 1 Comparison of the determined densities with non-equilibrium and equilibrium analysis. The density determined by the Kratky balance is $\rho = 1.055$ g/cm^3

MA (wt%) in the gradient	6	8	10	12	15
$\rho_{\text{nonequilibrium analysis}}$ [g/cm^3]	1.054	1.054	1.054	1.055	1.059
$\rho_{\text{equilibrium analysis}}$ [g/cm^3]	1.055	1.053	1.055	1.054	1.060

calculated. To get the signal of the pure sample, the signal of the gradient is subtracted from the original data. In order to get the density distribution rather than the radial distribution of the sample, the x-axis is converted from radius to density. The resulting curve is shown in the dashed line of Fig. 4. The value of the particle density resulting from nonequilibrium analysis is $\rho = 1.054$ g/cm^3. This is in even better agreement with the value from the Kratky balance ($\rho = 1.055$ g/cm^3) than the result from equilibrium analysis. In Fig. 4 both curves are plotted: by the equilibrium analysis (solid line) and from the nonequilibrium analysis (dashed line). The peak by the nonequilibrium analysis is slightly shifted compared with the peak by the equilibrium analysis, but this shift is within the error of the method. The widths of both peaks are equal.

The reliability of this type of analysis is again tested by analyzing all five gradients of the previous section at a nonequilibrium state. Figure 7 shows how the peak position is changing with the use of different amounts of metrizamide. The results of the analyses can be compared with the help of Table 1. The determined densities are, again, in excellent agreement with the value from the Kratky balance. The exception is again the experiment where 15 wt% metrizamide were used. The difficulty with this analysis is that the signal of the sample is very near the meniscus of the column. Therefore, the fitting of the gradient and the extrapolation was more difficult than for the other experiments. This problem can be seen at the baseline of the signal which is not as flat as it is for the other examples. But for all the other analyses very good results could be achieved by the nonequilibrium analysis of the

density gradient. We call this new type of nonequilibrium density gradients "dynamic density gradients".

Conclusions

The first part of the paper dealt with sedimentation velocity. It was shown that the particle size distribution of latex particles transparent enough to show fringes while using interference optics can be studied very well with the XL-I. Because there is no diffusion broadening of the band, the analysis of these systems with the "dc/dt"-program gives directly $g(s)$ distributions. Therefore, analyses at different times give always the same result, which is a good way to judge the goodness of the run. Moreover analyses at different times allow to determine between tiny amounts of particles and noise of the measurement. The second part of the paper concerned about density gradients. It was shown that a direct analysis of the data is possible by fitting the pure density gradient. This change of the refractive index is then transformed into absolute refractive index and further into density. As it is also possible to convert the axis of radial position into an axis of density, the real radial density distribution in the cell can be plotted. This procedure can be carried out in the equilibrium state of the gradient resulting in very precise values of the density of the sample. It can also be used at times where the gradient is not yet completely established. Even in this case excellent results can be achieved for the radial density distribution in the cell. We call these nonequilibrium density gradients "dynamic density gradients" like the H_2O/D_2O- or the water/percoll density gradients [6].

References

1. Schuster TM, Laue TM (eds) (1994) Modern Analytical Ultracentrifugation. Birkhäuser, Boston-Basel-Berlin
2. Mächtle W (1992) In Harding SE, Rowe AJ, Horton JC (eds) Analytical Ultracentrifugation in Biochemistry and Polymer Science, Ch 10. The Royal Society of Chemistry, Cambridge, UK, pp 147–175
3. Stafford III WF (1992) Analytical Biochemistry 203:295–301
4. Meselson M, Stahl FW, Vinograd J (1995) Proc. Nat. Acad. Sci. USA 43:581–588
5. Rickwood D (1992) Metrizamide. A Gradient Medium for Centrifugation Studies. Nyegraad & Co, Oslo, Norway
6. Mächtle W (1984) Colloid Polym Sci 262:270–282

Progr Colloid Polym Sci (1997) 107:166–171
© Steinkopff Verlag 1997

POLYMERS, COLLOIDS & SUPRAMOLECULAR SYSTEMS

Analytical ultracentrifugation as a tool in supramolecular chemistry: a feasibility study using a metal coordination array

D. Schubert
J.A. van den Broek
B. Sell
H. Durchschlag
W. Mächtle
U.S. Schubert
J.-M. Lehn

Prof. Dr. D. Schubert (✉)
J.A. van den Broek · B. Sell
Institut für Biophysik
Johann Wolfgang Goethe-Universität
Haus 74
Theodor-Stern-Kai 7
60590 Frankfurt am Main, Germany

H. Durchschlag
Institut für Biophysik und Physikalische
Biochemie
Universität Regensburg
93040 Regensburg, Germany

W. Mächtle
Kunststofflaboratorium
BASF AG
67056 Ludwigshafen, Germany

U.S. Schubert · J.-M. Lehn
Laboratoire de Chimie Supramoléculaire
Université Louis Pasteur
67000 Strasbourg, France

U.S. Schubert
Lehrstuhl für Makromolekulare Stoffe
Technische Universität München
85747 Garching, Germany

Abstract Self-assembling systems composed of organic molecules and metal ions represent one of the topics in modern supramolecular chemistry. Up to now, their physical characterization was essentially limited to crystals or to surface layers. Studies on solubilized systems, e.g., on their molecular mass distribution or association behavior, are rare. We have explored whether sedimentation equilibrium analysis in the Beckman Optima XL-A analytical ultracentrifuge can be successfully applied. A grid-like cobalt coordination array ("$[2 \times 2]$-Co(II)-grid") was used as a model compound.

The technical problem involved, concerning the chemical resistance of the cell components against organic solvents, was solved by using titanium centerpieces and polyethylene gaskets. Another potential problem, the high UV absorbance of the solvents, could be circumvented by measuring the absorbance versus radius profiles in the visible wavelength range, where "grids" and most other related compounds show absorption bands. Special efforts were made to solve the remaining problems: (1) the suppression of nonidealities in sedimentation behavior, which could prevent or greatly complicate a successful analysis of self-associating systems, and (2) the determination of the partial specific volume $\bar{v}$ of the complexes, in particular considering compounds either not available in sufficient quantities for density measurements or exhibiting low solubility. With respect to problem (1), solvent/salt systems were found in which the model compound apparently shows ideal sedimentation behavior. Procedures are suggested to overcome problem (2).

Key words Supramolecular chemistry – metal coordination array – self-association – sedimentation equilibrium analysis

Introduction

Certain organic molecules can associate, by self-assembly via noncovalent interactions, with other organic molecules and metal ions. The resulting assemblies are the subject of supramolecular chemistry, a "chemistry beyond the molecule" [1, 2]. The number and diversity of useful organic building blocks can be greatly increased by appropriate design. The supramolecular entities obtained have highly interesting chemical and physical properties and potential practical applications [2]. The field is in rapid development.

The problem of elucidating the structure of the assembled complexes turned out to be an intricate one, due to the weakness of the interactions involved. Of course, full

Progr Colloid Polym Sci (1997) 107: 166–171
© Steinkopff Verlag 1997

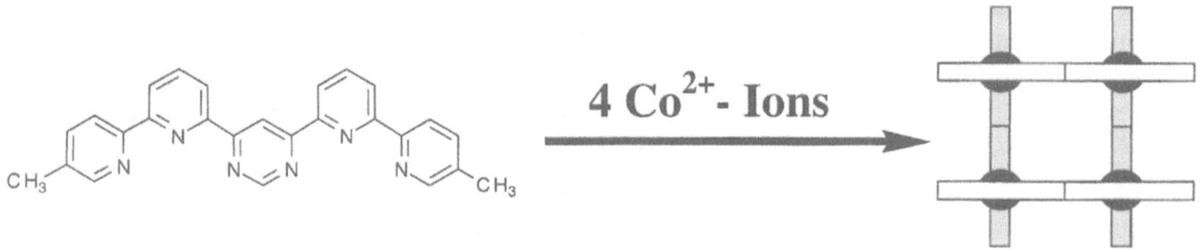

Fig. 1 Building blocks and reaction scheme of the Co coordination array [4]

structural information can be obtained from X-ray crystallography if crystals of sufficient quality are available. In other cases, interesting information may be extracted from surface layers of the material [3, 4]. The search for structural information on the *solubilized* supramolecular entities is apparently still in its infancy and employed up to now only a small number of techniques. NMR, a method of extreme usefulness for structure determination both in organic and biological chemistry, has contributed a number of important findings on supramolecular assemblies [2, 5]. Nevertheless, it is apparently of more limited use in supramolecular chemistry than in the other two fields, in particular when dealing with complexes which contain several copies of a single subunit [6]. However, not only structure but also the fundamental molecular properties like molar mass (or molar mass distribution) and association behavior have been determined in rare cases only, using mainly vapor pressure osmometry or electrospray mass spectrometry [6]. The latter technique has found several important applications [7, 8].

The physical characterization of complexes stabilized by weak, noncovalent interactions is a problem well-known from biochemistry, in particular with respect to the properties of the complexes in solution. Numerous methods are available. They include techniques for determining the molecular properties mentioned above: molar mass and association behavior (see, e.g., ref. [9]). Surprisingly, the classical and most frequently applied method for that purpose (not only in biochemistry but also in macromolecular chemistry), analytical ultracentrifugation, has not been used in supramolecular chemistry. This is even more surprising since those methods used till now for studying molar mass and self-association of solubilized supramolecular aggregates are certainly less potent. We have therefore undertaken to study the applicability of analytical ultracentrifugation to the field, in particular with respect to overcoming expected obstacles. As a model system, we have used a relatively simple supramolecular compound: a grid-like cobalt coordination array ("[2 × 2]-Co(II)-grid") consisting of four molecules of 4,6-*bis*(6-(2,2'-bipyridyl))pyrimidine and four Co(II) ions (plus appropriate counteranions), held together by noncovalent

interactions [4, 10] (Fig. 1). The usefulness of the compound as a model system stems from its expected low tendency towards self-association and its availability in amounts sufficient for conventional density measurements.

Materials and methods

Materials

The organic component of the Co coordination array, 4,6-*bis*(6-(2,2'-bipyridyl))pyrimidine, was synthesized by Stille-type carbon–carbon bond-forming reactions, using organo-tin intermediates [10]. The reaction of equimolar quantities of the molecules and Co(II) acetate in refluxing methanol exclusively leads to the formation of the tetranuclear complex [4, 10]. The complexes were isolated as hexafluorophosphate salts and were recrystallized twice from acetone/ether [4, 10]. The reagents used were purchased from Aldrich. For the ultracentrifuge experiments, stock solutions in acetone or nitromethane, with a sample concentration of 0.2 mg/ml, were prepared.

Analytical ultracentrifugation

Sedimentation equilibrium experiments were performed in a Beckman Optima XL-A ultracentrifuge, in connection with an An-60 Ti rotor. Titanium double-sector centerpieces of pathlength 12 mm (BASF AG, Ludwigshafen) and polyethylene gaskets were applied. Rotor speed was 40 000 rpm, rotor temperature 20 °C. Sample volume in most cases was 135 μl. Sedimentation equilibrium was reached after 5–20 h, depending on the solvent. The absorbance versus radius profiles $A(r)$ of the samples were recorded at a wavelength of 430 or 490 nm. The $A(r)$ data were evaluated assuming either ideal sedimentation behavior (using computer programs developed by Schuck [11, 12]) or the presence of nonidealities (applying a program from the Optima XL-A data analysis software). In the fits to the experimental data, the position of the baseline was treated as a free parameter. However, only those fits were accepted where the calculated baseline position did not deviate from zero by more than ± 0.006 absorbance units (which applied to the majority of fits).

Determination of solvent densities and of the partial specific volume of the "grids"

The densities of the pure solvents were taken from the literature [13]. The density increments due to added salt were determined by using a Paar DMA 02 density meter [14]. It was found that 100 mM NH_4PF_6, when added to acetone, increased solvent density by 0.0121 g/ml and, when added to nitromethane, by 0.0087 g/ml. The determination of the partial specific volume of the sample molecules, $\bar{v}$, was performed by two independent methods: (1) by measuring the density difference between sample solution and pure solvent in a Paar density meter [14]; and (2) by combining the results of sedimentation equilibrium experiments performed in solvents of different densities, which allows a simultaneous determination of $\bar{v}$ and molar mass [15].

Results and discussion

It is obvious that the compounds of supramolecular chemistry, with their molecular masses of a few kDa and their tendency for self-association, represent suitable and, at the same time, interesting subjects for sedimentation equilibrium analysis in the analytical ultracentrifuge. It is, however, apparent that several problems interfere with a successful determination of molar mass and of the state of association of these entities. One group of problems is of technical nature. Two additional problems are intrinsic ones which are frequently encountered in analytical ultracentrifugation.

Technical problems

The need for applying organic solvents to solubilize the supramolecular compounds poses a number of problems. One is connected with the strong UV absorbance of many organic solvents, two others are related to the action of the solvents on the cell components.

Many compounds of supramolecular chemistry are transparent in the visible wavelength range but absorb light in the UV, at wavelengths where also many of the usually applied solvents absorb. Frequently, solvent absorption exceeds absorption by the sample by orders of magnitude. With the present system, a Co coordination array, this problem does not arise, since the arrays show strong light absorption in the visible wavelength range also. In general, application of the recently introduced Optima XL-I ultracentrifuge, with an optical system sensitive to changes in refractive index, circumvents the problem. Other problems related to the solvent, however, persist (see below).

Most organic solvents have deterious effects on the centerpiece of the ultracentrifuge cell (even the Beckman Kel-F centerpieces will be unstable with many of the solvents of interest). The polymer chemists have long been faced with this problem. Those working for big chemical companies found the solution in manufacturing titanium centerpieces, as used in the present study. Solvent action on the centerpiece gaskets poses another problem. We have observed that all solvents used extract UV-absorbing material (in addition, possibly, metal ions) from commercial and noncommercial gaskets. The absorption seems to be caused by the softener used in foil production. The extracted material can be harmful in two ways: (1) it can contribute, in an uncontrolled way, both to the absorption and the refractive index of sample and reference, and (2) it may influence the properties of the compound under study. To minimize these effects, we have soaked the gaskets for several days in the solvent to be used in the run (with several solvent exchanges). In addition, we collected the absorbance versus radius profiles $A(r)$ at wavelengths where the unwanted material does not absorb.

Avoiding nonideal sedimentation behavior

It is generally accepted to be impossible to extract reliably, from sedimentation equilibrium experiments, more than 3–4 unknown parameters (except in special cases [11, 12]). Thus, when studying systems capable of self-association, it is highly important to avoid introducing virial coefficients as additional parameters, by choosing conditions where nonidealities in the sedimentation behavior are small enough to be neglected. For this purpose, we have studied the dependency of the sedimentation behavior of the supramolecular compound on ionic strength, in solutions containing varying concentrations of ammonium hexafluorophosphate. The Co coordination array used seemed to be particularly suited as a model system for investigating this problem, since it was thought to be stable and not to show a tendency towards ongoing self-association.

Salt-free solutions

Such solutions are known to promote nonideal sedimentation behavior of charged or partially charged molecules [16]. The sedimentation equilibrium profiles observed

Fig. 2 Plots of $\ln A$ versus r^2 for the Co coordination array in two different solvents: (a) in acetone, without salt (---; $\rho_0 = 0.7899$ g/ml) and with 25 mM NH_4PF_6 (—; $\rho_0 = 0.7929$ g/ml); (b) in nitromethane, without salt (---; $\rho_0 = 1.1371$ g/ml) and with 44 mM NH_4PF_6 (—; $\rho_0 = 1.1409$ g/ml). Sample concentration was approximately 3×10^{-5} M

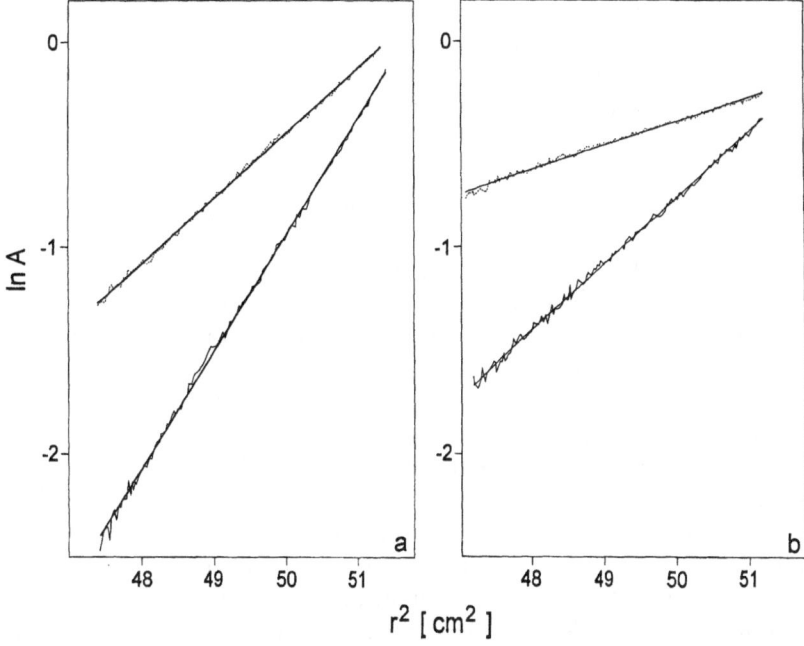

with the Co coordination arrays in salt-free solutions of acetone or nitromethane confirm this expectation: In both solvents, $\ln A(r^2)$ can be fitted by a straight line (Figs. 2a and b) the slope of which is, however, distinctly gentler than expected for a solution of the intact Co coordination array: Using a $\bar{v}$-value of 0.627 ml/g, as found by two different methods (see below), the relative molar mass of the arrays is obtained as 1740 (acetone) or 1122 (nitromethane), respectively, and thus significantly lower than the figure for the intact array ($M_r = 3061$). This seems to indicate that the counterions have in part dissociated from the arrays, thus leading to the so-called primary charge effect [16]. According to established theory [16], the equations describing the observed behavior are formally identical to the case of ideal sedimentation behavior. However, M_r is replaced by $M_r/(z + 1)$, where z is the number of ionizable sites on the compound under study. It follows that, in salt-free solutions of acetone or nitromethane, the Co coordination array has an effective, time-averaged charge of 0.7 and 1.7 elementary units. The difference in the results for the two solvents is probably due to differences in their polarity.

It should be noted that satisfactory fits to the experimental $A(r)$ distributions can also be obtained under the assumption that the arrays dissociate into building blocks of around one- and two-third of their initial molar mass. Since crystals grown in the absence of salt are built up from intact arrays [17], we consider this interpretation as extremely unlikely.

Ionic strengths of 10–50 mM

In the presence of low concentrations of NH_4PF_6, the samples apparently show ideal sedimentation behavior: this is demonstrated both by linear fits to the $\ln A(r^2)$ data (Fig. 2) and fits to the untransformed $A(r)$ data using a single exponential for ideal solutions (not shown). In both cases, the fits are of excellent quality, and the figure for $M_{eff} = M(1 - \bar{v}\rho_0)$, obtained from the slope of the $\ln A(r^2)$ fit or required for the best $A(r)$ fit, is close to that expected for the intact Co coordination array. Together with $\bar{v} = 0.627$ ml/g (see below), the data shown give $M_r = 3120$ (both in acetone and nitromethane), which deviates from the chemically determined figure by < 2%. This deviation is within the experimental error. It is thus clear that addition of 10–50 mM salt to the organic solvents used ensures ideal sedimentation behavior of the supramolecular compound under study.

Ionic strengths of 100–200 mM

It is to be expected that suppression of nonideal sedimentation behavior by addition of salt, as described above, persists at higher ionic strength. In accordance with this expectation, the $A(r)$ data collected at 100–200 mM NH_4PF_6 could be perfectly fitted without introducing virial coefficients. However, good fits were only obtained when, in the equations applied, terms characterizing oligomers of the array were considered. The size and relative amount of these oligomers, which in some experiments were also found at low or even zero ionic strength, were irreproducible up to now (see below).

Determination of the partial specific volume

Sedimentation equilibrium and velocity experiments on aqueous solutions of proteins generally use $\bar{v}$-values which were obtained by calculation, applying tabulated data for the density contribution of the different amino acid residues [18]. This procedure qualifies itself by the good agreement of calculated and measured data, the latter mostly obtained in a Paar digital density meter [14]. The possibility to use calculated $\bar{v}$-values can also be desirable for the complexes of supramolecular chemistry, in particular in the case where direct density measurement is difficult due to shortage of material or to limited solubility. It is clear, however, that also with these compounds the reliability of the figures calculated first has to be checked by comparison with experimental data. For the model system studied in this paper we have therefore performed two independent experimental determinations of $\bar{v}$, in order to establish a basis for subsequent investigations.

Determination of $\bar{v}$ by digital densimetry

The measurements were performed at 20 °C and sample concentrations of 3–12 mg/ml, with acetone or nitromethane as a solvent. Sample concentration was determined by weighing the dried sample. The results obtained, after extrapolating the data to zero sample concentration, were $\bar{v} = (0.629 \pm 0.013)$ ml/g in acetone and $\bar{v} = (0.625 \pm 0.009)$ ml/g in nitromethane (Fig. 3). Since both values agree within the experimental error, they can be combined into a mean of $\bar{v} = (0.627 \pm 0.010)$ ml/g. It should be noted, however, that omission of the highest data point in the nitromethane series reduces the corresponding figure to 0.612 ml/g.

Determination of $\bar{v}$ from sedimentation equilibrium experiments

The density measurements were performed at relatively high sample concentrations, where the complexes probably tend to aggregate. Since aggregation could, in principle, influence $\bar{v}$, we have also tried to extract $\bar{v}$ from sedimentation equilibrium data, i.e. from data on samples more dilute by a factor of approximately 100. For this purpose, M_{eff} values measured in the two solvents, which

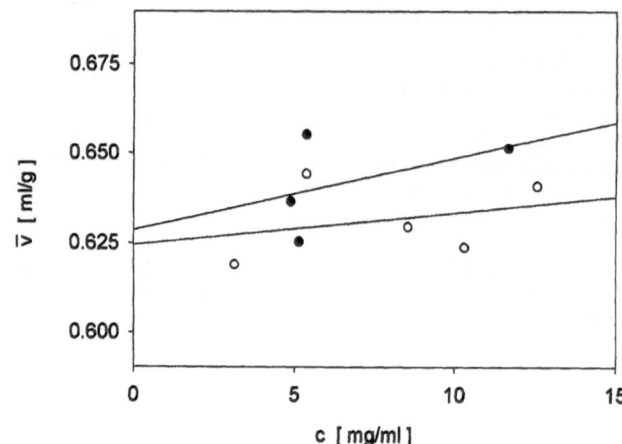

Fig. 3 Dependency of the partial specific volume of the Co coordination array, $\bar{v}$, on sample concentration. Solvent: (●) acetone, (○) nitromethane

greatly differ in density (0.79 vs. 1.14 g/ml), were combined to yield both M_r and $\bar{v}$ (this is an adaptation of a method by Edelstein and Schachman, in which the corresponding figures for proteins are calculated from sedimentation equilibrium data in H_2O/D_2O mixtures of differing composition [15]). The solutions contained enough NH_4PF_6 to ensure ideal sedimentation behavior of the samples. The results obtained were $\bar{v} = (0.627 \pm 0.010)$ ml/g and $M_r = 3120 \pm 150$. The former value is identical to that from density measurement. The deviation of the latter from the theoretical value is within the experimental error.

In principle, calculating $\bar{v}$ can be done in two ways [19]: (i) *ab initio*; (ii) performing a "calculus of differences", starting from the known $\bar{v}$-value of a closely related compound. Method (i) suffers at present from the relatively poor knowledge of the $\bar{v}$-increments of the building blocks of the compounds of supramolecular chemistry, as compared, e.g., with those of protein chemistry. This will lead to a much greater uncertainty in the results than in the case of proteins. Our original plan was to apply method (ii), using a closely related Co coordination array with a density known from X-ray crystallography [17] as a starting point. We did realize, however, that this density value included several contributions from the solvent used during crystallization; these contributions would require quite ambiguous corrections. In forthcoming studies, the $\bar{v}$-value for the Co coordination array determined by us may represent a suitable starting point.

Properties of the Co coordination array

The outcome of the present study confirms the fact that the Co coordination array can exist as a homogeneous entity in solution, with the same overall composition as in a crystal. In addition, our study has revealed another property of the compound which has been mentioned briefly above. In many experiments, part of the arrays showed strong self-association which we found difficult or impossible to reverse and which seemed to depend on the history of the sample. Up to now, we were unable to control this conversion. We have also observed that, even with arrays which finally were monomeric (e.g., the samples of Fig. 2), it took several days until the monomeric state was reached. A similar behavior was observed with another supramolecular compound unrelated to the arrays (Tziatzios, C. and Schubert, U.S., personal communication). Thus, controlling the state of association of the complexes is far from trivial. Considering the importance of the state of association of the compounds of supramolecular chemistry in many physical studies or planned applications (e.g., in nanotechnology [2,4]), its control by methods like analytical ultracentrifugation seems to be highly recommendable.

Conclusions

Typical questions which can be answered by analytical ultracentrifugation are important questions in supra-

molecular chemistry:

(1) Do the compounds really exist in solution?
(2) Do they show significant dissociation?
(3) Do they show significant self-association and (if yes) what type of self-association?

We have shown in this study that those problems which could prevent answering the above questions can be solved. The technical problems arising from the need to use organic solvents can be met by using titanium centerpieces and solvent extraction of the gaskets; nonideal sedimentation behavior of the samples can be remedied by adding salt, and practicable methods are available to determine the partial specific volume of the compounds. It follows that analytical ultracentrifugation, in particular sedimentation equilibrium analysis, is fully applicable to supramolecular chemistry. The problems described above, concerning the control of the supramolecular compounds' state of association, convincingly demonstrate that, in fact, there is a need for applying this technique.

Acknowledgements U.S.S. gratefully acknowledges financial support from Fonds der Chemischen Industrie and from Bayerisches Staatsministerium für Unterricht, Kultus, Wissenschaft und Kunst.

References

1. Lehn J-M (1988) Angew Chem 100:91
2. Lehn J-M (1995) Supramolecular Chemistry: Concepts and Perspectives. Verlag Chemie, Weinheim
3. Cohen SR, Weissbuch I, Popovitz-Biro R, Majewski J, Mauder HP, Lavi R, Leiserowitz L, Lahav M (1996) Israel J Chem 36:97
4. Schubert US, Volkmer D, Hanan GS, Lehn J-M, Hassmann J, Hahn CY, Waldmann O, Müller P, Baum G, Fenske D (1997) ISMRI 9, Symp Proc, in press
5. Baxter PNW, Lehn J-M, Fischer J, Youinou M-T (1994) Angew Chem 106:2432
6. Lawrence DS, Jiang T, Levett M (1995) Chem Rev 95:2229
7. Marquis-Rigault F, Dupont-Gervais A, van Dorsselaer A, Lehn J-M (1996) Chem Eur J 2:1395
8. Przybylsky M, Glocker MO (1996) Angew Chem 108:878
9. Cantor CR, Schimmel PR (1980) Biophysical Chemistry, Part II: Techniques for the Study of Biological Structure and Function. Freeman, San Francisco
10. Hanan GS, Schubert US, Volkmar D, Reviere E, Lehn J-M, Kritsakis N, Fischer J (1997) Can J Chem 75:169
11. Schuck P (1994) Prog Colloid Polym Sci 94:1
12. Schuck P, Legrum B, Passow H, Schubert D (1995) Eur J Biochem 230:806
13. Windholz M (ed) (1983) The Merck Index, 10th ed. Merck & Co, Rahway
14. Kratky O, Leopold H, Stabinger H (1973) Meth Enzymol 27:98
15. Edelstein SJ, Schachman HK (1967) J Biol Chem 242:306
16. Fujita H (1975) Foundations of Ultracentrifugal Analysis. Wiley, New York, 302
17. Hanan GS, Volkmer D, Schubert US, Lehn J-M, Baum G, Fenske D (1997) Angew Chem 109:1929
18. Durchschlag H (1986) In: Hinz H-J (ed), Thermodynamic Data for Biochemistry and Biotechnology. Springer, Berlin, 45
19. Durchschlag H, Zipper P (1994) Prog Colloid Polym Sci 94:20

Progr Colloid Polym Sci (1997) 107:172–179
© Steinkopff Verlag 1997

W. Borchard
H.M. Hinsken

The sedimentation velocity of a gelled polymer

Prof. Dr. W. Borchard (✉) ·
H.M. Hinsken
Angewandte Physikalische Chemie
der Gerhard-Mercator-Universität
Gesamthochschule Duisburg
47078 Duisburg, Germany

In memory of Prof. Dr. Rolf Haase†

Abstract In a centrifugal field a gel is a continuous system with a radial concentration gradient due to the volume force of the field and diffusion. In contrary to the known definition of the sedimentation coefficient of a gel where the movement of the gel meniscus is related to the driving force, the velocity of the center of mass of the polymer with respect to its acceleration has been introduced using irreversible thermodynamics. It is shown that there will be no induction period of this velocity although the meniscus does not move. From extrapolation to times where the selected rotor speed had just been reached, the sedimentation coefficient of the polymer of a κ-carrageenan/water gel could be determined. In addition to the sedimentation coefficient, the phenomenological and the diffusion coefficients were also calculated from the data. Differences between the sedimentation of a sol and a gel are pointed out and discussed.

Key words Centrifugal field – gel – sedimentation – diffusion – κ-carrageenan/water

Introduction

The sedimentation–diffusion equilibrium of a gel – a liquid mixed system showing up elasticity – has been treated in many papers [1–4]. With the results derived therein the thermodynamic and elastic properties of gels can be determined in centrifugal experiments if real continuous equilibria are established.

It is long ago that the first attempt was made to formulate the sedimentation velocity of a gel, which was considered to be the change of the radial position of the meniscus of the gel with time, where this meniscus is initially the interface between gel and vapor but becomes the interface between gel and solvent or solution [1]. This definition of the sedimentation velocity of a gel, which has been used also by other authors [5,6], is problematic because at low values of the centrifugal field there may be no movement of the meniscus of the gel although a concentration gradient will be established as soon as the external field is acting. The formation of concentration gradients inside a gel means that the mass of the crosslinked polymer in different volumes along the radial distance from the rotational axis has changed.

We want to show that the flux of the masses of the different components with time related to the field strength can be used to define a sedimentation coefficient of a component in a gelled system in complete analogy to the sedimentation coefficient of a component in a solution.

In recent papers it was shown that the movement of the meniscus of a gel is due to the conservation of the crosslinked polymer in the centrifugal cell if its concentration is raised at the cell bottom and decreased at the meniscus [2]. As the lowest polymer concentration possible corresponds to the swelling equilibrium of the gel, we state that this concentration has to be constant at the meniscus of the gel as soon as this interface starts to move to the bottom.

This can be considered to be the main difference between the sedimentation of a solution and a gel, because the concentration of an uncrosslinked polymer may go to zero close to the meniscus if the centrifugal field is high enough. This is not possible with a gel. As a consequence of this, the polymer concentration can easily be calculated inside the gel if both the concentration at the swelling equilibrium and the concentration gradients are known. Proceeding from this the boundary conditions for the transport equations may be easily formulated.

With the sedimentation coefficient defined in this manner there is no need to introduce the concept of an induction period of the sedimentation process [1,5,7].

The situation is schematically explained by means of a swelling-pressure–concentration diagram in Fig. 1. When introduced into the centrifugal cell, the partial density of the crosslinked polymer component in a binary system is given by ρ_2^0. If the gel is allowed to swell freely in the pure solvent the partial density will be $\rho_{2,s}$ at maximum saturation, which is the lowest polymer concentration without additional surface or volume forces acting on component 2. The coexistence curve $\pi_s(\rho_2)$ given by curve a in Fig. 1 tells us that a swelling pressure π_s has to act on the gel so that it may coexist with the pure solvent at concentrations $\rho_2 > \rho_{2,s}$. The swelling pressure is an osmotic pressure exerted on a gel where the gel surface acts as a semipermeable membrane. Thus the gel with the initial concentration ρ_2^0 will coexist with the pure solvent if it is under the swelling pressure $\pi_{s,1}$ which is identical with P_1.

In the centrifugal field we have a continuous equilibrium with a series of osmotically active pressure differences which were named partial hydrostatic pressures P_i [1]. At the angular rotational speed ω_1 inside the gel a concentration gradient is built up which is demonstrated by curve b. This curve has been constructed assuming that the concentration at the radial positions of the original meniscus (M,0) and the cell bottom (B) are as follows: $\rho_2(r_{M,0}) = \rho_2^0 - \Delta\rho_2$ and $\rho_2(r_B) = \rho_2^0 + \Delta\rho_2$, where $\Delta\rho_2$ is positive. At low angular speeds of the rotor the concentration changes $\Delta\rho_2$ are small, as is the osmotically active pressure P_2 at the bottom of the cell. The corresponding pressure at the meniscus has to be zero. We assume the same shape for curve b as for curve a. At sedimentation equilibrium $\Delta\rho_2$ increases with increasing rotational speeds. Therefore $\rho_2(r_{M,0})$ has to shift towards $\rho_{2,s}$, which is considered to happen at $\omega = \omega_2$ in this schematic diagram. In this case the osmotically active pressure at the cell bottom is raised to P_3. All sets of values of osmotically active pressures and concentrations are situated on the $\pi_s(\rho_2)$-curve a. Up to now the mean concentration is given by $\bar\rho_2 = \rho_2^0$. As soon as the rotational speed is increased further to ω_3, where we

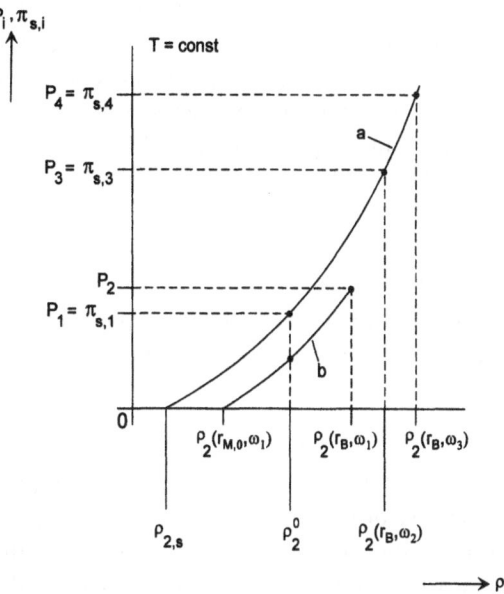

Fig. 1 Schematic swelling-pressure–concentration diagram of a crosslinked polymer; $\pi_{s,1}$ = swelling pressure, P_i = osmotically active pressures, ρ_2 = mass concentration of the polymer, $r_{M,0}$ = distance between meniscus of the gel and axis of rotation at time $t = 0$, ω_i = rotational speeds; see text

have $\omega_3 > \omega_2$, the osmotically active pressure at the bottom of the cell will become P_4 corresponding to the concentration $\rho_2(r_B, \omega_3)$. All indications of the rotational speed given in this section are not identical with those in the later equations. The concentration at the meniscus of the gel remains $\rho_2 = \rho_{2,s}$, so that inside the gel the concentration ρ_2 is given by $\rho_{2,s} \leqslant \rho_2 \leqslant \rho_2(r_B)$. It can be taken from Fig. 1 that the arithmetic mean $[\rho_{2,s} + \rho_2(r_B, \omega_3)]/2$ is larger than ρ_2^0 if the meniscus of the gel starts to move. Here we have

$$\bar\rho_2 = \rho_2^0 \frac{r_B^2 - r_{M,0}^2}{r_B^2 - r_{M,i}^2} \qquad (1)$$

which has been mentioned already by Svedberg [1] and Johnson [5]. As the original position of the meniscus of the gel, $r_{M,0}$, is smaller than any position of the meniscus $r_{M,i}$, as soon as it has moved, Eq. (1) describes the increase of the mean concentration of the crosslinked polymer in the gel for all equilibrium values $r_{M,i}$ with $r_{M,i} > r_{M,0}$ in a sector shaped cell.

Theory

In every volume element of a continuous system the average velocity of the particles of substances may differ so that there is a relative motion of the particles to each other which is differentiated from the motion by convection. The diffusion in its most general sense is the mean velocity of particles of kind k given by the vector $\mathbf{v}_k$ which is the velocity of particle k moving perpendicularly through the cell-fixed reference frame. The relative velocity of k is given by

$$\mathbf{v}_k - \mathbf{v} \, ,$$

where $\mathbf{v}$ is considered to be the velocity of the center of mass of the system [8]. The vector of the flux density of k, $\mathbf{J}_k$, may be defined as

$$\mathbf{J}_k \equiv \rho_k(\mathbf{v}_k - \mathbf{v}), \quad k = 1, 2, \ldots, N \, , \tag{2}$$

where ρ_k is the partial density or mass concentration of k. $\mathbf{J}_k$ is the flux density of k, defined as the moving mass of k per unit of time and area perpendicularly through the reference frame which moves with $\mathbf{v}$. The velocity of the center of mass $\mathbf{v}$ and the particle velocities $\mathbf{v}_k$ are interrelated by

$$\rho\mathbf{v} = \sum_k \rho_k\mathbf{v}_k \, . \tag{3}$$

We used the partial density ρ_k of k, because a gel is a huge particle where the molar mass is of an extremely high value and is not necessarily a well defined quantity.

Introducing the mass fraction w_k of k we get

$$\rho = \sum_k \rho_k \, , \tag{4}$$

$$w_k = \frac{\rho_k}{\rho} \, , \tag{5}$$

$$\sum_k w_k = 1 \, , \tag{6}$$

and

$$\mathbf{v} = \sum_k w_k\mathbf{v}_k \, . \tag{7}$$

Summing up over all particles k we obtain from Eq. (2) together with Eqs. (3)–(7)

$$\sum_k \mathbf{J}_k = 0 \, . \tag{8}$$

Equation (8) tells us that in a system of k particles there will be $k - 1$ independent fluxes.

The directly measurable flow of k relative to the cell, $\mathbf{J}_k^*$ is given by

$$\mathbf{J}_k^* = \rho_k\mathbf{v}_k \, . \tag{9}$$

$\mathbf{J}_k$ and $\mathbf{J}_k^*$ are interrelated by use of Eqs. (2), (3) and (9):

$$\mathbf{J}_k = \mathbf{J}_k^* - w_k \sum_k \mathbf{J}_k^* \, . \tag{10}$$

From this equation the relevant flux density $\mathbf{J}_k$ which is used in the thermodynamics of irreversible processes can be calculated. Only if the local center of mass does not change with time are the flux densities $\mathbf{J}_k$ and $\mathbf{J}_k^*$ identical. This means that convection does not occur.

In reality the flux densities $\mathbf{J}_k^*$ have to be determined separately – a consideration which may lead to serious experimental problems in determining the velocity of the center of mass in a certain small volume.

The following considerations are limited to a two particle or two component system consisting of a crosslinked nonelectrolyte polymer and the swelling agent. In principle it is easy to extend the description to more than two components or particles, as it has already been done for different reference systems and/or concentration variables [8,9].

The flux density $\mathbf{J}_k$ is proportional to the driving force $\mathbf{X}_k$, which is the acting force per mass of particle k. Both quantities have to vanish as the system approaches equilibrium. Thus we have for an incompressible system, where all partial specific volumes of particles k are constant,

$$\mathbf{J}_2 = \alpha_2\left[(1 - \tilde{V}_2\rho)\omega^2\mathbf{r} - \left(\frac{\partial\tilde{\mu}_2}{\partial\rho_2}\right)_{T,P}\left(\overrightarrow{\frac{\partial\rho_2}{\partial r}}\right)\right], \tag{11}$$

where α is the phenomenological coefficient, $\tilde{V}_2$ the partial specific volume of component 2 and $\tilde{\mu}_2$ the partial specific free enthalpy or the specific chemical potential of component 2. Introduction of the sedimentation coefficient of the crosslinked swollen polymer (gel), s_2, and also of the mutual diffusion coefficient, D,

$$s_2 \equiv \frac{\alpha_2}{\rho_2}(1 - \tilde{V}_2\rho) \tag{12}$$

and

$$D = \alpha_2\left(\frac{\partial\tilde{\mu}_2}{\partial\rho_2}\right)_{T,P} \, , \tag{13}$$

we obtain from Eqs. (11) and (12) in the limit $t \to 0$, when $\overrightarrow{(\partial\rho_2/\partial r)} \approx 0$:

$$\alpha_2 = \frac{\rho_2\mathbf{v}_2}{(1 - \tilde{V}_2\rho)\omega^2\mathbf{r}} \quad \text{or} \quad s_2 = \frac{d\mathbf{r}_s/dt}{\omega^2\mathbf{r}} \, . \tag{14}$$

It can be seen from this equation that the sedimentation coefficient of the polymer in the gel as defined by Eq. (12) is not the velocity of the meniscus but the velocity of the local center of mass of the gel component relative to the frame of the local center of mass of all components in the considered volume.

In the limit $t \to \infty$ we obtain from Eq. (11) with $\alpha_2 \neq 0$ that $\mathbf{J}_2$ and the bracketed term have to be zero. From a given value of α_2 which is determined by Eqs. (11) and (14), we are able to determine $(\partial\tilde{\mu}_2/\partial\rho_2)_{T,P}$ by measuring $\rho_2(r)$.

Using Eq. (13) the mutual diffusion coefficient D can be determined.

Regarding Eq. (14), the sedimentation coefficient s_2 is in principle a mobility which is thought to be characteristic of a gel of a given structure.

The same may be said concerning the diffusion coefficient. But it is known that both transport quantities are normally dependent on temperature, pressure and concentration, e.g., ρ_2.

It is possible to determine α_2, s_2 and D from a run at constant angular velocity, $\omega = \omega_1$. If after attainment of the sedimentation–diffusion equilibrium the angular velocity is raised again instantaneously to $\omega = \omega_2$, where we have $\omega_2 > \omega_1$, it is evident from Eq. (11) that the second term in brackets is not equal to zero because there is a constant concentration gradient. As soon as this angular velocity $\omega = \omega_2$ is reached, the first term has to be changed correspondingly. This procedure may be repeated up to the highest value ω_n.

In decreasing the angular velocity rapidly from the highest chosen value back to $\omega = \omega_{n-1}$ there will be a return of the movement of the center of mass of the network component due to the diffusion term. If $\omega_{n-1} = 0$, back diffusion is the *only* driving force for the velocity of the center of mass of component 2. From these considerations it is possible to determine s_2 and D of a gel slab or of local parts in it, if the concentration–distance curve $\rho_2(r)$ can be measured.

Experimental

The fine powdered κ-carrageenan, which was investigated, included 11.3% water by weight. It is extracted from seaweeds and is in reality a polyelectrolyte with sulfate groups within the chains and potassium ions as counterions. The potassium content was determined to be 10% by weight. Calculation of the sulfate content of the regular chain in moles shows it to be roughly the same as the potassium content. No additional salt was added. Therefore we treated the system as a non-electrolyte system.

The gels were prepared in the range of 1–5% as described in detail before [4]. The lower limit is set by the critical gel concentration for network formation, the higher limit by the optical density. The soluble parts of the polymer were not removed by dialysis because of their negligible amount. After the gels had been filled in the titanium cells as hot solutions, they were allowed to cool down and reform in the refrigerator. The gel/vapor meniscus, which is parabolic in the 4° ultracentrifugal sector cell, was cut by a scalpel to get a roughly plane surface.

The vacuum chamber of the AUC was cooled down to 10 °C for three days before the cells were placed in the six hole rotor of the Model E which is equipped with a schlieren optical system. The first speed selected was 9000 rpm (called ω_1 later on in the text). The first pictures were taken immediately after the selected speed was reached (2.5 min) by means of a video camera and a frame grabber which is embedded in a personal computer. These are the reference pictures for the five cells filled with gels. The further pictures were taken after different time intervals (increasing with elapsing time). After three weeks the gradients were assumed to be constant. The next higher speed (ω_2), 12 000 rpm, was selected and held for another three weeks, pictures being taken from time to time. Afterwards the speed was increased to 15 000 rpm (ω_3) and held for the same time. At this speed an exclusion of solvent could be observed in all the gels and the consequent gel/solvent meniscus began to move towards the bottom of the cell.

In the next steps the speed was lowered again to the earlier chosen values (first to ω_2, then to ω_1) sufficient time being allowed to elapse for attainment of the revised equilibrium concentration gradient from that pertaining at the higher speed.

Each picture was overlaid with the corresponding reference picture and the area between the two schlieren lines was integrated stepwise. For the calculation of the local polymer concentration from the derivative of the refractive index, the area of the calibration cell for the chosen schlieren angles (40° and 60°) and the refractive index increment of the system has to be known.

Results and discussion

Some typical schlieren patterns are presented in Figs. 2–4 for a κ-carrageenan/water gel with a polymer concentration of 4.5% by wt at 10 °C [10]. In Fig. 2 the first picture was taken after an acceleration time of 3 min, within which the centrifugal speed was changed from zero to 9000 rpm. It can be seen that there is a steep increase of the refractive index gradient at the meniscus of the gel and also at the bottom, although a change of the concentration may not have occurred. In the whole gel a small value of (dn/dr) is observed. It can be differentiated from the base line which can be constructed from the signals in the reference holes and in the air above the gel.

We do not yet know the reason for the positive and negative values of (dn/dr), but it may be due to the occurrence of a stress-optical effect as the gel column is compressed in the direction of the external field. During the application of a pressure the gel becomes anisotropic, which may lead to a birefringence of the passing light. As there is no acting pressure at the meniscus, the effect might be provoked by a slight friction of the gel at the walls of

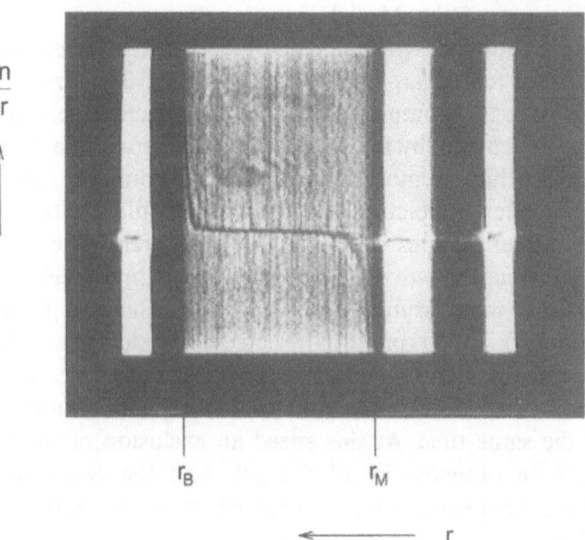

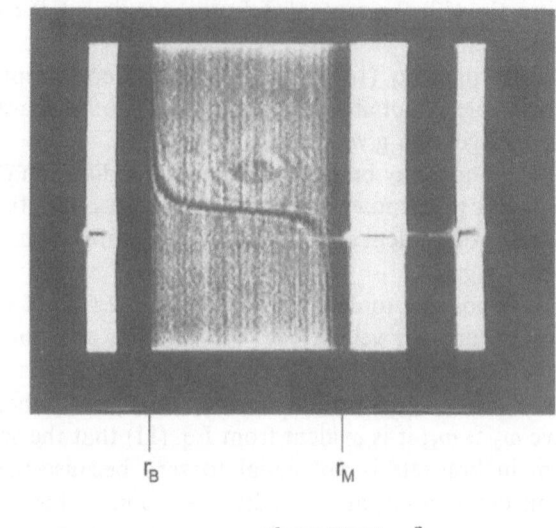

Fig. 2 Original schlieren pattern of a κ-carrageenan/water gel with a polymer concentration of 4.5% by weight at 10 °C taken 3 min after starting the run at 9000 rpm and a schlieren angle of 40°; r = radial distance, M = meniscus, B = bottom, (dn/dr) = gradient of the refractive index

Fig. 4 Schlieren pattern of the same gel as in Figs. 2 and 3 taken after 20 d at 9000 rpm, 21 d at 12 000 rpm and 21 d at 15 000 rpm; see text

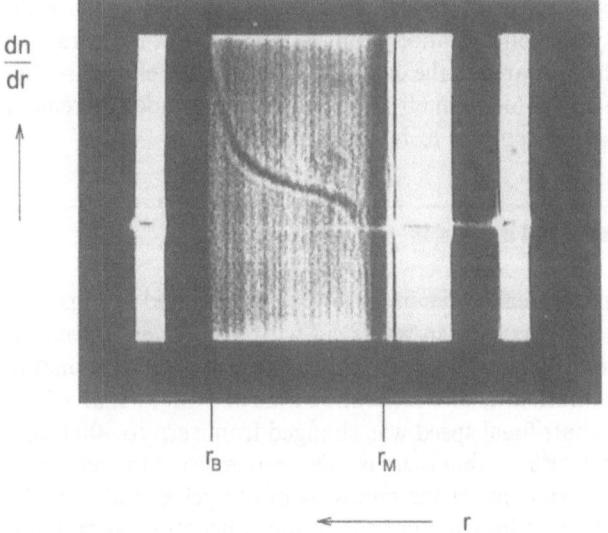

Fig. 3 Schlieren pattern of the same gel as that in Fig. 2 but taken after 20 d at 9000 rpm and 21 d at 12 000 rpm; see text

Fig. 2 have been subtracted from all following schlieren patterns, because the appearance of the "additional" gradients are independent of the changes of concentration in which we are interested. Figure 3 is an original schlieren pattern which was taken after a run at 9000 rpm for 20 d (days) and at 12 000 rpm for 21 d. The increase of the refractive index gradient is mainly due to the change in polymer concentration which increases from the meniscus to the bottom. In Fig. 4, which was recorded after further centrifugation at 15 000 rpm for 21 d, the movement of the meniscus can be seen for the first time. At this speed the gel exudes solvent because the polymer concentration of the swelling equilibrium, which is the lowest possible one, has been established at the meniscus.

The time to reach an equilibrium value with these gel columns was several weeks, and hence the experiments were extremely time consuming. The movement of the meniscus for weaker gels with lower polymer concentrations was observed at lower speeds of the rotor.

The schlieren patterns have been transferred to diagrams of $\rho_2(r)$ versus r by

$$\frac{d\rho_2}{dr} = \frac{d\rho_2}{dn}\frac{dn}{dr}, \tag{15}$$

the cell. Alternatively, the treatment of the gel surface by a scalpel may be responsible for the effects at the meniscus r_M, which remain nearly unchanged at higher fields. Because of these considerations, we have taken the schlieren pattern shown in Fig. 2 as a reference for all patterns at 9000 rpm. This means that the gradients (dn/dr) versus r in

where $(dn/d\rho_2)$ is the refractive index increment, which has been taken from the literature [11].

Integration of the concentration gradient curves, $(d\rho_2/dr)$ versus r, yields the corresponding polymer concentration distribution, an example of which is presented for the gel with a polymer content of 4.5% by wt at 10 °C

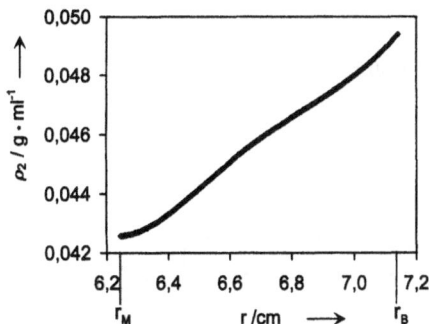

Fig. 5 Partial density ρ_2 in g ml^{-1} of a gel with a polymer content of 4.5% by weight as a function of radial distance after 16 d at 9000 rpm; indices M and B refer to the meniscus and bottom of the gel

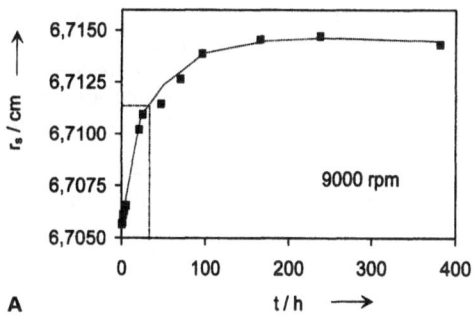

A

Fig. 6A Time dependence of the center of mass of the polymer r_s for a κ-carrageenan/water gel with a polymer content of 4.5% by weight centrifuged at a rotational speed of 9000 rpm

in Fig. 5. For the run at 9000 rpm and a time of 16 d it can be seen that the partial density ρ_2 is, to a good approximation a linear function of r.

The center of mass of the polymer was calculated for different sedimentation times from partial density/radial distance curves $\rho_2(r)$. The radial distance of the center of mass of the polymer $r_s = |\mathbf{r_s}|$ is given by

$$r_s = \frac{\int_{r_M}^{r_B} \rho_2(r)\, r^2 dr}{\int_{r_M}^{r_B} \rho_2(r)\, r\, dr} \cdot \frac{\sin\alpha}{\text{arc}\,\alpha} \qquad (16)$$

where α is the radiant of half of the angle of the sector. The factor $(\sin\alpha/\text{arc}\,\alpha)$ for a sector of $4°$ is given by 0.9998, which is essentially unity. For the calculations following Eq. (16) the $\rho_2(r)$-curve was fitted by means of a linear or quadratic regression curve.

The coordinates of the center of mass are plotted versus time for the run at 9000 rpm in Fig. 6A. It can be seen that the movement of the center of mass of the physically crosslinked polymer is only about 0.053 mm in 25 h in the linear parts at the beginning of the curve, which has been represented in an enlarged scale in Fig. 6B. After 250 h the $r_s(t)$-curve levels off due to attainment of sedimentation–diffusion equilibrium. Such attainment of an equilibrium state has to be proved later by an experiment at the same speed (9000 rpm), but starting from a previous equilibrium position at a higher speed.

From Eq. (14) the sedimentation coefficient of the gel in the limit $t \to 0$ was determined by integration in complete analogy to the treatment of the sedimentation of solutions [9], resulting in the expression

$$\ln\left(\frac{r_s}{r_{s,0}}\right) = \omega^2 s_2 t, \qquad (14a)$$

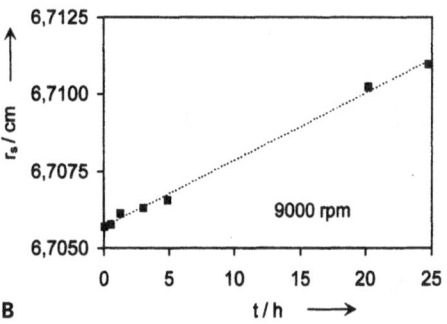

B

Fig. 6B Enlargement of the marked area in Fig. 6A

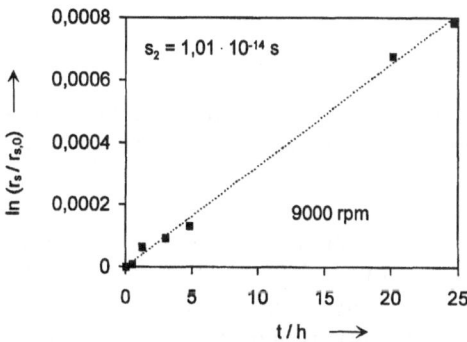

Fig. 7 Logarithmic plot of the center of mass of the polymer r_s relative to the extrapolated value $r_{s,0}$ at time $t = 0$ from the linear plot in Fig. 6B; s_2 is the sedimentation coefficient of the crosslinked polymer in the gel; see text

where r_s is the center of the polymer mass at time t and $r_{s,0}$ that at $t = 0$. The value $r_{s,0}$ has been determined by extrapolating the linear part of the $r_s(t)$-curve, because of the effects mentioned above. This was also the case for all first values which have been determined directly after a change of the rotational speed. A plot of the values of $\ln(r_s/r_{s,0})$ versus t is shown in Fig. 7, where all values up to

Table 1 Transport coefficients of a thermoreversible κ-carrageenan/water gel with a polymer content of 4.5% by weight; ω = rotational speed in rotations per minute (rpm), s_2 = sedimentation coefficient of the polymer in the gel in seconds (s), α_2 = phenomenological coefficient of the polymer in $s\,g\,ml^{-1}$, D = mutual diffusion coefficient of the gel in $cm^2\,s^{-1}$

ω[rpm]	$s_2[10^{-13}\,s]$	$\alpha_2[10^{-15}\,s\,g\,ml^{-1}]$	$D[10^{-10}\,cm^2\,s^{-1}]$
9000	0.101	1.0	3.8
12 000	0.136[a]	1.3	3.7
12 000	0.126[b]	1.2	3.4

[a] and [b] refer to Eqs. (18c) and (18d)

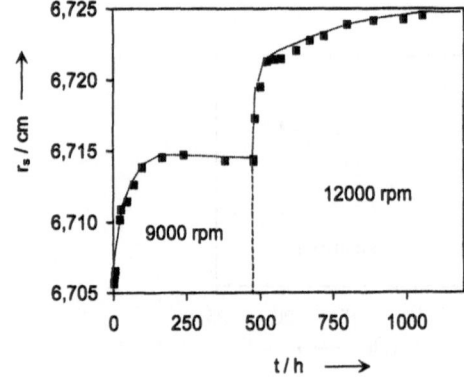

Fig. 8 Overview of the motion of the center of mass of the polymer of the gel at 9000 and 12 000 rpm; see text

25 h are on a straight line. From its slope a sedimentation coefficient in the center of mass frame is given by $s_2 = 1.01 \times 10^{-14}\,s = 0.101S$ ($1S = 10^{-13}\,s$), a result also reported in Table 1.

In the same way the data are treated when the rotational speed was raised to 12 000 rpm. The change of the center of mass of the polymer as a function of time is presented in Fig. 8. The curve is similar to the movement of r_s at 9000 rpm. For the evaluation of the data we have taken two different procedures.

(a) Regarding Eq. (11) we noted that at sedimentation–diffusion equilibrium the flux density $\mathbf{J}_2 = \rho_2\mathbf{v}_2$ has become zero, and that the bracketed term has to also vanish. Combination of the resultant expression with Eq. (12) for the situation with $\omega = \omega_1 = 9000$ rpm, $r_s = r_{s,1}$ leads to the relationship:

$$\rho_2 s_2 \omega_1^2 r_s = \alpha_2\left(\frac{\partial \tilde{\mu}_2}{\partial r}\right)_{T,P} . \tag{17}$$

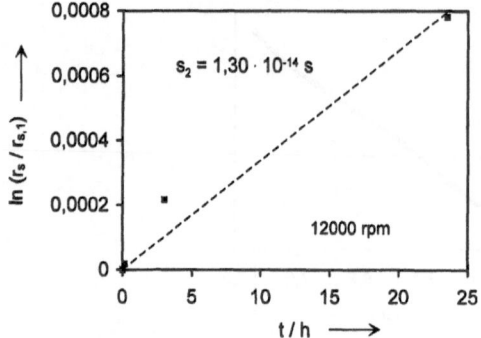

Fig. 9 Logarithmic plot of the relative center of mass of the polymer $(r_s/r_{s,1})$ versus time at 12 000 rpm; s_2 is the sedimentation coefficient of the cross-linked polymer in the gel; see Fig. 7 and text

If now the rotational speed is instantaneously changed from ω_1 to ω_2 we derive from Eqs. (11), (12) and (17)

$$\rho_2 \mathbf{v}_2 = s_2 \rho_2 (\omega_2^2 \mathbf{r}_s - \omega_1^2 \mathbf{r}_{s,1}) \tag{18a}$$

or

$$s_2\,dt = \frac{d\mathbf{r}_s}{\omega_2^2 \mathbf{r}_s - \omega_1^2 \mathbf{r}_{s,1}} \tag{18b}$$

$\mathbf{r}_{s,1}$ is the equilibrium position of $\mathbf{r}_s$ at ω_1 and also for $t = 0$ at the new rotational speed $\omega = \omega_2 = 12\,000$ rpm.

For short times we have $r_s \approx r_{s,1}$, and hence integration of Eq. (18b) leads to

$$\frac{\ln(\mathbf{r}_s/\mathbf{r}_{s,1})}{\omega_2^2 - \omega_1^2} = s_2 t . \tag{18c}$$

(b) If $\mathbf{r}_{s,1}$ is taken to be constant and $\mathbf{r}_s$ varies distinctly during the experiment we get by integration of Eq. (18b):

$$f(\omega, r_s) = \ln\frac{\omega_2^2 \mathbf{r}_s - \omega_1^2 \mathbf{r}_{s,1}}{\mathbf{r}_{s,1}(\omega_2^2 - \omega_1^2)} = \omega_2^2 s_2 t . \tag{18d}$$

From the slopes of the plots on the left-hand sides of Eqs. (18c) or (18d) versus t the sedimentation coefficient s_2 is obtained under the conditions mentioned.

From Fig. 8 the first point at the new speed at 12 000 rpm has been omitted. The linear position of $\mathbf{r}_{s,1}(t)$ during the first 23 h has been extrapolated to the time when the new speed was reached to get $\mathbf{r}_{s,1}$. The plot of $\ln(\mathbf{r}_s/\mathbf{r}_{s,1})$ versus t where $r_s = |\mathbf{r}_s|$ and $r_{s,1} = |\mathbf{r}_{s,1}|$ is linear (Fig. 9). Additionally, $f(\omega, r_s)$ from Eq. (18d) is plotted versus t in Fig. 10. The sedimentation coefficients calculated from the slopes of the lines in both figures are very similar and, within the range of accuracy, identical with the result at 9000 rpm. All sedimentation coefficients are gathered in Table 1. The conclusion is that the sedimentation coefficient as it has been defined in this paper has a value which is about 10-fold lower than those values found in the corresponding non-gelling solutions.

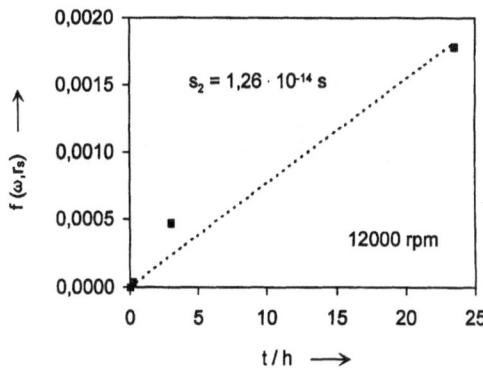

Fig. 10 Dependence of the function $f(\omega, r_s)$ as defined by Eq. (18d) upon time of centrifugation at 12 000 rpm; s_2 is the sedimentation coefficient of the crosslinked polymer in the gel; see text

With Eq. (12) the phenomenological coefficient α_2 of the polymer component in the gel has been calculated

using $\tilde{V}_2 = 0.51$ ml g^{-1} from [11] and the density–concentration relationship from [4].

Starting from Eqs. (11)–(13) and the condition for sedimentation–diffusion equilibrium ($\mathbf{J}_2 = 0$), we get immediately

$$D = \frac{s_2 \omega_i^2 r_s \rho_2}{(\mathrm{d}\rho_2/\mathrm{d}r)} . \tag{19}$$

If we choose the mean values of the concentration gradient in the gel column $\mathrm{d}\rho_2/\mathrm{d}r$ at the rotatioinal speeds we obtain a mean value of D for the gel which is the same at 9000 and 12 000 rpm as expected. That D is nearly three orders of magnitude lower than known values of polymer solutions agrees with results of Rehage, who measured diffusion coefficients of chemically crosslinked polystyrene gels in organic solvents [12].

Acknowledgements We thank the Deutsche Forschungsgemeinschaft (DFG) and the Max-Buchner-Stiftung for financial support of the project.

References

1. Svedberg T, Pedersen KO (1940) In: Ostwald W (ed) Die Ultrazentrifuge, Handbuch der Kolloidwissenschaft, Vol. VII. Steinkopff-Verlag, Dresden and Leipzig, pp 26–29
2. Borchard W (1991) Progr Colloid Polym Sci 86:84
3. Cölfen H, Borchard W (1991) Progr Colloid Polym Sci 86:102
4. Hinsken HM, Borchard W (1995) Colloid Polym Sci 273:913
5. Johnson P (1971) J Photographic Sci 19:49
6. Bohonek J, Spühler A, Ribeaud M, Tomka 1 (1976) In: Cox J (ed) Academic Press, London, p 37
7. Johnson P, Matcalfe JC (1967) Eur Polym J 3:423
8. Haase R (1963) Thermodynamik der irreversiblen Prozesse. D. Steinkopff, Darmstadt, p. 239 ff.
9. Fujita H (1962) Mathematical Theory of Sedimentation Analysis. Academic Press, New York, p 9 onwards
10. Hinsken HM Thesis in preparation
11. Snoeren THM (1976) In: Veeman H, Zonen BV (eds) Kappa-Carrageenan, A study on its Physico-chemical Properties, Sol–gel Transition and Interaction with Milk Proteins.
12. Rehage G (1959) Sympos über Makromoleküle, Wiesbaden, II A 15

Progr Colloid Polym Sci (1997) 107:180–188
© Steinkopff Verlag 1997

H.G. Müller

New contributions of analytical ultracentrifugation to the investigation of dispersions

Dr. H.G. Müller (✉)
Bayer AG
ZF-FPP, E-41
51368 Leverkusen, Germany

Abstract The particle size distribution of dispersions is an important parameter for viscosity, gloss and opacity and can be determined with great precision by analytical ultracentrifugation.
A method using this technique has been developed at Bayer [1] and is now in widespread use. It was recently found to be the most satisfactory method in a recent ringtest of techniques for the submicron range. The high performance of this method is seen from the fact that a sample containing nine components can be analysed exactly. Its high resolving power is proved by a baseline separation of monodisperse components, differing by 10% of diameter only. To detect low quantities of small particles, measurements are performed at different concentrations and then coupled, yielding a high degree of precision. This method has been successfully applied to competitive growth experiments in emulsion polymerization and other areas.

Key words Analytical ultracentrifugation – particle size distribution – competitive growth – nano particles – ultrafine particles

Introduction

The ultracentrifugation method developed by Scholtan and Lange [1] and since widely adopted [2] for the determination of particle size distributions is one of the best methods available for use in the submicron particle size range. In a recent ringtest [3] of available methods it was found to give the best results.

Samples containing nine individual components were correctly analysed (Fig. 1). The high performance of this method is underlined by the fact that it has not been necessary to couple measurements of different cells and different concentrations for this result, but this analysis has been conducted in one cell in one run.

The baseline separation of two components, differing in diameter by only 10%, is further evidence of the high resolving power of the method (Fig. 2). It should be mentioned here, that the physical basis of this method is Stokes' law and the application of the light scattering theory of Mie, which is valid for absorbing and nonabsorbing particles, as has been described in [1].

As can be seen from Fig. 3, which shows an extremely fine cadmium sulphide dispersion of just 1 nanometer diameter, the method is also useful for nanoparticles. The density of CdS is 4.82 g/ml, the measurement has been conducted by Schlieren-optics on a model E.

These examples show the improvements in the method when it was refined in 1987 [4]. The method has recently been adopted by Prof. Antonietti at the Max-Planck Institute for colloid and interface research in Berlin-Teltow.

Given the fact that it is already a very powerful tool is there any need for the method to be developed still further? The answer to this is to be found in the demands placed on

Progr Colloid Polym Sci (1997) 107:180–188
© Steinkopff Verlag 1997

Fig. 1 Result of a test mixture of 9 calibration latices measured in one cell in one run

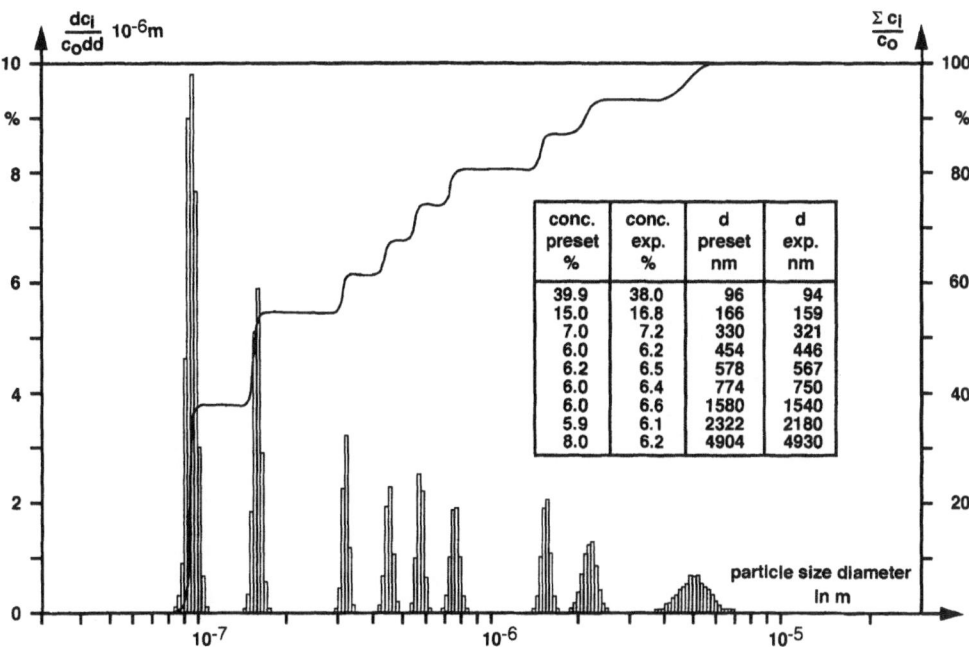

conc. preset %	conc. exp. %	d preset nm	d exp. nm
39.9	38.0	96	94
15.0	16.8	166	159
7.0	7.2	330	321
6.0	6.2	454	446
6.2	6.5	578	567
6.0	6.4	774	750
6.0	6.6	1580	1540
5.9	6.1	2322	2180
8.0	6.2	4904	4930

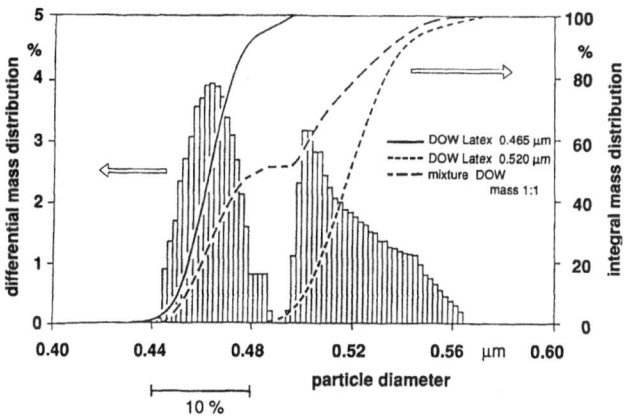

Fig. 2 Resolving power of the auc method for particle size distributions

the method. One of the most important of these is the transition to seed technology in latex production. In terms of colloid chemistry, a key requirement for the use of this technology is that no new particles form upon growth of the seed and that the seed is free of ultrafine particles. An analysis of these ultrafine and possibly new particles means determining a small number of fine particles and a large number of coarse ones in one sample.

The challenge

This aim initially appeared unattainable using the ultracentrifugation method since small particles are known to cause disproportionately less turbidity than large ones, so that they have a disproportionately small effect on measurement.

The solution

The "magnifying glass"

The first stage in solving the problem was – in addition to the usual measurement of the distribution at a dilution of approx. $100 \times - 1000 \times$ – to carry out a second measurement at a concentration up to $10 \times$ higher (i.e., at a dilution of only $10 \times -100 \times$).

With this approach the large particles will lie outside of the measuring range and only the fine particle fraction of the sample will be measured. These fine particles are then integrated into the measurement at normal concentration and the ratio of concentrations of the two measurements permits the correct weighting of a fine particle fraction possibly included in the measurement.

This further development in the method for determining particle size distribution has taken place at the same time at our laboratory and somewhere else [5]. It has been used to investigate competitive particle growth in emulsion polymerization [6] and formed the basis for the development of novel kinetic models in this field, recently published [7]. The aim was to simulate such processes.

Figure 4 shows an example of competitive particle growth, but with only some of the chosen samples shown.

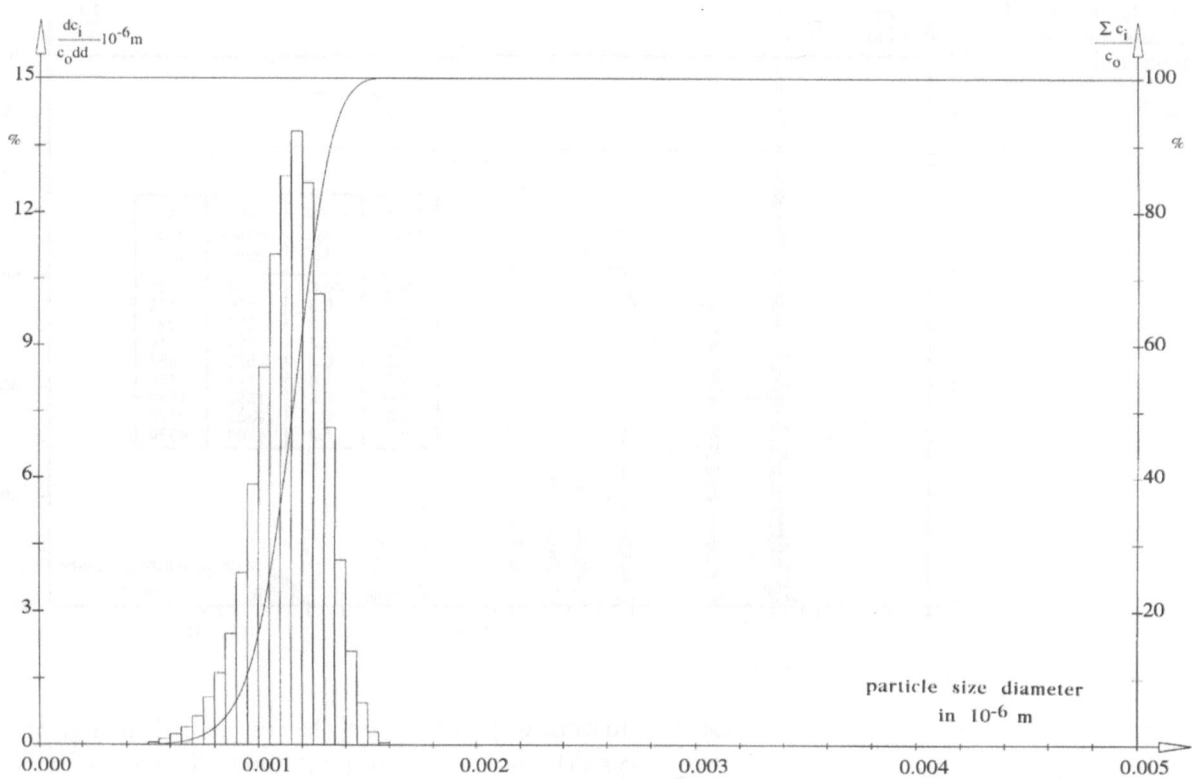

Fig. 3 Particle size distribution of nano particles, here a CdS sample

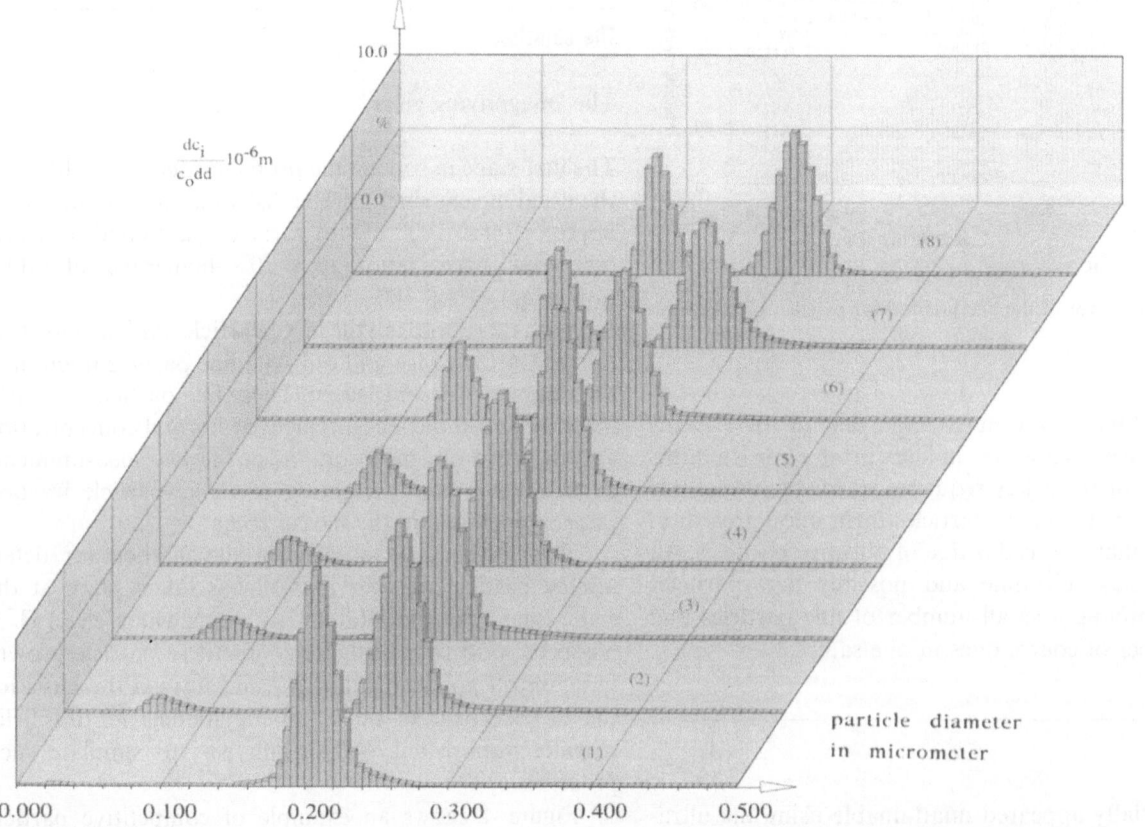

Fig. 4 Particle size distributions of an emulsion polymerization with competitive growth. The conversion increases with increasing number in brackets

Progr Colloid Polym Sci (1997) 107:180–188
© Steinkopff Verlag 1997

The total particle number was calculated from each of these distributions using Eq. (1), except for the very first of these, which results from an addition of the original distributions of both components. Equation (1) reads

$$N = \frac{6 \times 10^{21} m_p}{\pi \rho_p d_{VN}^3} \qquad (1)$$

where $m_p = c' \rho_{H_2O}/(100 - c')$ is the mass of polymer per ml aqueous phase, d_{VN} the diameter at the mean volume of number distribution, ρ_p the particle density and c' the percent by weight of latex (w/w) in the undiluted latex sample.

Equation (1) yields the particle number per ml aqueous phase, without taking account of the concentration of monomers, emulsifiers and salts. This total number was entered in Fig. 5 for all the samples investigated in this series as a function of conversion.

The high constancy of the total particle number not only shows that the polymerization process proceeded without the formation of new particles but also just how precise the determination of particle size distribution is using the "magnifying glass" method. It must be remembered here that the particle number is related to the third power of the particle diameter (cf. Eq. (1)).

In addition to the analysis for competitive particle growth, this method has also been a powerful tool in other areas [8,9].

1. The confirmation that the particle number is constant when using an external seed for the carboxylated SBR latices, allowing the particle diameter to be specifically adjusted, was an essential requirement for the switch to seed technology within Bayer's Rubber Group. One cause of continuous product fluctuations was therefore eliminated.

2. The "magnifying glass" method plays a decisive role in the basic investigations of the colloidal properties of aqueous 2-component polyurethane systems [10].

A characteristic example of the structure of such dispersions is presented in Fig. 6 and shows the change over time of the colloidal structure of such a reacting system. Two different particle size fractions are shown side by side. The small particles, between 0.02 and 0.1 μm in size, are polyol particles swollen with polyisocyanate. The large particles, of the order of 1 μm are polyisocyanate droplets which are emulsified by polyol particles.

This "magnifying glass" method initially appeared to be the answer to the problem of reliably determining

Fig. 5 Total number of particles for the distributions in Fig. 4 as a function of conversion

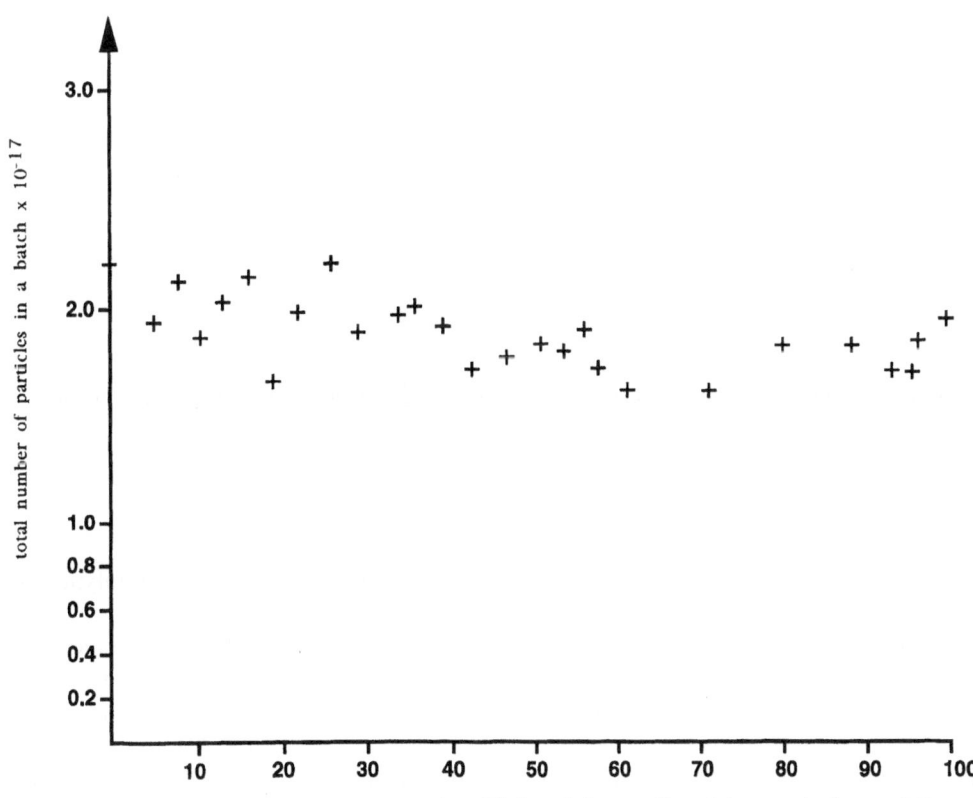

total number of particles in a batch × 10^{-17}

conversion (%) in relation to the total amount of monomer

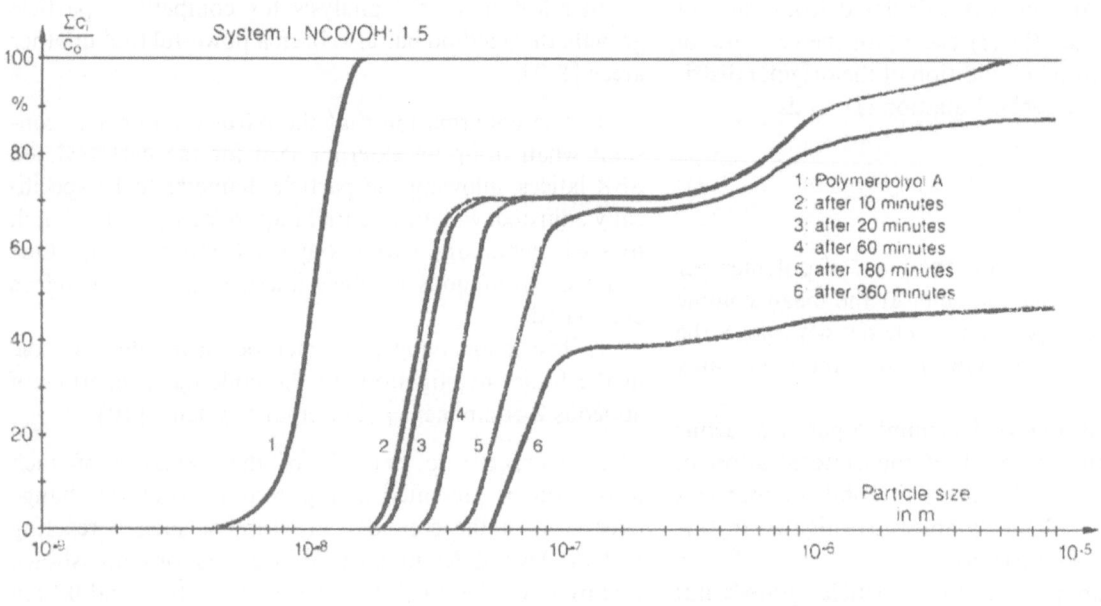

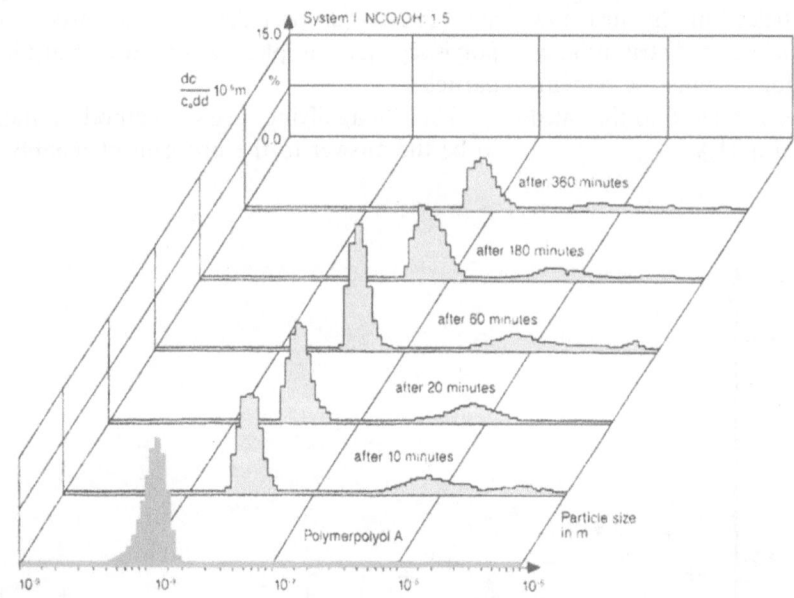

Fig. 6 Investigation of the time dependent colloidal structure of an aqueous 2-K-PUR system made from a polyol (first distribution) and a polyisocyanate

a low small-particle size fraction in the presence of a high large-particle size fraction, a requirement deriving from seed polymerization. However, a bimodal latex (Fig. 7) was derived from a seed latex (Fig. 8) which seemed to have a monomodal distribution despite the use of the "magnifying glass" approach. This raised the question of whether small particles were in fact responsible for the disruptive effect.

The "double magnifying glass"

In a further development of the approach referred to above it was decided to use the highest possible concentration, i.e., the original sample. It became soon obvious that the pelleting of the main mass of the disperse component in an analytical cell at a starting concentration of approx. 40% by weight would present difficulties since a large part of

Progr Colloid Polym Sci (1997) 107:180–188
© Steinkopff Verlag 1997

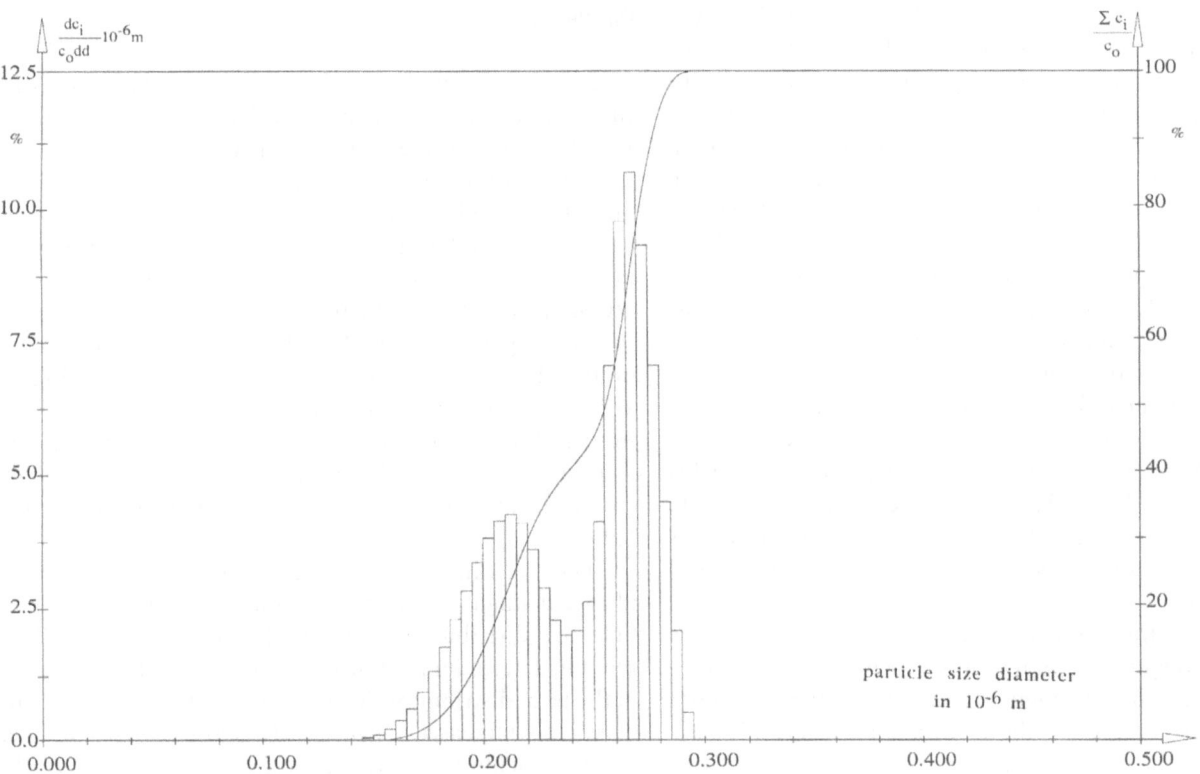

Fig. 7 Latex made by a seed polymerization from seed latex in Fig. 8

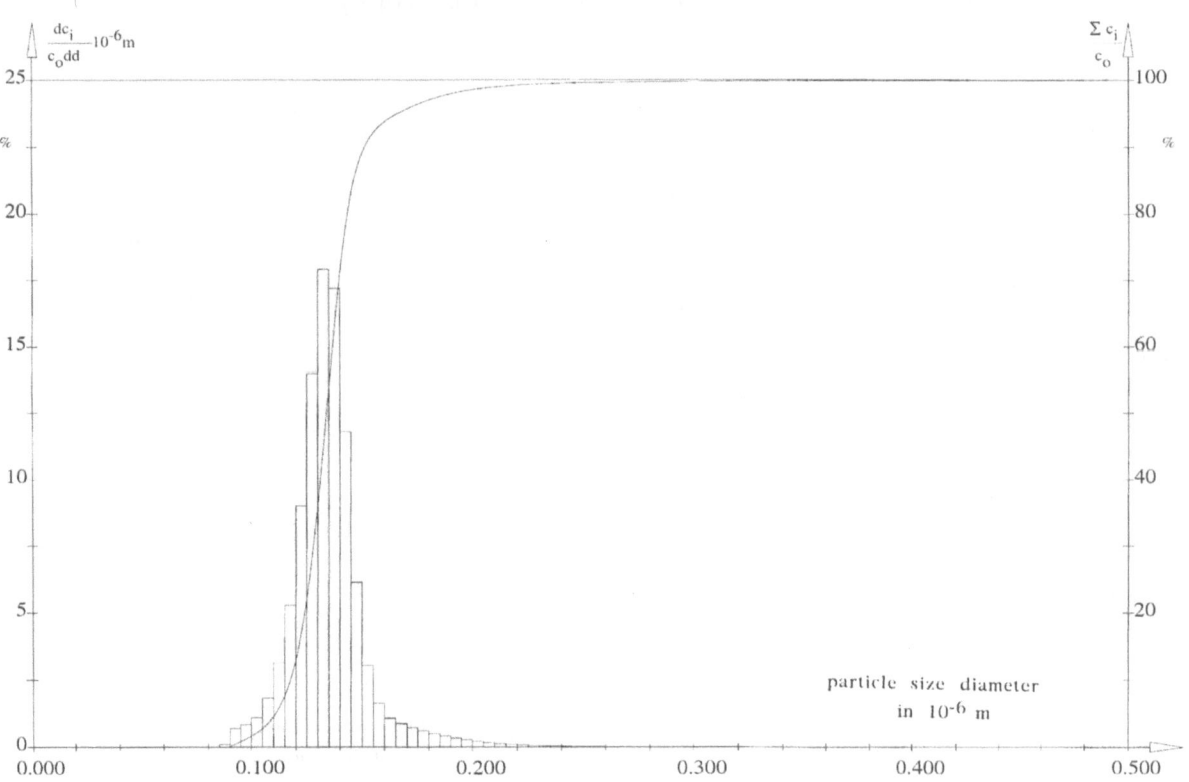

Fig. 8 Seed latex for latex in Fig. 7, distribution resulting from a measurement with "magnifying glass"

the cell volume would be filled with latex solids. The seed latex sample was therefore pre-centrifuged at the original concentration in a preparative ultracentrifuge and the aqueous phase then analysed in an analytical ultracentrifuge cell in terms of particle size distribution. The conditions for preparative ultracentrifugation were chosen so that particles with a diameter of 0.1 μm and above are sedimented (this calculation assumes a strongly diluted dispersion). Figure 9 shows the results for the seed latex. It can be seen that this seed latex actually has an ultrafine component with a d_{50} value of 31 nm.

To counter the objection that the "double magnifying glass" approach would inevitably lead to the discovery of such fine particles because of the methodology used, a study was set up using seed latex which yields a monomodal latex. The results are shown in Fig. 10. It can be seen that the corresponding analysis for normal seed latex shows only a fine fraction of the overall distribution.

The mass fraction of the smallest particle component is 0.4% by weight. This value can be determined by establishing the solids content of the supernatant or more precisely by evaluating the area of the "schlieren" peak after prior calibration. Figure 11 gives a summary of this study.

Discussion

A comparison of the results shown in Figure 11 with the example from the competitive particle growth in Fig. 4 shows that the fine components of the two starting materials are similar both in terms of mass fraction (0.4% and 0.5%) and diameter (31 and 25 nm), but the end distributions differ in that the individual components are clearly separated from one another in Fig. 4 but are already growing together in Fig. 11. This is because the starting particles in Fig. 11 are already at a much more advanced stage of growth than those in Fig. 4. Whereas the diameter of the large particles in Fig. 4 grows from 210 to 310 nm – corresponding to a 3.2-fold growth in particle volume – the particle diameter in Fig. 11 increases from 130 to 270 nm, a 9-fold increase in particle volume.

These results show the exceptional performance of the ultracentrifugation method for the determination of particle size distributions. This is the very first instance of reliably determining a fine fraction of just 0.4% by weight of the total distribution. The use of the "double magnifying glass" has increased sensitivity 10-fold over the previous method, yielding an overall increase in sensitivity of the method for fine fractions by a factor of 100.

Fig. 9 Ultrafine particles in the distribution of latex in Fig. 8, detected by use of the "double magnifying glass"

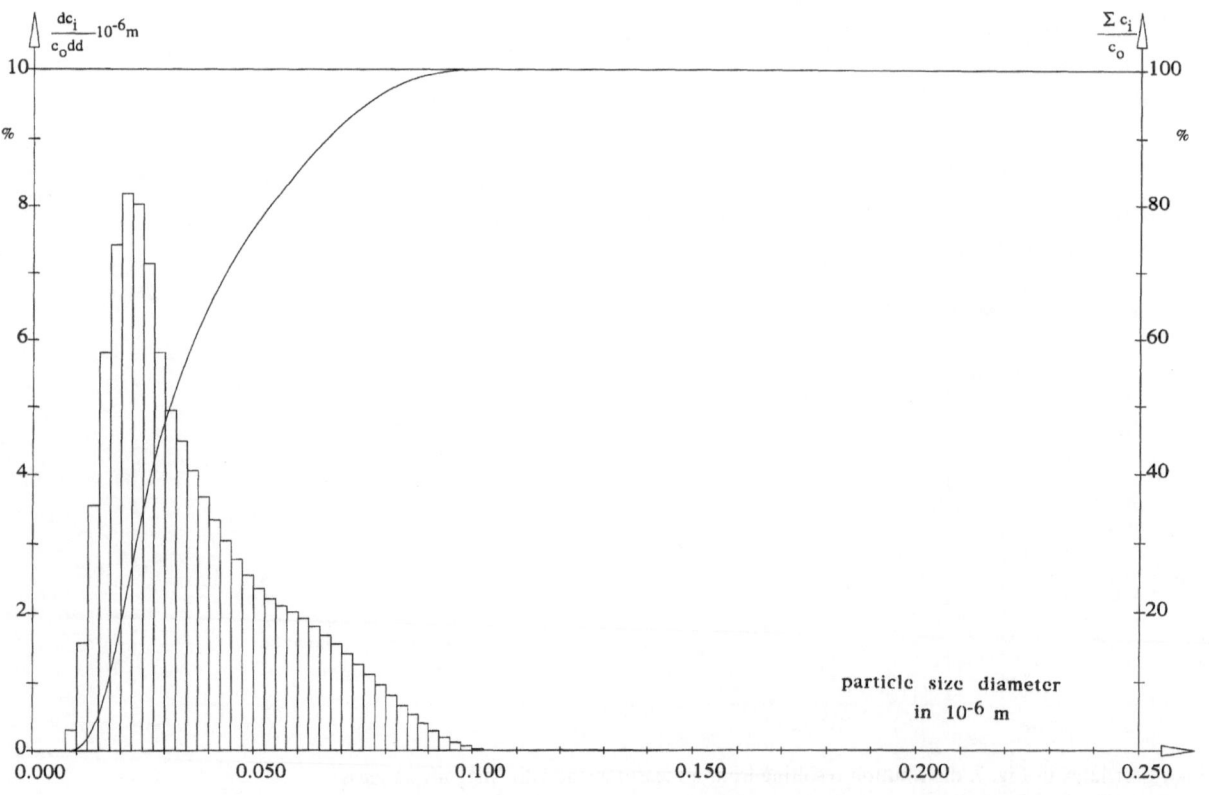

Progr Colloid Polym Sci (1997) 107:180–188
© Steinkopff Verlag 1997

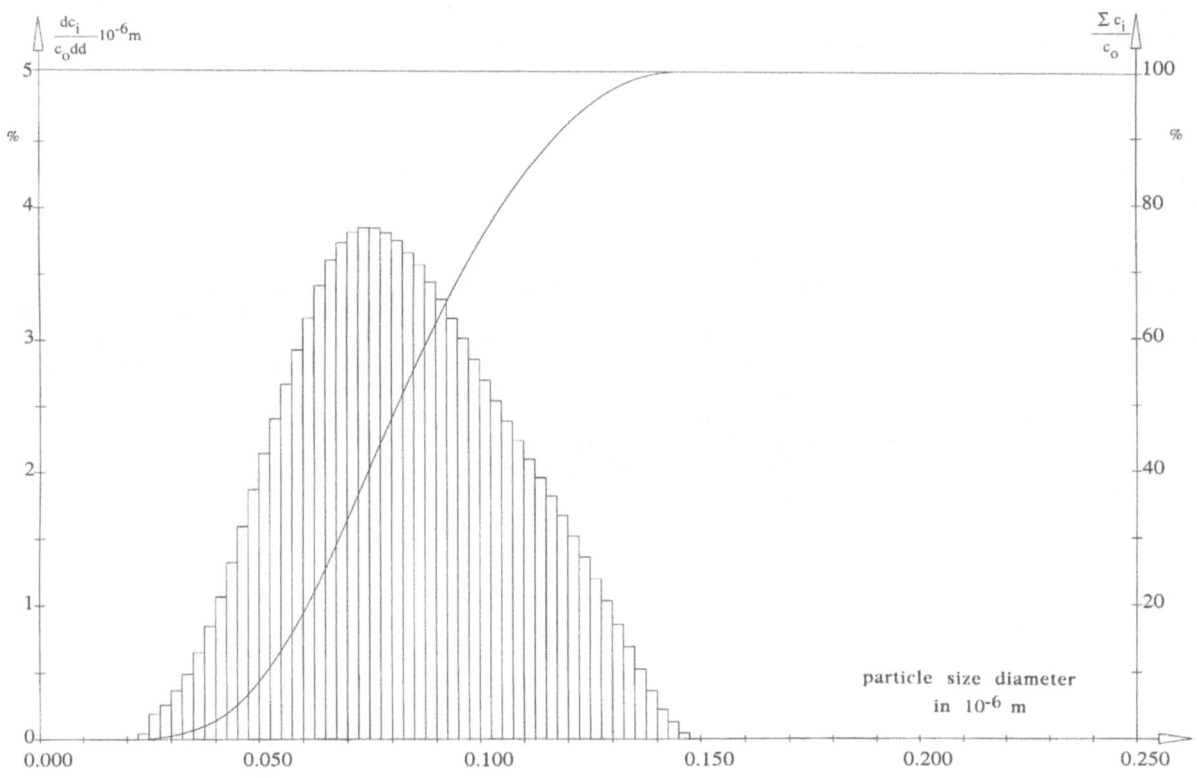

Fig. 10 Looking for ultrafine particles in the distribution of a usual seed latex with the "double magnifying glass"

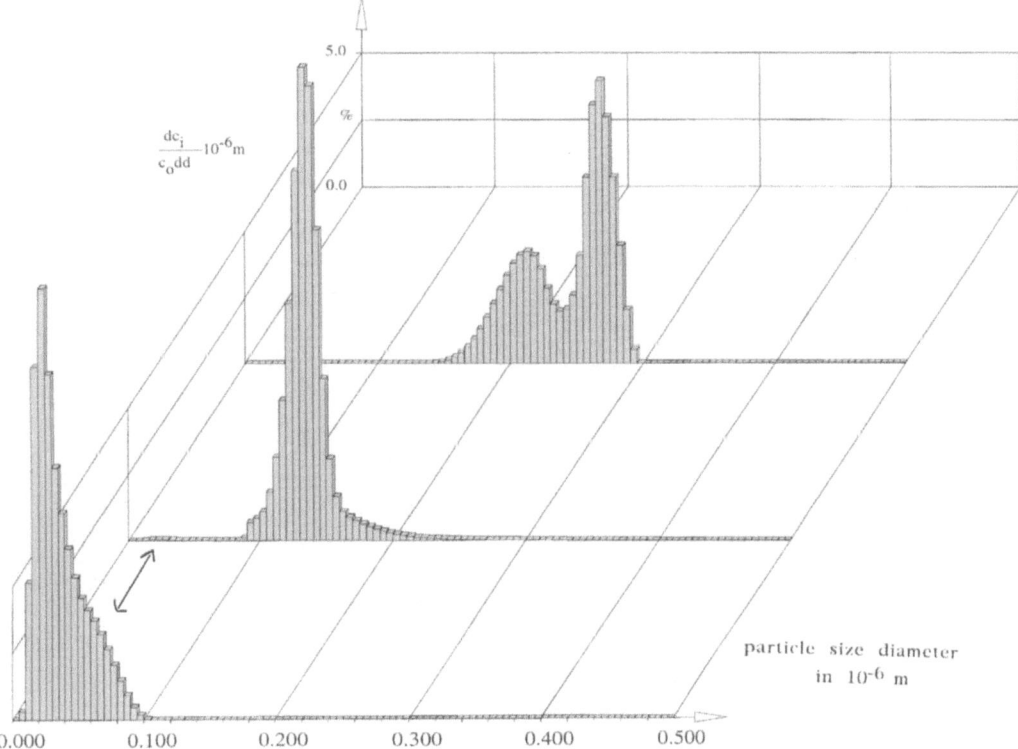

Fig. 11 Overall view for understanding the bimodal distribution in Fig. 7

Moreover, neither soap titration nor any other method allowed the smallest particle size fraction of the fine latex to be established. This example also shows how important it is in seed technology to have a seed which is free of fractions of ultrafine particles, and illustrates the problems which even very low quantities of such "impurities" can cause.

References

1. Scholtan W, Lange H (1972) Kolloid-ZuZ Polymere 250:782–796
2. Mächtle W (1984) Makromol Chem 185:1025–1039
3. Lange H (1995) Part Part Syst Charact 12:148–157
4. Müller HG (1989) Colloid Polym Sci 267:1113–1116
5. Mächtle W (1988) Die Angew Makromol Chem 162:35–52
6. Vanderhoff JW, Vitkuske JF, Bradford EB, Alfrey T (1956) J Polym Sci 20: 225–234
7. Bachmann R, Pallaske U, Schmidt A, Müller HG (1995) In: Dechema Monographs Vol. 131. VCH Verlagsgesellschaft, Weinheim, pp 65–74
8. Grunder R, Kim YS, Ballauff M, Kranz D, Müller HG (1991) Angew Chem 103: 1715–1717
9. Müller HG, Schmidt A, Kranz D (1991) Progr Colloid Polym Sci 86:70–75
10. Probst J, Müller HG, Jürgens E (1992) Poster contribution to the Gordon research conference on polymer colloids, Irsee, Germany

Progr Colloid Polym Sci (1997) 107:189–192
© Steinkopff Verlag 1997

GELS, EMULSION AND DISPERSIONS

Determination of hydrodynamic radius: a comparison of ultracentrifuge methods with dynamic light scattering

P.M. Budd
R.K. Pinfield
C. Price

Dr. P.M. Budd (✉) · R.K. Pinfield · C. Price
Department of Chemistry
University of Manchester
Manchester M13 9PL, United Kingdom
E-mail: Peter.Budd@man.ac.uk

Abstract Hydrodynamic radius, r_h, may be determined from the limiting diffusion coefficient, D_0, from the limiting sedimentation coefficient, s_0, or, using simple models for the concentration dependence, from the sedimentation coefficient, s, at a finite concentration. For water/AOT/heptane water-in-oil microemulsions, values of r_h determined from s are shown to agree well with those obtained from s_0 or D_0 determined using an analytical ultracentrifuge, and with literature values from dynamic light scattering. For several aqueous polyurethane dispersions, sedimentation velocity demonstrated the presence of small species with $r_h < 5$ nm and medium-sized species with r_h in the range 10–17 nm, as well as much larger aggregates. For these complex dispersions, dynamic light scattering behavior was dominated by the large aggregates and analysis by the CONTIN method was not able to detect all the species present.

Key words Hydrodynamic radius – microemulsion – polyurethane dispersion – sedimentation velocity – diffusion – light scattering

Introduction

The hydrodynamic radius, r_h, for a colloidal particle dispersed in a continuous medium may be evaluated from the limiting mutual diffusion coefficient, D_0, utilizing the Stokes–Einstein relationship

$$r_h = \frac{kT}{6\pi\eta D_0},\qquad(1)$$

where k is the Boltzmann constant, T the absolute temperature and η the viscosity of the medium. Values of r_h may also be obtained from the limiting sedimentation coefficient, s_0, using

$$r_h = \left\{\frac{9\eta s_0}{2(\rho_D - \rho_M)}\right\}^{1/2},\qquad(2)$$

where ρ_D and ρ_M are the densities of the dispersed phase and the medium, respectively. With some systems it is difficult or inappropriate to extrapolate to infinite dilution, in which case r_h can in principle be evaluated from the sedimentation coefficient, s, at a finite concentration using

$$r_h = \left\{\frac{9\eta s}{2(\rho_D - \rho_M)} \times f(\phi)\right\}^{1/2},\qquad(3)$$

where $f(\phi)$ is a function of the volume fraction of the dispersed phase, ϕ [1]. A number of expressions for $f(\phi)$ have been proposed. For concentrated dispersions of solid particles, Barnea and Mizrahi [2] obtained the semi-empirical equation

$$f(\phi) = \frac{(1 + \phi^{1/3})\exp(5\phi/3(1 - \phi))}{(1 - \phi)^2}.\qquad(4)$$

For systems in which $\phi < 0.1$, a theoretical study by Batchelor [3] gave

$$f(\phi) = \frac{1}{1 - 6.55\phi}.\qquad(5)$$

In recent years, dynamic light scattering has been widely employed for the determination of diffusion coefficients, and hence of hydrodynamic radii. Analytical ultracentrifugation can also provide hydrodynamic radii, either from diffusion coefficients measured by synthetic boundary methods, or from sedimentation coefficients derived from sedimentation velocity data [4]. In this contribution, results from ultracentrifuge methods are compared with those from dynamic light scattering for two different types of colloidal system: (1) water/Aerosol OT/heptane water-in-oil microemulsions and (2) aqueous polyurethane dispersions.

Microemulsions form readily when water is dispersed in an organic medium such as n-heptane, with the double-tailed anionic surfactant dioctylsulphosuccinate (Aerosol OT, AOT) as dispersant. There have been a number of studies of water/AOT/alkane microemulsions by ultracentrifuge [1, 5–9] and other [10–12] methods and the size of the microemulsion droplets has been found to depend on the ratio $R = $ [water]/[AOT]. In the present work, microemulsions were used as a model system for assessing the applicability of Eqs. (1)–(5).

Aqueous polyurethane dispersions are used in the coatings industry. Their composition may be complex, making particle size analysis difficult. In the present work, a comparison is made between sedimentation velocity data, analyzed using Eq. (3) with Eq. (5), and dynamic light scattering data, analyzed by the CONTIN method, for dispersions prepared in our laboratory.

Experimental

Microemulsions

AOT (Sigma Chemical Company) was dissolved in n-heptane (BDH Analar Grade) and an appropriate quantity of distilled water added. With a little shaking, clear microemulsions formed. Microemulsions were prepared with $[AOT] = 0.05$ mol dm^{-3} for $R = 20$, 25 and 50, and at higher values of [AOT] for $R = 25$. Ultracentrifuge experiments were performed at 23°C. For data analysis, the viscosity and density of heptane were taken as 4.09×10^{-4} kg m^{-1} s^{-1} and 684 kg m^{-3}, respectively. The density of the dispersed phase was estimated from the data of Mathews and Hirschhorn [5].

Aqueous polyurethane dispersions

The preparation of aqueous polyurethane dispersions, in which the polymer incorporates acid groups, is described elsewhere [13]. N-methyl-2-pyrrolidone (NMP) is present as a cosolvent. The dispersions are referred to by a code (e.g. R1.6/5.0/20, where R1.6 indicates the method of preparation, 5.0 indicates the proportion of acid groups and 20 relates to the amount of NMP cosolvent in the initial dispersion). In the context of the present study, the significant feature of these dispersions is that they are complex fluids containing more than one species. Sedimentation velocity and dynamic light scattering experiments were performed at 20°C. Polyurethane dispersions were diluted to concentrations in the range 1–8 wt% solids for sedimentation studies and to 0.1 wt% solids for light scattering. Densities for dispersions and solvents were determined using a Paar DMA 60 digital density meter.

Analytical ultracentrifugation

A Beckman Model E Analytical Ultracentrifuge was employed with Schlieren optics. For diffusion measurements a capillary-type synthetic boundary centrepiece was used, and for sedimentation velocity measurements either a double-sector aluminum-filled epon or a single-sector aluminum centerpiece was used. Diffusion coefficients were evaluated by the height–area method and sedimentation coefficients from the rate of movement of the maximum in the Schlieren image [4].

Dynamic light scattering

For light scattering studies, polyurethane dispersions were diluted to 0.1 wt% solids with filtered (Millipore 0.22 μm) solvent and were clarified by repeated filtration (Millipore 0.45 μm). Dynamic light scattering measurements were carried out at an angle of 90° and a temperature of 20°C using an Otsuka DLS-700 photon correlation spectrometer fitted with a 5 mV He–Ne laser (wavelength = 632.8 nm). Distributions of hydrodynamic radii were evaluated by the linear regularization CONTIN method [14].

Results and discussion

Microemulsions

Table 1 gives sedimentation and diffusion coefficients for microemulsions determined using the Analytical Ultracentrifuge. For $R = 25$, D_0 was evaluated as 9.4×10^{-11} m^2 s^{-1} and s_0 as 69×10^{-13} s. Application of Eqs. (1) and (2), respectively, give a consistent value of $r_h = 5.6$ nm. This is in good agreement with the dynamic light scattering results of Aveyard et al. [11], as can be seen in Fig. 1,

Table 1 Sedimentation and diffusion coefficients for water/AOT/heptane microemulsions at 23°C

$R = \dfrac{[\text{water}]}{[\text{AOT}]}$	[AOT] $[\text{mol dm}^{-3}]$	$s\ [10^{-13}\ \text{s}]$	D $[10^{-11}\ \text{m}^2\ \text{s}^{-1}]$
20	0.05	33.9	—
25	0.05	46.0	8.91
25	0.10	35.1	7.63
25	0.15	27.9	7.54
50	0.05	130	—

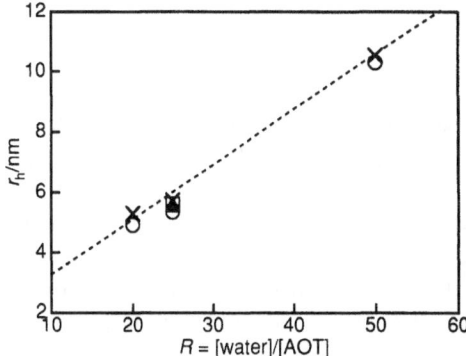

Fig. 1 Dependence of r_h on R for water/AOT/heptane microemulsions from dynamic light scattering studies [11] (- - - -), and values of r_h determined from measurements in the ultracentrifuge of D_0 (▲), s_0 (□) and of s at [AOT] = 0.05 mol dm^{-3} using Eq. (4) (×) and Eq. (5) (○).

which illustrates the dependence of r_h on R derived from numerous light scattering experiments.

Figure 1 also shows hydrodynamic radii calculated from s at a finite concentration, [AOT] = 0.05 mol dm^{-3}, using Eqs. (3)–(5), for $R = 20, 25$ and 50. At the concentrations studied, Eqs. (4) and (5) give similar results. Good

agreement is seen with the relationship from dynamic light scattering and, for $R = 25$, with the results from ultracentrifuge data extrapolated to infinite dilution.

Aqueous polyurethane dispersions

Table 2 summarizes results from sedimentation velocity experiments on a range of aqueous polyurethane dispersions. On centrifugation of the dispersions there was an initial decrease in turbidity of the suspension, accompanied by a buildup of sediment at the bottom of the cell. This indicates the presence of a large, rapidly sedimenting component which could not be quantified under the experimental conditions employed. Further centrifugation showed there to be additional components present in the system. In most cases, a fast boundary was observed in the Schlieren image, with s in the range 20–100×10^{-13} s, followed by a slow boundary, with s of 7.0×10^{-13} s or less. Sedimentation velocity thus indicates the presence of at least three distinct species in most dispersions, which may be interpreted for these systems as being unimers, micelles and aggregates. For a sample (R1.6/5.0/20) studied at relatively high concentration and high rotor speed, the slow boundary was itself observed to be bimodal. The mean density of the dispersed phase in the dispersions was evaluated from density measurements as 1.166 ± 0.005 g cm^{-3}. Table 2 gives hydrodynamic radii calculated using Eq. (3) with Eq. (5). It can be seen that there is reasonable agreement between results obtained at different concentrations and at different rotor speeds.

Light scattering behavior is dominated by the largest species present in a dispersion. Figure 2 shows distributions of hydrodynamic radii determined using CONTIN

Table 2 Sedimentation coefficients and hydrodynamic radii from sedimentation velocity of aqueous polyurethane dispersions

Dispersion	Conc. [wt% solids]	Speed [rev min^{-1}]	Slow boundary s $[10^{-13}\text{s}]$	Slow boundary r_h [nm]	Fast boundary s $[10^{-13}\text{s}]$	Fast boundary r_h [nm]
R1.6/5.0/10	1.0	40 000	2.9	2.9	37.9	10.3
R1.6/5.0/20	1.0	20 000	7.1	4.5	37.7	10.3
R1.6/5.0/20	4.0	20 000	6.9	4.9	40.9	11.9
R1.6/5.0/20	7.95	40 000	5.7, 2.2[a]	5.3, 3.3[a]	20.8	10.1
R1.6/5.0/30	1.0	10 000	—	—	43.8	11.1
R1.6/5.0/30	1.0	40 000	5.7	4.0	38.7	10.4
R1.6/7.0/25	1.0	20 000	4.9	3.7	78.1	14.8
R1.6/7.0/30	1.0	40 000	3.1	2.9	34.8	9.9
R1.6/9.0/30	1.0	20 000	1.9	2.3	—	—
75N1.6/5.0/30	1.0	20 000	—	—	102.3	17.0
75N1.6/9.0/30	1.0	20 000	—	—	92.9	16.2

[a] Bimodal Schlieren peak

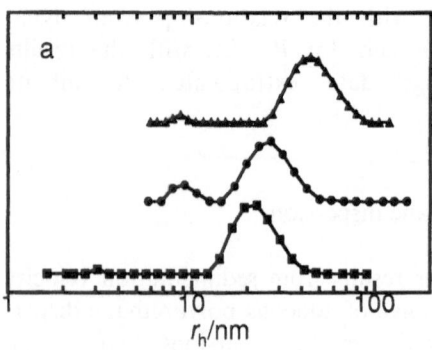

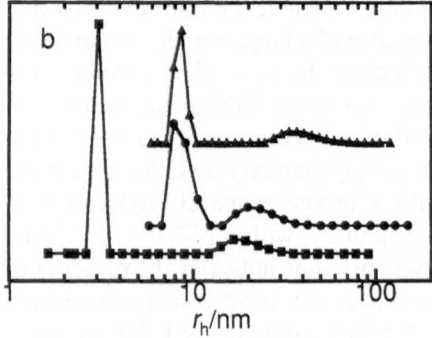

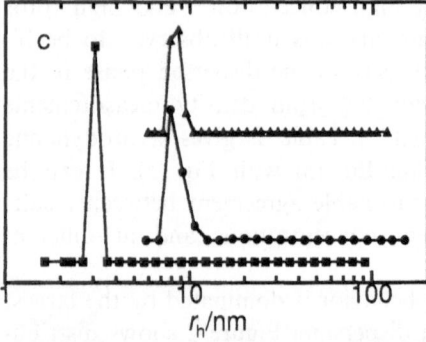

Fig. 2 Distributions of hydrodynamic radii for aqueous polyurethane dispersions determined by dynamic light scattering using CONTIN: (a) intensity distribution, (b) weight distribution and (c) number distribution for R1.6/5.0/10 (■) R1.6/5.0/20 (●) and R1.6/5.0/30 (▲).

Table 3 Hydrodynamic radii from CONTIN analysis of dynamic light scattering data for aqueous polyurethane dispersions (0.1 wt% solids)

Dispersion	Low-intensity peak r_h [nm]	High intensity peak r_h [nm]
R1.6/5.0/10	3.1	23.5
R1.6/5.0/20	9.1	26.9
R1.6/5.0/30	8.5	42.7
R1.6/7.0/25	—	22.6
R1.6/7.0/30	8.4	26.0
R1.6/9.0/30	5.2	20.8
75N1.6/5.0/30	34.3	73.8
75N1.6/9.0/30	—	68.7

for a series of dispersions (R1.6/5.0) prepared with different amounts of NMP co-solvent. In each case, in the intensity distribution (Fig. 2a) a major peak is seen with $r_h > 10$ nm, and a minor peak at smaller sizes. However, in the weight distribution (Fig. 2b) it can be seen that the peak at smaller sizes represents the majority of material by mass. In the number distribution (Fig. 2c) the peak at large size is insignificant.

Table 3 gives hydrodynamic radii corresponding to the maxima in the intensity distribution. For the R1.6/5.0 series of dispersions, a comparison with Table 2 suggests that the CONTIN peak at smaller sizes corresponds to the slow boundary in sedimentation velocity for R1.6/5.0/10 and to the fast boundary for R1.6/5.0/20 and R1.6/5.0/30. For the dispersions studied, CONTIN is able at best only to resolve two components.

In conclusion, it is seen that when dealing with complex fluids containing multiple species, dynamic light scattering results may be misleading. A simple analysis of sedimentation velocity data, such as that employed here, can provide useful complementary information about the size distribution.

References

1. Hwan R-N, Miller CA, Fort T (1979) J Coll Interf Sci 68:221
2. Barnea E, Mizrahi J (1973) Chem Eng J 5:171
3. Batchelor GK (1972) J Fluid Mech 52:245
4. Budd PM (1989) In: Allen G, Bevington JC, Booth C, Price C (eds) Comprehensive Polymer Science, Vol 1, Chap 10. Pergamon, Oxford, 199
5. Mathews MB, Hirschhorn E (1953) J Coll Sci 8:86
6. Eicke H-F, Rehak J (1976) Helv Chim Acta 59:2883
7. Zulauf M, Eicke, H-F (1979) J Phys Chem 83:480
8. Robinson BH, Steytler DC, Tack RD (1979) J Chem Soc Faraday Trans I 75:481
9. Oldfield C, Freedman RB, Robinson BH (1996) J Chem Soc Faraday Trans I 92:73
10. Robinson BH, Toprakcioglu C, Dore JC, Chieux P (1984) J Chem Soc Faraday Trans I 80:13
11. Aveyard R, Binks BP, Clark C, Mead J (1986) J Chem Soc, Faraday Trans I 82:125
12. Aveyard R, Binks BP, Mead J, Clint JH (1988) J Chem Soc Faraday Trans I 84:675
13. Pinfield RK (1996) PhD Thesis, University of Manchester
14. Provencher SW (1982) Computer Phys Comm 27:213

Progr Colloid Polym Sci (1997) 107:193
© Steinkopff Verlag 1997

AUTHOR INDEX

Progr Colloid Polym Sci (1997) 107:194
© Steinkopff Verlag 1997